Solid State Lasers

New Developments and Applications

NATO ASI Series

Advanced Science Institutes Series

A series presenting the results of activities sponsored by the NATO Science Committee, which aims at the dissemination of advanced scientific and technological knowledge, with a view to strengthening links between scientific communities.

The series is published by an international board of publishers in conjunction with the NATO Scientific Affairs Division

A	**Life Sciences**	Plenum Publishing Corporation
B	**Physics**	New York and London
C	**Mathematical and Physical Sciences**	Kluwer Academic Publishers
D	**Behavioral and Social Sciences**	Dordrecht, Boston, and London
E	**Applied Sciences**	
F	**Computer and Systems Sciences**	Springer-Verlag
G	**Ecological Sciences**	Berlin, Heidelberg, New York, London,
H	**Cell Biology**	Paris, Tokyo, Hong Kong, and Barcelona
I	**Global Environmental Change**	

Recent Volumes in this Series

Volume 313 —Dissociative Recombination: Theory, Experiment, and Applications
edited by Bertrand R. Rowe, J. Brian A. Mitchell, and André Canosa

Volume 314 —Ultrashort Processes in Condensed Matter
edited by Walter E. Bron

Volume 315 —Low-Dimensional Topology and Quantum Field Theory
edited by Hugh Osborn

Volume 316 —Super-Intense Laser–Atom Physics
edited by Bernard Piraux, Anne L'Huillier, and Kazimierz Rzążewski

Volume 317 —Solid State Lasers: New Developments and Applications
edited by Massimo Inguscio and Richard Wallenstein

Volume 318 —Relativistic and Electron Correlation Effects in Molecules and Solids
edited by G. L. Malli

Volume 319 —Statics and Dynamics of Alloy Phase Transformations
edited by Patrice E. A. Turchi and Antonios Gonis

Volume 320 —Singular Limits of Dispersive Waves
edited by N. M. Ercolani, I. R. Gabitov, C. D. Levermore, and D. Serre

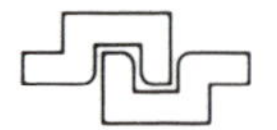

Series B: Physics

Solid State Lasers

New Developments and Applications

Edited by

Massimo Inguscio

Department of Physics and
European Laboratory for Nonlinear Spectroscopy, (LENS)
University of Florence
Florence, Italy

and

Richard Wallenstein

University of Kaiserslautern
Kaiserslautern, Germany

Plenum Press
New York and London
Published in cooperation with NATO Scientific Affairs Division

Proceedings of a NATO Advanced Study Institute on
Solid State Lasers: New Developments and Applications,
held August 31–September 11, 1992,
at Elba Island, Tuscany, Italy

NATO-PCO-DATA BASE

The electronic index to the NATO ASI Series provides full bibliographical references (with keywords and/or abstracts) to more than 30,000 contributions from international scientists published in all sections of the NATO ASI Series. Access to the NATO-PCO-DATA BASE is possible in two ways:

—via online FILE 128 (NATO-PCO-DATA BASE) hosted by ESRIN, Via Galileo Galilei, I-00044 Frascati, Italy

—via CD-ROM "NATO Science and Technology Disk" with user-friendly retrieval software in English, French, and German (©WTV GmbH and DATAWARE Technologies, Inc. 1989). The CD-ROM also contains the AGARD Aerospace Database.

The CD-ROM can be ordered through any member of the Board of Publishers or through NATO-PCO, Overijse, Belgium.

Library of Congress Cataloging-in-Publication Data

NATO Advanced Study Institute on Solid State Lasers: New Developments and Applications (1992 : Elba, Italy)
Solid state lasers : new developments and applications / edited by Massimo Inguscio and Richard Wallenstein.
p. cm. -- (NATO ASI series. Series B, Physics ; v. 317)
"Published in cooperation with NATO Scientific Affairs Division."
Includes bibliographical references and index.
ISBN 0-306-44598-0
1. Solid-state lasers--Congresses. 2. Laser spectroscopy--Congresses. 3. Quantum electronics--Congresses. I. Inguscio, M. II. Wallenstein, Richard. III. North Atlantic Treaty Organization. Scientific Affairs Divisi. IV. Title. V. Series.
TA1705.N38 1994
621.36'61--dc20 93-38226
CIP

ISBN 0-306-44598-0

A Division of Plenum Publishing Corporation
233 Spring Street, New York, N.Y. 10013

Printed in the United States of America

PREFACE

This volume contains the lectures and seminars presented at the NATO Advanced Study Institute on "Solid State Lasers: New Developments and Applications" the fifteenth course of the Europhysics School of Quantum Electronics, held under the supervision of the Quantum Electronics Division of the European Physical Society. The Institute was held at Elba International Physics Center, Marciana Marina, Elba Island, Tuscany, Italy, August 31 - September 11, 1992.

The Europhysics School of Quantum Electronics was started in 1970 with the aim of providing instruction for young researchers and advanced students already engaged in the area of quantum electronics or wishing to switch to this area from a different background. Presently the school is under the direction of Professors F.T. Arecchi and M. Inguscio, University of Florence, and Prof. H. Walther, University of Munich, and has its headquarters at the National Institute of Optics (INO), Florence, Italy. Each time the directors choose a subject of particular interest, alternating fundamental topics with technological ones, and ask colleagues specifically competent in a given area to take the scientific responsibility for that course.

Past courses were devoted to the following topics:

1. 1971: Physical and Technical Measurements with Lasers
2. 1972: Nonlinear Optics and Short Pulses
3. 1973: Laser Frontiers: Short Wavelength and High Powers
4. 1974: Cooperative Phenomena in Multicomponent Systems
5. 1975: Molecular Spectroscopy and Photochemistry with Lasers
6. 1976: Coherent Optical Engineering
7. 1977: Coherence in Spectroscopy and Modern Physics
8. 1979: Lasers in Biology and Medicine
9. 1980: Physical Processes in Laser Material Interactions
10. 1981: Advances in Laser Spectroscopy
11. 1982: Laser Applications to Chemistry
13. 1987: Instabilities and Chaos in Quantum Optics
14. 1989: Applied Laser Spectroscopy

The objective of the ASI on "Solid State Lasers: New Developments and Applications" was to bring together young researchers and top level scientists in the fields where advanced research and interdisciplinary applications of solid state lasers are being made. A major aim was to illustrate the basic aspects of the new, advanced laser sources as well as the rapidly developing technical applications. It was shown how fascinating and useful can be the transfer of high technology techniques from highly specialized laboratories to a more general use.

The scientific organization of the course was taken care of by F.T. Arecchi, University of Florence, R.I. Byer, Stanford University, M. Inguscio, Director of the ASI, University of Florence, R. Wallenstein, University of Kaiserslautern, and H. Walther, University of Munich and MPI for Quantum Optics.

Lectures started with the description of new concept resonators including special designs for high brightness operation. Also, configurations for mode locked operation were discussed as well as monolithic designs and quantum confined resonators.

Basic properties and new developments in solid state laser materials were covered also in relation to crystal growth problems and giving emphasis to materials for tunable operation and nonlinear generation.

Essentially all solid state lasers were presented in a series of exhaustive lectures. The contributions to the present volume include diode pumped solid state lasers, high power diode lasers, rare earth ion lasers, Yb: Yag or YLF based lasers, color center lasers and fibre lasers. Extension of the spectral coverage by means of optical parametric generation and harmonic generation is also illustrated, while the contribution of solid state lasers to the production of X-ray radiation is discussed.

Techniques for coherent improvement and frequency control are discussed in particular in conjunction with semiconductor diode lasers.

Selected examples in atomic and molecular spectroscopy, including frequency metrology and environmental analysis, can also be found.

There were 16 invited lecturers, 6 seminars speakers and 59 other participants at the Institute. They came from three main different research areas: solid state physics, optics and quantum electronics, atomic and molecular physics.

A novelty was constituted by the official participation of researchers from Central and East Europe countries, according to the new rules stated by NATO. Also in this case the participation was very active and a significant exchange of information was possible.

The success of the NATO-ASI was also helped by the charming atmosphere of the small village in the Elba Island. We could also take advantage of the logistic support of the Elba International Physics Center and we would like to thank Mrs. Antonella Sapere secretary of the Center.

We wish to express our appreciation to the NATO Scientific Affairs Division, whose financial support made the Institute possible. We also acknowledge the contribution of the Consiglio Nazionale delle Ricerche (CNR).

The National Science Foundation contributed with two travel grants for USA participants. Additional funding has been provided by NATO national officer for the participation of students from Turkey and Portugal.

We wish to thank Mrs. Maria B. Petrone of INO, who, as secretary of the Europhysics School of Quantum Electronics, significantly helped in the organization of this course.

Together with the participants, we are grateful to Anna Chiara Arecchi for her competent and enthusiastic assistance during the course. We remember with great pleasure the Quartet (Renzo Pelli - Flute, Warwick Lister - Violin, Anne Lokken - Viola and Ursula Koenig - Violoncello) whose performance of music by Mozart and Haydn was particularly impressive.

M. Inguscio

R. Wallenstein

CONTENTS

RESONATOR DESIGNS FOR HIGH BRIGHTNESS SOLID-STATE LASERS

G. Cerullo, S. De Silvestri, V. Magni, O. Svelto

Centro di Elettronica Quantistica e Strumentazione Elettronica del CNR
Dipartimento di Fisica del Politecnico
Piazza L. da Vinci 32 - 20133 Milano, Italy

1. INTRODUCTION

The generation of diffraction limited beams of high power or energy represents a research field of great interest for many scientific and industrial applications. Laser materials of large volume are available and design procedures for optical resonators with large cross-sectional area fundamental modes, that can exploit the laser medium, are a demanding problem. Stable resonators operating on the fundamental TEM_{00} mode produce high quality diffraction limited beams; however, the transverse mode dimension and, as a consequence, the extracted energy from the gain medium are generally small. On the other hand, unstable resonators can sustain fundamental modes of wide cross-section, but the output beam quality and divergence can be degraded by the spatial profile of the output beam.

In this work we consider two resonator designs for the generation of diffraction limited beams of high power, which have been applied to Neodymium lasers. The first concerns dynamically stable resonators of low misalignment sensitivity for continuous wave (c.w.) TEM_{00} lasers with thermal lensing in the active medium. Two active media are considered, namely Nd:YAG and Nd:YLF. The second resonator design concerns unstable resonators with variable reflectivity output mirrors of super-gaussian reflectivity profile, which have been applied to pulsed Nd:YAG lasers of low and high average power.

2. STABLE RESONATORS FOR C.W. TEM_{00} MODE OPERATION

Multimode laser beams with an average power up to the kilowatt can presently be generated with Nd:YAG lasers. On the contrary, the output power is reduced to few tens of watts when a stable resonator operating in the TEM_{00} mode is used, mainly because of the small overlapping volume of the TEM_{00} mode with the active material. To increase the output power with stable resonators, a large volume TEM_{00} mode is required. However, if appropriate design criteria are not applied, the resonators modes and output power become dramatically sensitive to small perturbations and to the mirror misalignment. The problem is further complicated by the lens effect produced in the laser

Solid State Lasers: New Developments and Applications
Edited by M. Inguscio and R. Wallenstein, Plenum Press, New York, 1993

rod by pumping. In fact for Nd:YAG the rod dioptric power can reach considerable values even at moderate lamp input power levels (e.g. about 4-5 m^{-1}) and plays a fundamental role in determining the laser performances. In the section 2.1 we present simple guidelines for the design of resonators with an internal variable lens, that represents the pumped rod, in order to obtain large fundamental mode volume and minimum misalignment sensitivity. We also point out the main differences for Nd:YAG and Nd:YLF resonator design. In section 2.2 we present the experimental results obtained with those two active materials.

2.1. RESONATOR DESIGN FOR Nd:YAG AND Nd:YLF LASERS

The dioptric power 1/f of the thermal lens induced in the laser rod is proportional to the input pump power[1,2]:

$$1/f = (k/\pi a^2)P_{in} \tag{1}$$

where a is the rod radius and k is a constant depending on the opto-mechanical properties of the laser rod and on the pumping efficiency. A resonator containing a variable lens presents, as function of the dioptric power 1/f of the lens, two ranges of optical stability, called stability zones, of the same width $\Delta(1/f)$. The spot size, w_3, of the TEM_{00} mode (half width at $1/e^2$ of the maximum of the gaussian intensity profile) on the rod reaches, inside each stability zone, a minimum where the resonator is said to be "dynamically stable", since it is insensitive, to the first order, to fluctuations of the variable lens. The value w_{30} of the spot size at this point is related to the width of the stability zone by the relationship[3,4]:

$$w_{30}^2 = (2\lambda/\pi) / \Delta(1/f) \tag{2}$$

The spot size w_{30} can be assumed to be representative of the spot size inside the whole stability zone since the spot size varies slowly with the rod dioptric power, except near the stability limit, where it diverges. Equation (2) shows that the fundamental mode volume inside the laser rod ($\propto w_{30}^2$) is inversely proportional to the dioptric power range for which the resonator remains stable. The resonator sensitivity against mirror misalignment[5] in one of the stability zones, called zone II, is much higher than in the other and diverges near one of the stability limits; it is therefore crucial to design the resonator so that it works in the other zone, called zone I. The stability zone of low misalignment sensitivity is characterized by the fact that, at one of its limits, the intracavity lens images the end mirrors on each other.

Combining Eqs. (1) and (2) one obtains the range of input power ΔP_{in} for which the resonator is stable[6]:

$$\Delta P_{in} = (2\lambda/k)(a/w_{30})^2 \tag{3}$$

The quantity $(a/w_{30})^2$ is the ratio between the cross sections of the rod and of the TEM_{00} mode. For an optimized resonator the rod should be the limiting aperture. Since for good TEM_{00} operation the ratio a/w_{30} must be in the range between 1.2 and 2, a/w_{30} is approximately a constant and independent of the rod size[7,8]. Therefore, from Eq. (3) the input power range for which the resonator is stable depends only on the parameter k, that is related to the properties of the laser medium and to the pumping efficiency. Since the pump cavity is usually well optimized it follows that ΔP_{in} can be assumed as a figure of merit of the solid state material. Assuming a/w_{30} = 1.5 to 1.8 for Nd:YAG one obtains using the data reported in the literature ΔP_{in} = 300-500 W. The situation is more favourable for Nd:YLF where ΔP_{in} = 1.9-3.1 kW for light polarized parallel to the c axis of the crystal and even greater than 6.5 kW for light polarized perpendicular to the c axis. For a given value of the spot size w_{30}, a factor of two in the pump power stability range can be gained by designing the resonator with the stability zones joined[3,4].

The small ΔP_{in} in Nd:YAG has a first practical drawback on the fact that it is a small fraction of the input power (usually 6-10 kW) and therefore requires a careful setting of the input power in order to obtain and maintain laser action. A more serious drawback comes from the stress induced thermal birefringence, which gives rise to two different values of focal length for polarization of the beam along the radial direction, f_r, and along the tangential one, f_t. The ratio between these two values[1,2] is $f_t/f_r \approx 1.2$, while the mean dioptric power, $1/f' = (1/f_r + 1/f_t)/2$, is with a good approximation proportional to the pump power entering the lamp[2]. The thermal birefringence can split the pump power stability regions for radial and tangential polarizations, for large fundamental mode volumes or high input powers, to such an extent that they do not overlap. For example, if a resonator is designed to have spot size w_{30} = 1.8 mm, the width of the stability zone is 0.2 diopters. On the other hand, assuming in Eq. (1) $k/\pi a^2 = 0.6\ m^{-1}\cdot kW^{-1}$ as appropriate for a 6.4 mm diameter rod[2], when the thermal lens dioptric power for tangential polarization is 3 diopters (i.e., at about 5 kW pump power) that for radial polarization is 3.6 diopters, therefore the two stability regions for different polarizations do not overlap and the laser cannot work. Therefore, for laser operation the width of the stability zone must satisfy the following condition:

$$\Delta\frac{1}{f} = \frac{2\lambda}{\pi w_{30}^2} > \frac{1}{f_r} - \frac{1}{f_t} \approx \frac{0.18}{f'} \approx 0.18\frac{k}{\pi a^2}P_{in} \tag{4}$$

Assuming again $k/\pi a^2 = 0.6\ m^{-1}\cdot kW^{-1}$, Eq. (4) yields:

$$w_{30}^2\ P_{in} < 6.3\ mm^2\cdot kW \tag{5}$$

This result means that the maximum achievable fundamental mode volume in a Nd:YAG rod ($\propto w_{30}^2$) at high power is limited. As an example, at an input power of 10 kW the spot size must be less than 0.8 mm. For Nd:YLF crystal the situation is completely different since the thermal induced radial birefringence is negligible. In addition the gain depends on the polarization direction of the light with respect to the c axis of the crystal, therefore the electromagnetic radiation inside the cavity is naturally polarized. The Nd:YLF being anisotropic, different values of dioptric power are measured for the two different beam polarizations corresponding to the 1047 transition (electric field parallel to the c axis) and 1053 transition (electric field orthogonal to the c axis). Due to astigmatism two values of dioptric power are measured for each polarization in two orthogonal planes. For each polarization the astigmatism of the thermal lens can be enough to produce a complete separation of the two stability zones for the light in the planes parallel and perpendicular to the optical axis[9], but this effect can be compensated for by using cylindrical optics.

In the following we present simple design criteria for Nd:YAG and Nd:YLF lasers. The general resonator configuration for Nd:YAG laser is shown in Fig. 1(a). The optimized resonators for fundamental mode operation can be easily designed according to the following procedure[4,8]. One has first to set the value of the dynamically stable mode spot size w_{30} in the rod, which usually should correspond to the maximum value consistent with large mode volume for high extraction efficiency and low losses caused by the rod aperture. This value automatically sets the width of the stability zone according to Eq. (2). Then, one requires that the laser works in the low misalignment sensitivity zone. The spot size w_{30} is completely determined by the distance L_2 and the radius of curvature R_2. The distance L_1 is determined assuming that at the upper stability limit the laser rod images a mirror into the other. The radius of curvature of the second mirror R_1 does not influence the spot size w_{30} and the position of the stability zone. The radius R_1 can be chosen to minimize the resonator misalignment sensitivity. In the case of Nd:YLF lasers the general resonator configuration is shown in Fig. 1(b). The small values of the thermal focal length for Nd:YLF require the presence of a spherical lens f_S to allows operation in the region with low misalignment sensitivity (zone I). The presence of astigmatism can be easily compensated by inserting in the resonator a cylindrical lens f_C. The resonator design criteria are similar to those for Nd:YAG lasers. The distance L_2 and the radius of curvature R_2 determine the spot size w_{30} in the rod. The relative position of spherical and cylindrical lenses allow the resonator to operate in zone I and also guarantees a superposition between the two stability zones in the two orthogonal planes. Note that in the case of Nd:YAG the differences in focal length for different polarization cannot be compensated for by simple optical systems.

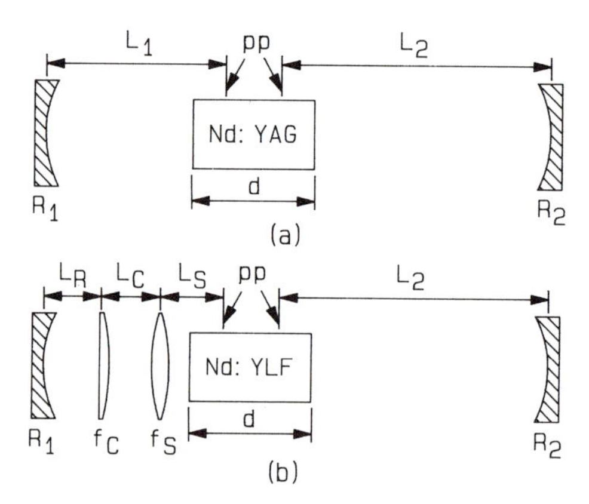

Figure 1. (a) Resonator configuration for Nd:YAG laser: mirrors of radii of curvature R_1 and R_2 are placed at a distance L_1 and L_2 from the rod principal planes (pp) located at a distance d/2n from the rod faces (d is the rod length and n is its refractive index). (b) Resonator configuration for Nd:YLF laser: f_S and f_C are respectively a spherical and cylindrical lens, L_S and L_2 are the distances respectively of the spherical lens and the mirror with R_2 radius of curvature from the rod principal planes.

2.5. EXPERIMENTAL RESULTS

The experimental work was done using Nd:YAG and Nd:YLF rods with flat faces 104 mm in length by 6.4 mm in diameter. The crystals were pumped by a 6 mm $\times$ 100 mm Krypton arc lamp in an elliptical gold plated pot. The maximum pump power entering the lamp was about 7.5 kW. For the purpose of comparison the Nd:YAG and Nd:YLF rods were tested in a multimode resonator: the maximum output power at 7 kW input power were 120 W and 80 W for Nd:YAG and Nd:YLF respectively. The beam quality was characterised through the M^2 factor[10].

For Nd:YAG, many different resonators using various combinations of radii of curvature of the mirrors were designed and tested[11]. Laser action could not be achieved until the design spot size in the rod w_{30} was reduced to be less than about 1 mm, (according to the rod radius a spot size of 2 mm should give the optimum filling factor). This result is in substantial agreement with the theoretical discussion presented in the previous section: for spot size larger than about 1 mm the splitting of the stability zones for radial and tangential polarization prevents the laser from working. Since in general the output power increases by increasing the input power and the mode volume, but, on the other hand, the optical stability range decreases and the birefringence effects become stronger an optimum tradeoff should exist between pump power level and TEM_{00} spot size. To find the optimum we tested several resonators designed for spot sizes in the rod ranging from 0.7 and 1 mm. Two groups of resonators were tested: the first one with distinct stability zones, the second with joined stability zones. All resonators were operated without any internal aperture except that of the rod. The output mirror reflectivity was 80% for all resonators. The behavior of the output power as a function of the input power is similar for the resonators of the two groups. The laser exhibits two thresholds, both determined by the transition between the regimes of optical stability and instability of the resonator. For resonators with distinct stability zones the range of input power within which the laser works varies between about 1 kW (spot size w_{30} = 1 mm) and 2 kW (spot size w_{30} = 0.7 mm), in agreement with the theoretical predictions. Accordingly, for resonators with joined stability zones the pump power stability range was measured to be nearly double. For resonator with distinct stability zones M^2 decreases as the spot size is increased and reaches values close to the diffraction limit (M^2 = 1) for spot sizes of 1 mm, where the output power of about 20 W has been measured. For spot sizes of 0.7 and 0.8 mm it also decreases as the upper stability limit is moved towards high values. In the case of joined stability zones the values of M^2 decrease as the spot size is increased up to 0.95 mm, where it reaches values close to one and the output power approaches 40 W.

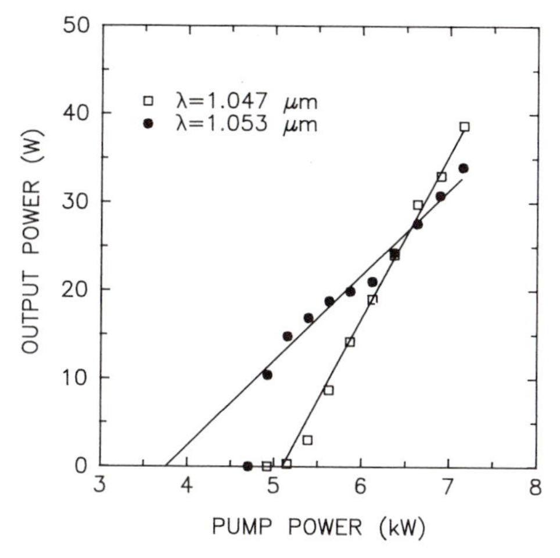

Figure 2. Input-output curves for an optimized TEM_{00} resonator for Nd:YLF laser. The resonator parameters [see Fig.1(b)] are: for λ = 1047 nm f_C = 1018 mm, f_S = 220 mm, L_R = 173 mm, L_C = 107 mm, L_S = 180 mm, L_2 = 888 mm and for λ = 1053 nm f_C removed, f_S = 183 mm (tilted at an angle of 11°), $L_R + L_C$ = 198 mm, L_S = 212 mm, L_2 = 784 mm; the radii of curvature R_1 = 500 mm, R_2 = -200 mm of the two end mirrors are the same for both wavelength.

For Nd:YLF a set of optimized resonators has been designed[9] for both 1053 nm and 1047 nm corresponding to light with the electric field polarized perpendicular and parallel respectively to the c crystallographic axis, which is orthogonal to the rod axis. On account of the difference in the thermal lens for the polarization directions corresponding to the 1053 nm line and to the 1047 line, the optimized resonators are optically stable only for a specific wavelength. Therefore no tuning nor polarizing elements are required to select the operating wavelength and polarization. The first set of experiments was carried out at 1047 nm. Various resonators using different combinations of cylindrical and spherical lenses were designed according to the measured values of Nd:YLF dioptric power to achieve astigmatism compensation at the highest lamp power. For values of spot size in the rod greater than 1.65 mm, a beam with a clean gaussian profile was measured. Figure 2 shows a typical input-output curve for w_{30} = 1.8 mm, which was found to be the optimum value. With the resonator described in Fig. 1(b) an output power of 40 W was obtained with 7 kW pump power, with a reflectivity of the output mirror of 80%, which was the optimum value. The output power long term stability was measured to be better than 2%. Note that the threshold is determined by the optical stability of the resonator, which depends on the pump power level. At an input power of 7kW, corresponding to 40 W output power, the thermal astigmatism of the rod was adequately compensated and the output beam had a circular cross-section. The shape of the output beam profile and the M^2 measurement, which yielded at the maximum output power the value 1.12, confirmed the beam to be essentially diffraction limited. A second set of experiments was carried out at 1053 nm. Using an output coupler reflectivity of 90% a maximum TEM_{00} output power of 35 W was obtained.

3. UNSTABLE RESONATORS WITH VARIABLE-REFLECTIVITY OUTPUT COUPLERS

A usual technique to generate diffraction limited laser beams of high power relies on the use of unstable resonators. This solution is well suited especially for lasers having active medium of large cross-section and high gain, since the electromagnetic field provides a wide filling of the active material, giving an efficient coupling between the optical mode and the gain medium. Besides this advantage, however, there are some drawbacks. In traditional unstable resonators the output beam is extracted around the edge of a totally reflecting mirror and presents a typical doughnut shape with highly pronounced fringes produced by diffraction wavelets originating at the mirror edge. Furthermore, owing to the mode crossing phenomenon[12], the transverse mode discrimination may be critical and the resulting beam quality is significantly reduced compared to that of stable resonators operating in the TEM_{00} mode. The low beam quality also affects the far field profile, where a remarkable fraction of the energy is spread in side lobes around the central peak, worsening the divergence. An unstable resonator with a transmitting coupler with a radially variable reflectivity profile can overcome these shortcomings. By using these mirrors, the hole in the near-field disappears, leaving a dip whose depth and width depend on the resonator configurations, the diffraction fringes are greatly reduced, due to the absence of the mirror sharp shape, and the losses degeneracy is removed, giving greater transverse mode discrimination. As a consequence, the near-field is more uniform, with a significant reduction of high frequency components, and the mode purity is increased; these effects are also reflected in the far-field, where the energy fraction present in side lobes around the central peak is greatly reduced.

The availability of variable reflectivity mirrors whose parameters can be tuned independently is extremely important in the practical realization of a number of different design specifications. The super-gaussian mirrors which have been recently introduced for Nd:YAG lasers[13,14], besides allowing large mode volume, can also satisfy this requirements. The reflectivity profile R(r) of a super-gaussian mirror is expressed by:

$$R(r) = R_0 \exp[-2(r/w_m)^n] \qquad (6)$$

where r is the radial coordinate, R_0 is the peak reflectivity, w_m the mirror spot size and n the super-gaussian order. A thin film vacuum deposition technique has been developed[15] which allows the fabrication of multidielectric mirrors whose reflectivity profile is under control. This method is based on the shadowing effect of a fixed mask with circular aperture, placed between the crucible and the substrate. The mirrors consists of a glass substrate covered with a double layer anti-reflection coating over which a third layer of high refractive index material has been deposited. The thickness of the upper layer is radially decreasing from about $\lambda/4$ to zero, following a bell-shaped profile that provides for the super-gaussian reflectivity.

In section 3.1 we present general design criteria for unstable resonators with super-gaussian reflectivity mirrors and present some application to low frequency pulsed Nd:YAG lasers. In section 3.2 we present the design criteria of unstable resonator with super-gaussian mirror for high average power Nd:YAG laser, where the rod thermal lensing cannot be neglected.

3.1. GENERAL DESIGN CRITERIA

For an unstable resonator with super-gaussian mirrors the eigenvalue diffraction equation cannot be solved in closed form. However, under condition for validity of geometrical optics the optical modes and losses can be easily calculated from a self consistency equation[13]. For a super-gaussian resonator the fundamental mode is a super-gaussian function of the same order as that of the reflectivity profile. The field amplitude $u_0(r)$ incident on the super-gaussian mirror is given by:

$$u_0(r) = A \exp[-(r/w_i)^n] \qquad (7)$$

where A is an arbitrary constant and w_i is given by:

$$w_i = w_m (M^n - 1)^{1/n} \qquad (8)$$

where M is the geometrical magnification after one round trip. The round trip power losses L are given by:

$$L = 1 - R_0/M^2 \qquad (9)$$

independently of the super-gaussian order n. A comparison of the results obtained with geometrical optics and those by numerical solution of the Fresnel diffraction integral in cylindrical coordinates has been performed for several configurations. The agreement between geometrical and diffractive optics is satisfactory provided that approximately $n < 15$, $M > 1.5$ and the geometrical mode intensity at the aperture edge is less than 5% of the peak. The output beam profile transmitted through a super-gaussian mirror is no longer super-gaussian and presents a central dip if $R_0M^n > 1$, as often occurs in cases of practical interest. Therefore the near field presents a hollow whose depth and width increases with the order n. In the far field the hollow results in side lobes which are fairly negligible for $n < 10$.

Simple guidelines can be applied to the design of super-gaussian resonators[16]. The mode spot size should be optimized for good balancing between efficient filling of the gain medium and low beam perturbations caused by diffraction at the aperture of the active material. Regarding the choice of the geometrical magnification a trade-off has to be achieved among the mode intensity discrimination ratio ($1/M^2$), the output losses and the angular misalignment sensitivity proportional to $M/(M-1)^2$. The choice of the peak reflectivity R_0 is not only related to the output losses, but also determines the shape of the output beam; the near-field profile is maximally flat if the additional constraint $R_0M^n = 1$ is satisfied.

Unstable resonators with super-gaussian mirrors have been applied to a pulsed Nd:YAG laser with a rod 75 mm long with diameter 6.3 mm, operating at a repetition rate of 1 Hz. The schematic of the unstable resonator with a magnification M = 1.8 is shown in Fig. 3. Figure 4 shows the experimental output energies as a function of the input energy with super-gaussian mirrors of orders n = 2.8, 5, 9, and 35. For any value of n the output beam profile closely agreed with that calculated by Eqs. (7) and (8), the beam divergence was close to the diffraction limit, and the far field profile agreed very well with the calculated Fourier transform of the theoretical near field. The same laser equipped with a stable multimode resonator made up by two concave mirrors of 8 radius of curvature provides an energy of 570 mJ about 1.3-1.5 times greater than that of super-gaussian resonators with an output beam divergence 10 times the diffraction limit.

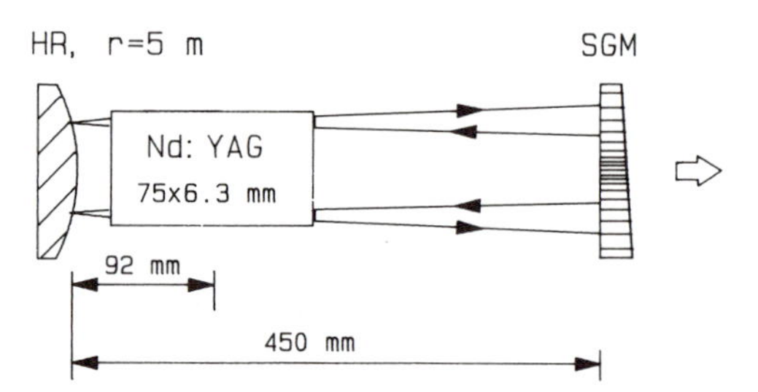

Figure 3. Schematic of the unstable resonator configuration with super-gaussian (SG) reflectivity mirror for low gain pulsed Nd:YAG laser.

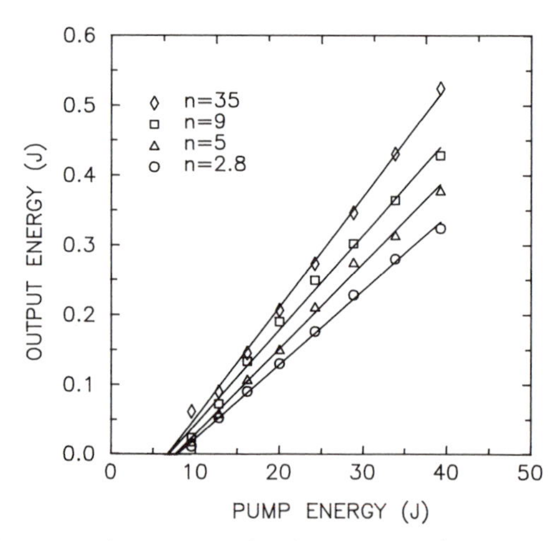

Figure 4. Output energies versus input energy for the super-gaussian resonator of Fig. 3 for different mirror orders n =2.8, 5, 9, 35; the mirror spot sizes were 1.82, 1.92, 2.03, 2.25 mm respectively.

3.2. UNSTABLE RESONATOR FOR HIGH AVERAGE POWER Nd:YAG LASER

In recent years Nd:YAG lasers with output power exceeding one kilowatt have been developed and are now available[17,18]. In these lasers stable multimode resonators are typically used and the output beam is characterized by a divergence 30-100 times greater than the minimum limit imposed by the diffraction. On account of the strong thermal lensing of Nd:YAG rods and of its dependence upon pumping power, however, the application of unstable resonators to high average power Nd:YAG lasers is not trivial. A novel unstable resonator configuration, called rod-imaging using a super-gaussian graded reflectivity mirror as output coupler has been recently developed. This resonator allows to tame rod thermal lensing so as to obtain high output power, nearly diffraction limited beams and low misalignment sensitivity.

In high power Nd:YAG lasers the rod thermal lensing strongly influences all the characteristics of the resonator. The requirements to be satisfied for a practical application of unstable resonators to Nd:YAG lasers can be summarized as follows: (i) low round-trip magnification (M) to avoid over-coupling; (ii) limited variation of magnification, and hence of the losses, with pump power; (iii) mode matching to the rod cross section, so as to ensure efficient power extraction without excessive internal losses and mode distortion due to diffraction at the rod aperture; (iv) limited variation of the mode size in the rod with the pump power; (v) control of the position of possible mode focal points inside the resonator to avoid damaging of optical components; (vi) low misalignment sensitivity.

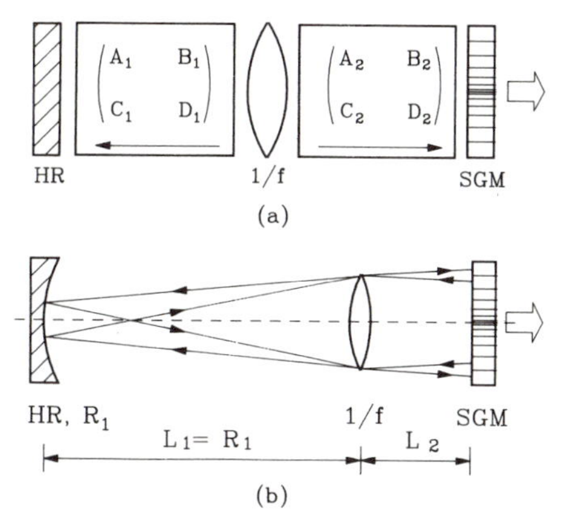

Figure 5. (a) Resonator containing a lens of dioptric power 1/f, representing the pumped rod, and arbitrary optical systems described by the ray matrices. HR is a totally reflecting mirror and SGM a super-gaussian output mirror. The arrows indicate the direction to which the matrices refers. (b) Rod-imaging unstable resonator with indicated the geometrical mode contour.

We consider a general unstable resonator, represented with two plane mirrors that enclose a lens of variable dioptric power, 1/f, sandwiched between two optical system described by the corresponding ray matrices, as shown in Fig. 5(a). If the end mirrors were curved, they should be intended to be resolved in a plane mirror and a lens (included in the matrix) of focal length equal to the mirror radius of curvature. We assume that the output is taken from a graded reflectivity mirror of super-gaussian profile (SGM), whereas the second mirror is uniform and totally reflecting (HR). The geometrical optics provides an adequate description of the resonator modes. If we denote by

$$\begin{vmatrix} A & B \\ C & D \end{vmatrix} \tag{10}$$

the one way matrix describing the ray propagation from mirror HR to mirror SGM, the boundaries between the stable and unstable region are found for the values of 1/f for which one of the matrix element vanishes. For any resonator four stability limits defining two stability zones are thus found. In correspondence to the stability limits the magnification is $M = \pm 1$ and the mode of the resonator in the planes of the end mirrors is constituted either by spherical waves focused on the mirror surfaces or by plane waves. In the real resonator with curved end mirrors these plane waves correspond to spherical waves converging into the mirror center of curvature. According to the previous considerations, different configurations of unstable resonators can be chosen. Let's now examine the mode behaviour in correspondence to each of these limits. When $A = 0$ the mode is focused on the super-gaussian output mirror (SGM) and the spot size on the lens diverges. This solution has to be discarded since the mode is not determined by the super-gaussian mirror, but by the rod aperture. When $B = 0$ the mode is focused on both mirrors and for the same reason also this choice must be abandoned. When $C = 0$ the mode on the mirrors is constituted by plane waves; the drawback of this configuration is that the misalignment sensitivity of the resonator diverges. For $D = 0$ the mode is focused on mirror HR and is a plane wave on mirror SGM: therefore the mode spot size remains finite on the super-gaussian mirror and on the rod. Close to $D = 0$ the magnification is negative ($M < -1$); this corresponds to a low misalignment sensitivity (negative branch unstable resonator). In conclusion, a configuration near the stability limit $D = 0$ appears to be a promising solution. Two other requirements are that mode spot size in the rod and magnification vary within acceptable values for a given range of rod dioptric powers. Minimum variation in the spot size is reached when the two waves propagating in opposite directions inside the resonator have the same dimension in the variable lens. This is obtained when the matrix element D_1 [see Fig. 5(a)] vanishes so that the left arm of the resonator makes an image of the rod onto itself. This resonator configuration has been named "*rod-imaging*".

We have implemented[19] a rod-imaging resonator for a laser head housing, in a ceramic cavity, a 152.4×6.35 mm Nd:YAG rod pumped by two flashlamps. The schematic of the resonator is shown in Fig. 5(b): it is made by a concave totally reflecting mirror with 170 mm radius of curvature placed at 170 mm from the rod

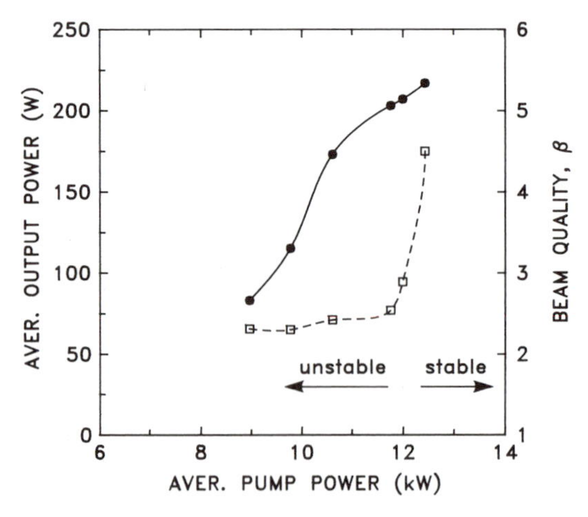

Figure 6. Average output power (solid line, left scale) and beam quality factor β (dashed line, right scale) as a function of the average input power obtained with a Nd:YAG laser using a rod imaging resonator.

principal plane to obtain the rod-imaging condition and by a plane super-gaussian mirror (n = 5) with peak reflectivity R_0 = 0.5 and spot size w_m = 1.92 mm (L_2 = 170 mm). The stability limit D = 0 is reached when the rod focal length is exactly equal to the radius of the rear mirror. The resonator magnification starts from M = -5.83 when the rod is not pumped and reaches M = -1 at the stability limit. The laser was pumped at a repetition rate of 15 Hz by rectangular pulses of 230 kW peak power. Figure 6 shows the behaviour of the average output power as a function of the average input power as obtained by varying the pulse duration. In correspondence to the transition into the stable regime, which occurs at 12 kW input power, no particular feature is noticeable on this curve. The beam quality is evaluated by the parameter $\beta = D\theta_d / \lambda$, where D is the diameter (full width at $1/e^2$) of the near field profile and θ_d is half the angle (half width at $1/e^2$) of the far field profile. The behaviour of this figure of merit as a function of the pump power is also shown in Fig. 6. It can be seen that the beam quality parameter, β, remains constant up to 12 kW input power, i.e. until the resonator is unstable, and almost corresponds to the diffraction limit. As the resonator enters the stable region the laser operates in multimode regime and the divergence rapidly increases. The fact that the beam divergence remains low and that the output power monotonically increases with the pump power are both important features for the applications as well as the observed stability of the output power with time. These features are attributed to the special resonator design. To fully appreciate the significance of the previous results, they should be compared with those obtained with a traditional stable multimode resonator typically used for this class of lasers. At 10.5 kW input pump power the stable resonator (made by two flat mirrors placed symmetrically at 200 mm from the rod principal planes) generates a beam of 280 W (average) with β = 39, whereas the rod-imaging super-gaussian resonator generates a beam of 186 W with β = 2.2: the brightness of the beam therefore increases from 6.6 $MW \cdot cm^2 \cdot sr^{-1}$ to 1400 $MW \cdot cm^2 \cdot sr^{-1}$.

4. CONCLUSIONS

We have presented a review of our recent theoretical and experimental results in the field of optical resonators for diffraction limited lasers of high power. In particular, two resonator configurations have been considered, namely: stable resonator for c.w. TEM_{00} operation with thermal lensing effect in the active material and unstable resonators with radially variable-reflectivity mirrors of super-gaussian profile. The theoretical results for c.w. TEM_{00} lasers have led to the design of large mode volume, dynamically stable resonators optimized for minimum misalignment sensitivity. The application of the design procedures to c.w. Nd:YAG and Nd:YLF lasers has led to the generation of diffraction limited beams of 40 W in both cases. These are presently the best results obtained for c.w. operation using a single rod. For unstable resonators we have demonstrated that the use of super-gaussian mirrors allows to calculate with straightforward procedures the fundamental transverse mode and to exploit efficiently the active medium. The combination of a super-gaussian mirror with a new unstable resonator configuration (rod-imaging) has led to the generation of an almost diffraction limited beam of 200 W average power from a Nd:YAG laser.

Acknowledgments

The present paper has been partially supported by the National Research Council (C.N.R.) of Italy under the "Progetto Finalizzato" on Electro-optical Technologies.

REFERENCES

1. J.D.Foster and L.M.Osterink, Thermal effects in a Nd:YAG laser, *J. Appl. Phys* 41:3656(1970).
2. W.Koechner. "Solid State Laser Engineering," (III Edition) Springer-Verlag, Berlin (1992) chap. 7.
3. V.Magni, Multielement stable resonators containing a variable lens, *J. Opt. Soc. A* 4:1962(1987).
4. V.Magni, Resonators for solid-state lasers with large-volume fundamental mode and high alignment stability, *Appl. Opt.* 25:107(1986).
5. S.De Silvestri, P.Laporta, and V.Magni, Misalignment sensitivity of solid-state laser resonators with thermal lensing, *Opt. Commun.* 59:43(1886)
6. S.De Silvestri, P.Laporta, and V.Magni, Pump power stability range of single mode solid-state lasers, *IEEE J. Quantum Electron.* QE-23:1999(1987).
7. J.Dembowsky and H.Weber, Optimal pin hole radius for fundamental mode operation, *Opt. Commun.* 42:133(1982).
8. S.De Silvestri, P.Laporta, and V.Magni, 14-W continuous wave mode-locked Nd:YAG laser, *Opt. Lett.* 11:785(1986).
9. G.Cerullo, S.De Silvestri, and V.Magni, High efficiency, 40 W cw Nd:YLF laser with large TEM_{00} mode, *Opt. Commun.* 93:77(1992).
10. M.W.Sasnet, *in:* "The Physics and Technology of Laser Resonators," Adam Hilger, Bristol (1989), p. 132-134.
11. G.Cerullo, S.De Silvestri, V.Magni, and O.Svelto, Output power limitations in CW single transverse mode Nd:YAG lasers with a rod of large cross-section, *Opt. Quantum Electron.* (1993) in press.
12. A.E.Siegman. "Lasers," Oxford University Press, Oxford (1986), p. 874.
13. S.De Silvestri, P.Laporta, V.Magni, and O.Svelto, Solid-state laser unstable resonators with tapered reflectivity mirrors: the supergaussian approach, *IEEE J. Quantum Electron.* QE-24:1172(1988).
14. S.De Silvestri, P.Laporta, V.Magni, O.Svelto, and B.Majocchi, Unstable resonators with super-gaussian mirrors, *Opt. Lett.* 13:201(1988).
15. G.Emiliani, A.Piegari, S.De Silvestri, P.Laporta, and V.Magni, Optical coatings with variable reflectance for laser mirrors, *Appl. Opt.* 28:2832(1989).
16. S.De Silvestri, V.Magni, O.Svelto, and G.Valentini, Lasers with super-gaussian mirrors, *IEEE J. Quantum Electron.* QE-26:1500(1990).
17. C.L.M.Ireland, Kilowatt power levels extend industrial-YAG processing capability, *Laser Focus* 24:49(1988).
18. T.Yamada, S.Nishimura, S.Yoshida, Y.Fujimori, and K.Ishikawa, Multikilowatt continuously pumped Nd:YAG laser, *in:* "Conference on Laser and Electro-optics Technical Digest Series," Opt. Soc. Am., Washington (1988).
19. V.Magni, S.De Silvestri, and O.Svelto, Rod-imaging super-gaussian unstable resonator for high power solid-state lasers, *Opt. Commun.* 94:87(1992).

DEFINING AND MEASURING LASER BEAM QUALITY

A. E. Siegman

Edward L. Ginzton Laboratory
Stanford University
Stanford, California 94305–4085

ABSTRACT

This lecture will review a very useful approach that has recently been developed for defining the transverse beam quality as well as other propagation parameters of arbitrary real laser beams. We will also describe a convenient new instrument for measuring the propagation parameters of high-power laser beams, and summarize the results of beam quality measurements made on various types of laser devices.

INTRODUCTION

One of the most important properties of a laser oscillator is the highly collimated or spatially coherent nature of the laser's output beam. The degree to which a real laser beam approaches perfection in this regard is often referred to as the "beam quality" of the output beam, sometimes phrased as the "times diffraction limited" or TDL value of the beam. The spatial beam quality of the output beam, since it determines how tightly the laser beam can be focused, or how little it will spread in propagating to the far field, is a critical parameter in the practical performance of a laser in a wide range of practical applications, as well as a critical design parameter in designing optical beam trains and focusing systems for such a laser. Despite the great practical importance of understanding how a given laser beam will diverge in the far field or focus in the near field, however, reliable information on laser beam parameters, derived from careful measurements on real laser devices, has often been lacking in the past. Indeed there has been little agreement on measurement methods for characterizing transverse laser beam properties, and even less standardization on how the basic concept of "laser beam quality" should be rigorously defined, much less measured.

In this lecture we summarize a useful approach that has been developed in recent years for defining the transverse beam quality and other propagation parameters of an arbitrary real laser beam, including beams with significant astigmatism or asymmetry. We will also describe a convenient new instrument, one of several that have recently been developed for measuring the beam quality and other optical propagation parameters of high-power laser beams. Finally we will show a number of experimental examples illustrating how such beam quality measurements can give significant insight into the oscillation physics of many different types of lasers, as well as predicting their performance in practical laser applications.

Solid State Lasers: New Developments and Applications
Edited by M. Inguscio and R. Wallenstein, Plenum Press, New York, 1993

ANALYTICAL FORMULATION

It has of course long been known that laser oscillators using stable resonators will typically oscillate in lowest and perhaps higher-order Hermite-gaussian (or possibly Laguerre-gaussian) transverse modes; and it is also well recognized that the lowest-order or TEM_{00} mode from such a stable cavity provides a more or less ideal beam profile both for analytical and for many practical purposes. Because of this, the observation of a gaussian intensity profile in the output beam from an experimental laser is often taken as a sign that the laser is indeed producing a high-quality beam. It must be realized, however, that this conclusion is by no means always valid. Figure 1 of this lecture shows for example a "no-gaussian gaussian" beam profile—that is, a highly gaussian appearing beam intensity mode which in fact contains no TEM_{00}

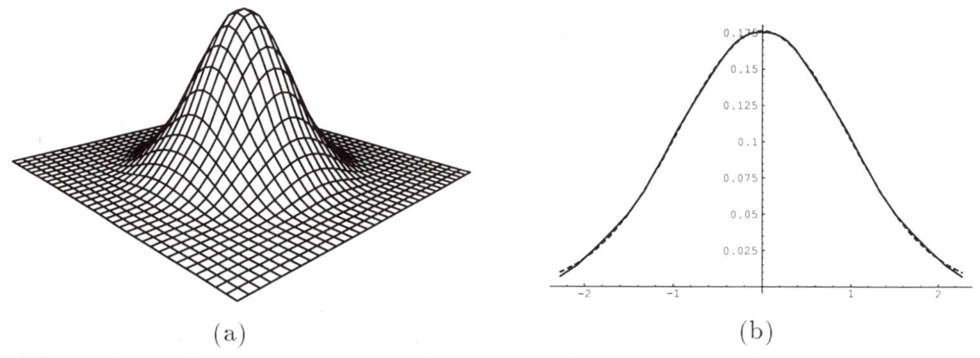

Figure 1. (a) Portrait of a "no-gaussian gaussian" beam profile. This computer-generated beam profile is highly gaussian in character, yet contains no TEM_{00} mode content at all. (b) Cross-section of the same beam profile.

mode component at all. Instead the profile shown is made up entirely of an incoherent superposition of five higher-order Laguerre-gaussian modes, including 45% of the LG01* mode, plus 17% of the LG10, 20% LG11*, 11.5% LG20, and 6% LG21*. The M^2 value of this mixture (to be defined later on) is approximately 3.1, i.e., the beam as shown is approximately 3.1 times diffraction limited as compared to a true gaussian beam profile.

To develop a more general and useful way of characterizing the propagation characteristics and the beam quality of arbitrary laser beams, including beams from unstable-resonator lasers and other lasers which may not employ stable cavities at all, we can first of all recall that the gaussian spot size $w(z)$ which provides the transverse scale factor for any family of Hermite-gaussian or Laguerre-gaussian free-space modes expands with distance in free space in the form

$$w^2(z) = w_0^2 + \left(\frac{\lambda}{\pi w_0}\right)^2 (z - z_0)^2 = w_0^2 \left[1 + \left(\frac{z - z_0}{z_R}\right)^2\right] \tag{1}$$

where $z_R \equiv \pi w_0^2/\lambda$ is the Rayleigh range of the Hermite-gaussian modes. The free-space propagation of a gaussian beam is therefore fully characterized by its waist spot size w_0 and its waist location z_0, with the far-field angular spread being given by $w(z) = \lambda/\pi w_0$. The propagation of such a gaussian beam through any complex paraxial optical system can also be fully characterized using these same two parameters plus a simple ray-matrix or ABCD-matrix formalism.

To extend this gaussian-beam result into a more general formalism for describing arbitrary and possibly highly nongaussian real laser beams, we can also note that the spot size w as conventionally defined for a gaussian beam is related to the transverse variance σ^2 or standard deviation σ of a TEM_{00} beam by $w = 2\sigma$. Taking this as a clue we then propose that the "real-beam spot sizes" W_x and W_y for any arbitrary nongaussian beam at any plane z should be defined by $W_x(z) \equiv 2\sigma_x(z)$ and $W_y(z) \equiv 2\sigma_y(z)$ where $\sigma_x(z)$ and $\sigma_y(z)$ are the time-averaged standard deviations of x and y evaluated on the real beam intensity profile $I(x, y, z)$ at that plane z. Given this definition it is possible to prove in several different ways that the axial variations of these real-beam spot sizes for any arbitrary, nongaussian, either fully coherent or partially incoherent, single-mode or multimode, smooth or distorted real beam in free space will be given by

$$W_x^2(z) = W_{0x}^2 + M_x^4\left(\frac{\lambda}{\pi W_{0x}}\right)^2 (z - z_{0x})^2 = W_{0x}^2\left[1 + \left(\frac{z - z_{0x}}{z_{Rx}}\right)^2\right] \tag{2}$$

and

$$W_y^2(z) = W_{0y}^2 + M_y^4\left(\frac{\lambda}{\pi W_{0y}}\right)^4 = W_{0y}^2\left[1 + \left(\frac{z - z_{0y}}{z_{Ry}}\right)^2\right] . \tag{3}$$

In other words, to within the paraxial degree of approximation the free-space propagation of the transverse spot sizes $W_x(z)$ and $W_y(z)$ for any real laser beam will be determined by the waist spot sizes W_{0x} and W_{0y} and waist locations z_{0x} and z_{0y}, almost exactly like the gaussian parameters w_0 and z_0 for a gaussian beam.

The important addition to this formalism is, however, that the propagation of the real beam, in addition to being dependent on the inverse waist size, also depends on the so-called "beam quality factors" M_x^2 and M_y^2 in the appropriate transverse direction. In fact, the far-field beam widths in the two transverse directions for the real beam are given by

$$W_x(x) \simeq M_x^2\left(\frac{\lambda}{\pi W_{0x}}\right) \quad \text{and} \quad W_y(z) \approx M_y^2\left(\frac{\lambda}{\pi W_{0y}}\right) \tag{4}$$

in the large-z limits $z \gg z_{Rx}$ and z_{Ry}. The beam quality factor M^2 is thus, in a physical interpretation, a measure of the near-field times far-field "space-beamwidth product" of the arbitrary real beam normalized to the space-beamwidth product for an ideal TEM_0 gaussian. One can in fact show that for an optical beam, just as for a radio-frequency signal pulse or a quantum wave packet, the minimum uncertainty product based on second moments always occurs for a gaussian beam profile or gaussian wave function. The beam quality factor as we have defined it thus always has $M^2 \geq 1$, with $M^2 = 1$ only for the ideal TEM_{00} gaussian beam. Since the far-field beam spread is now given by $M^2 \times (\lambda/\pi W_0)$ in each transverse coordinate, the M^2

value can serve as a rigorous definition of the "times diffraction limited" (TDL) value for an arbitrary real beam compared to a TEM_{00} gaussian beam. As an alternative way of viewing this, we can note that the Rayleigh ranges z_{Rx} and z_{Ry} introduced in the final terms of Eqs. (2) and (3) are given by

$$z_{Rx} \equiv \frac{\pi W_{0x}^2}{M_x^2 \lambda} \quad \text{and} \quad z_{Ry} \equiv \frac{\pi W_{0y}^2}{M_y^2 \lambda} \tag{5}$$

A beam with a given waist size W_0 and a beam quality factor M^2 thus has a shorter Rayleigh range and therefore diverges more rapidly than would an ideal gaussian with the same waist size. (Some people find it helpful to interpret this as saying that the beam diverges due to diffraction as if its wavelength λ were increased to an effective value of $M^2\lambda$.)

Over and above the far-field beam spread aspects, Eqs. (2) and (3) also say that the axial propagation of the spot sizes $W_x(z)$ and $W_y(z)$ for any arbitrary laser beam is fully characterized at every plane z by the six parameters W_{0x}, z_{0x} and M_x^2 in the x transverse direction and W_{0y}, z_{0y} and M_y^2 in the y direction. One can further show that these quadratic propagation equations will in fact be valid for any choice of perpendicular x and y axes in the transverse plane. The parameter values W_{0x}, z_{0x}, M_x^2, and so on will, however, have the widest separation and will give the most significant description of the beam if the x and y coordinates are rotated to the principal axes of the beam in question, that is, the axes in which the cross moment $\overline{xy}$ over the beam intensity profile is zero. The most general real beam possessing only simple or ordinary astigmatism can then be characterized by its "waist asymmetry" ($W_{0x} \neq W_{0y}$), its "conventional astigmatism" ($z_{0x} \neq z_{0y}$), and its "divergence asymmetry" ($M_x^2/W_{0x} \neq M_y^2/W_{0y}$). A beam of this type, which is generally called a "simple astigmatic" or "orthogonal" beam, is thus fully characterized, at least to second order, by 7 parameters, namely the 6 beam parameters described above plus the angular orientation of the principal axes with respect to the laboratory x, y coordinates. The most general possible laser beam is the so-called "general astigmatic" or "nonorthogonal" beam, in which the phase fronts and the intensity profiles are both astigmatic but with different principal axes, so that both sets of principal axes can twist or rotate about the z axis as the beam propagates. Such a beam still rigorously obeys Eqs. (2) and (3) in any fixed laboratory frame, but requires 10 independent second moments to fully describe its propagation and rotation. Nonplanar ring resonators can in fact produce output beams which are of this general astigmatic or nonorthogonal or twisted character, and devices which can measure all ten beam propagation parameters are being developed.

EXPERIMENTAL MEASUREMENT TECHNIQUES

The analytical results given in the preceding section, in particular the quadratic propagation formulas for the beam variances $\sigma_x^2(z)$ and $\sigma_y^2(z)$, have been buried in the optics literature for many years. Their utility for measuring and characterizing laser beam quality has only been recognized in recent years, however, stimulated in large part by the development of various commercial instruments for measuring beam intensity profiles and beam quality factors on real laser beams. An essential feature here is that measurements need to be made only of (time-averaged) beam intensity profiles or intensity moments, with much more difficult phasefront measurements not being required. Indeed to determine the 3 basic parameters W_0, z_0 and M^2 in each transverse direction, in principle one needs only to measure the second moments of the beam intensity profile $I(x, y, z)$ at three selected values of z along the beam and then fit the resulting spot sizes to the quadratic expressions in Eqs. (2) and (3). As a practical matter, of course, measurements at a large number of z values are usually desirable, with several measurements made close to the waist locations and several more far from the waist, to obtain good accuracy in the fits.

Beam profile measurements are readily made using charge-coupled-device (CCD) arrays or similar intensity measuring instruments, particularly if pulsed lasers are to be measured. As a practical matter, however, especially since the analytical results are neatly separable in the x and y transverse coordinates, independent knife-edge measurements in the x and y directions provide an even more convenient method for determining the integrated intensity profiles and beam moments in these two di-

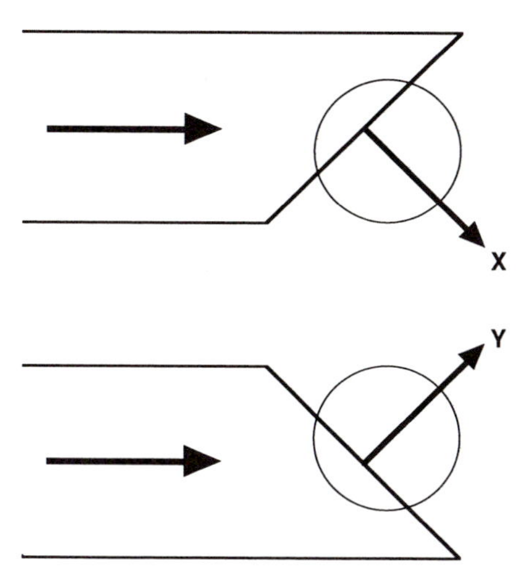

Figure 2. Knife-edge measurement method showing that a beam profile can be measured in both the x and y directions with one direction of motion by using $\pm 45^\circ$ knife edges.

rections, at least for cw laser beams. Note that the first derivative of a measured knife-edge profile along, say, the x direction gives the intensity profile $I(x, y)$ automatically integrated along the y direction, and similarly for the x-integrated y profile. Moreover, both the x and y profiles can be independently measured with a single direction of motion using $\pm 45^\circ$ knife edges as shown in Fig. 2. (Addition of a third knife edge oriented perpendicular to the direction of motion even permits one to determine the $\overline{xy}$ value and hence the principal axes.) Figure 3 shows the essential elements of a commercial instrument, the Coherent "ModeMaster", which makes use of this approach. A number of other instruments using other kinds of sampling methods, including point samplers for high-power beams and CCD arrays for electronic and single-shot pulsed profile measurement, are also commercially available.

Measurements of laser beam quality using this type of instrument on a variety of real laser beams have proven to be very useful, not only for predicting the effectiveness of laser beams for practical applications such as materials processing or nonlinear optical experiments, but also for aligning lasers for best performance in the laboratory or on the production line, and for studying the basic physics and the modal properties of laser devices. As one example of this, Fig. 4 shows in the lower plot the power output of a commercial argon-ion laser operated at three different discharge currents as a function of the intracavity mode-control aperture size. For all of the operating points shown in this plot the beam profile appears, at least on cursory inspection, to be a good quality gaussian beam. The upper plot shows, however, the M^2 value as monitored in real time for the same laser under the same conditions. One point that is immediately evident is that for the smaller aperture sizes the mode-control aperture

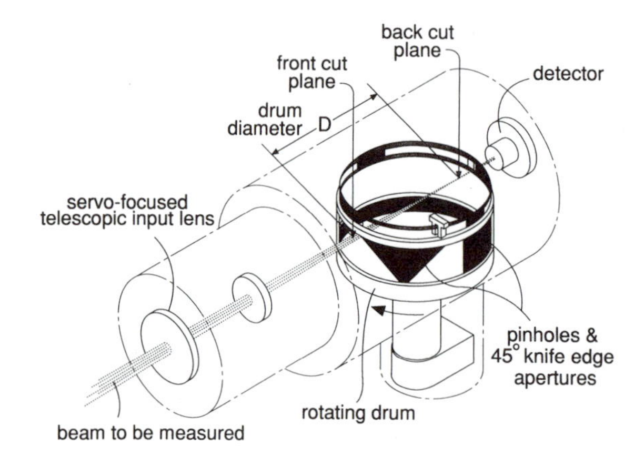

Figure 3. Schematic diagram of the "ModeMaster" beam quality and beam parameter measuring instrument.

is small enough to cut into the TEM_{00} mode of the stable resonator to a significant extent, causing diffraction effects which increase the beam quality factor to $M^2 \simeq 1.3$ rather than the theoretical value of 1.0 for the TEM_{00} mode. The M^2 value then improves toward $M^2 \approx 1.0$ for larger apertures, up to a point where the M^2 value suddenly deteriorates sharply for larger apertures. This sudden deterioration is a clear indication of the onset of a higher-order "donut" mode in the laser. The observed output profile of the laser beam remains almost unchanged and quasi gaussian in character as the laser goes through this transition point, however, with no matching discontinuities evident in the corresponding power output versus aperture curves. The effectiveness of this laser beam in materials processing or nonlinear optics experiments, or the calibration factor for beam intensity in a scientific experiment, would be very significantly altered by the sudden change in the M^2 value. Very similar results can be obtained using solid-state lasers as well.

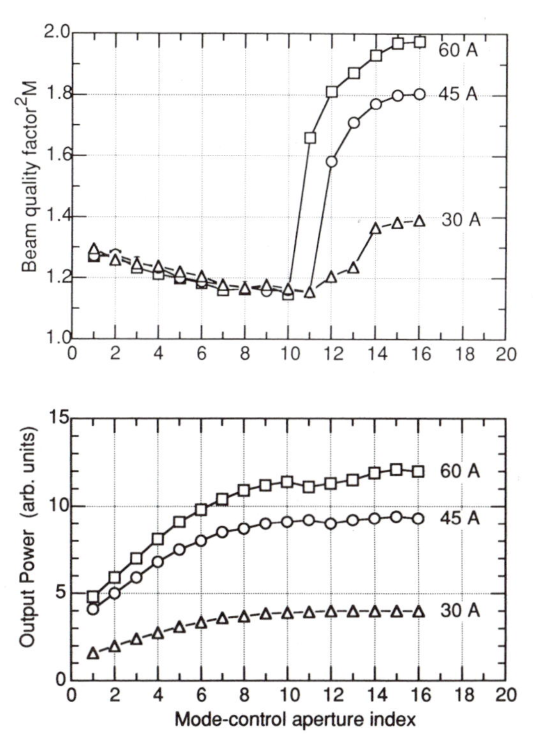

Figure 4. Measurements of total power output and beam quality factor M^2 for a commercial argon laser as a function of the index number for a series of intracavity mode-control apertures with aperture diameters ranging from $\approx$ 180 microns for the lowest index to $\approx$ 400 microns for the largest index value.

To confirm that the sharp transition points in the M^2 curves of Fig. 4 represent higher-order mode onset, computer simulations of a similar laser system were carried out using a recently developed "PARAXIA" software package for desktop computers, with results as shown in Figs. 5 and 6. In these calculations two independent optical waves were circulated around inside a stable laser cavity using Huygens integral in the manner originally pioneered by Fox and Li. A simulated gain sheet was also added at one end of the laser cavity, with the initial gain profile homogeneously saturated by the incoherent superposition of the two optical waves. For the cases shown in Fig. 5, below the break point or knee in each of the curves one of the two waves evolved into a slightly distorted gaussian profile, while the amplitude of the other wave dropped down to the numerical noise level of the calculation. For larger aperture values, however, the second wave evolved into a slightly distorted LG01* or "donut" profile whose relative amplitude increased with increasing aperture size. As shown in Fig. 6, this higher-order mode, when added incoherently to the initial mode, served primarily to broaden slightly the apparent gaussian profile of the overall time-averaged beam. Without the measured M^2 data shown in the top part of Fig. 3 it would be very difficult to observe the onset of the higher-order mode in this situation.

Similar but more detailed results for the behavior of high-power CO_2 lasers and of dye lasers as a function of cavity adjustment and of pump power beam quality have also been measured. In general, real-time measurements of beam quality factor permit lasers to be adjusted for optimum stability and optimum beam brightness (which is not in general the same as maximum power output) both in the laboratory and on the production line.

The measured results in this case result from the onset of independently oscillating higher-order Hermite-gaussian (HG) modes in the laser. Suppose that the time-averaged beam intensity profile in this situation is written in the form

$$I(x,y) = \sum_n \sum_m |c_{nm}|^2 |\tilde{u}_{nm}(x,y)|^2 \tag{6}$$

where the functions $\tilde{u}_{nm}(x,y)$ are nm-th order HG modes and the $|c_{nm}|^2$ are the fractional powers in each mode. (We are assuming that each transverse mode oscillates independently at a separate oscillation frequency here, so that the intensity is an incoherent superposition of the mode intensities.) The beam quality factors for the multimode beam are then given by the weighted sums

$$M_x^2 = \sum_n \sum_m (2n+1)|c_n|^2 \quad \text{and} \quad M_y^2 = \sum_n \sum_m (2m+1)|c_n|^2 \tag{7}$$

The limiting case of a pure donut mode corresponds to $|c_{00}|^2 = 0$ and $|c_{10}|^2 = |c_{01}|^2 = 1/2$, leading to $M_x^2 = M_y^2 = 2$, as indicated by the limiting values in Figs. 4 and 5. We should perhaps emphasize again, however, that the analytical formulas of Eqs. (2) and (3) are completely general and apply equally well to unstable resonators or to badly distorted or aberrated beams as well, with no underlying assumptions as to stable-cavity or HG mode components.

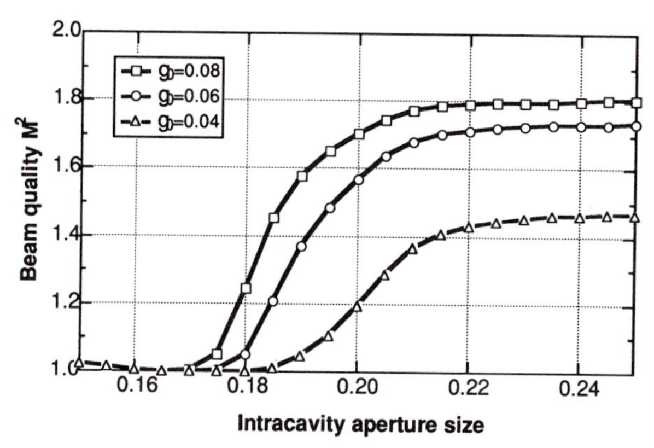

Figure 5. Computer simulations similar to the experimental results in Fig. 4 , using a Fox-and-Li calculation procedure in radial coordinates for a stable laser cavity with two circulating waves and a saturable gain sheet pasted on one end mirror. Only one of the circulating waves builds up to oscillation for aperture diameters below the break point in each of the M^2 curves.

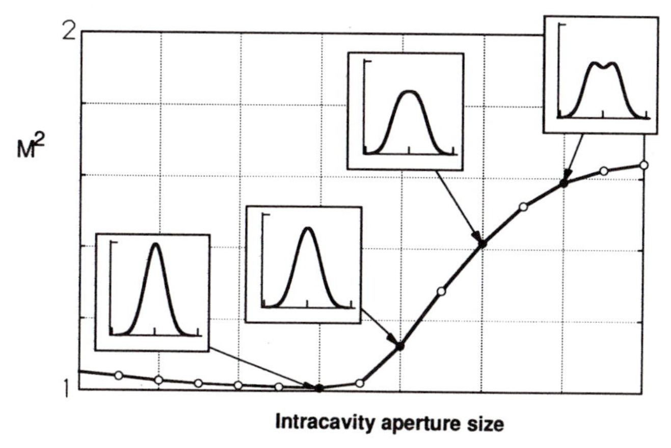

Figure 6. Illustration of the overall time-averaged beam profiles resulting from superposition of the two oscillating modes in a computer simulation like Fig. 5.

EFFECTS OF PHASE ABERRATIONS ON BEAM QUALITY

In addition to perturbation effects or multimode oscillations in laser devices themselves, phase aberrations in an external optical system can also lead to degraded transverse beam quality for a laser beam passing through the system. Elementary aberrations such as spherical aberration, coma, and astigmatism may be present, for example, in beam-expanding telescopes, focusing lenses, and other optical elements used to collimate or focus a laser beam. Thermal effects can cause significant spherical aberrations as well as thermal focusing in lenses, output windows and other optical elments used with high-power lasers. Thermal blooming can also produce significant spherical aberration for a high-power beam passing through an absorbing atmosphere or other medium. Finally, deviations from the paraxial approximation to Snell's law can produce significant spherical aberration for a curved laser wavefront exiting from a high-index dielectric medium, as for example in wide-stripe unstable-resonator diode lasers.

It is important, therefore, to know how seriously a given degree of spherical aberration will degrade the transverse beam quality of a given laser beam. As one example of this we can consider a laser beam with an initial beam quality M_{x0}^2 passing through a one-dimensional quartic or spherically aberrating element with a quartic phase aberration of the form $\exp[-j(2\pi/\lambda)(c_4 x^4)]$. The degradation in beam quality caused by this aberrating element is then given by

$$M_x^2 = \sqrt{[M_{x0}^2]^2 + [M_{xq}]^2} \tag{8}$$

where the added quartic contribution M_{xq}^2 is given by

$$M_{xq}^2 = \frac{16\pi\beta_x}{\lambda}\, c_4\, \overline{x^4}\ . \tag{9}$$

In this expression $\overline{x^4}$ is the fourth moment of x on the beam intensity profile $I(x,y)$, and $\beta_x = [(\,\overline{x^2}\,\overline{x^6} - \overline{x^4}^2)/\,\overline{x^4}^2]^{1/2}$ is a dimensionless factor of order $\simeq 0.5$ which depends on the exact shape of the laser beam profile. Figure 7 shows an example of how the beam quality of an initially near-gaussian beam deteriorates on passing through a 10 cm focal length plano-convex lens, as a function of beam size at the lens, for the lens used in both right and wrong orientations.

BEAM QUALITY OF LASER ARRAYS

As another and perhaps somewhat surprising example of the utility of the beam quality factor, we can examine what happens when the beam quality concept is applied to in-phase and out-of-phase laser arrays. Phased arrays are often suggested as a means of obtaining higher laser power by using multiple array elements while also obtaining good beam quality by maintaining all of the array elements in phase. Figure 8 shows, for example, the near-field amplitude profile and the far-field intensity profile for both an out-of-phase waveguide mode or phased-array supermode (left-hand column), such as often occurs in laser diode arrays, and also for the same near-field profile with alternating array elements brought into phase using binary phase plates (right-hand column). As expected the in-phase array produces a single on-axis far-field lobe rather than the double-lobed beam produced by the alternating-phase array.

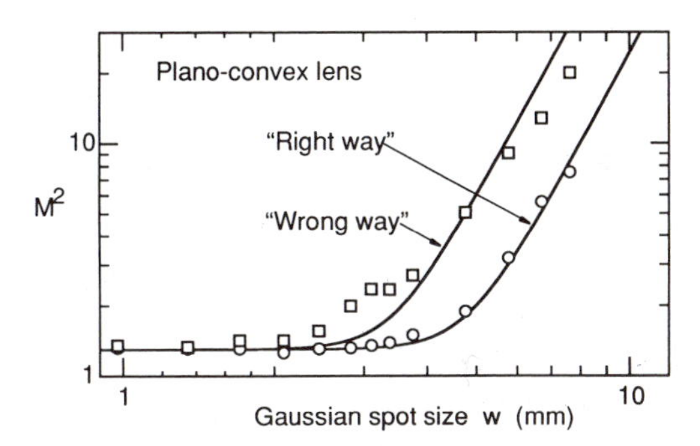

Figure 7. Experimental results for the beam quality degradation produced by spherical aberration in a thin plano-convex lens as a function of the gaussian beam spot size of the beam passing through the lens. The solid curves are obtained from the theoretical expressions of this lecture.

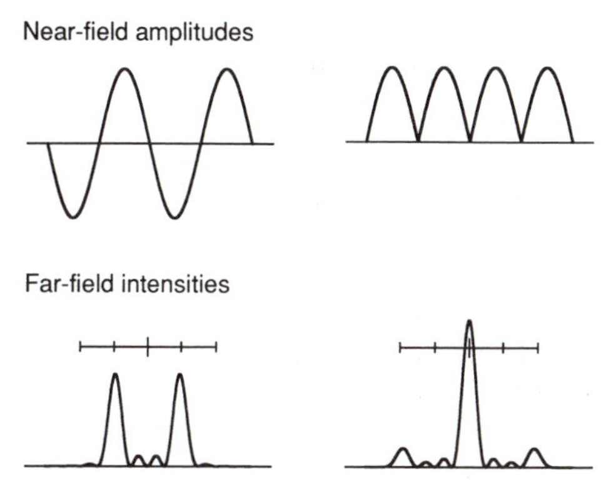

Figure 8. Near-field E-field profiles and associated far-field intensity profiles, for an asymmetric waveguide or diode-array mode with 4 alternating half cycles (left-hand column) and for the same mode with the alternating half-cycles brought into phase using a binary plate (right-hand column). The markers on the far-field profiles are at 1 and 1.5 standard deviations from the centerline.

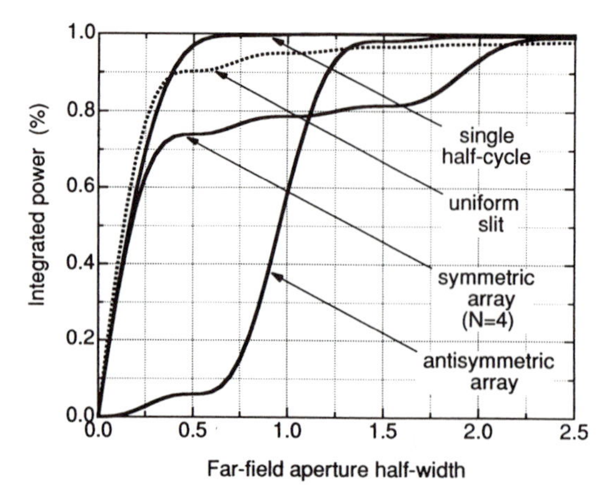

Figure 9. Integrated power versus far-field aperture half-width (or far-field angular half-width) for the two beam profiles shown in Fig. 7, plus similar curves for a uniformly illuminated slit and a single sinusoidal half-cycle near-field pattern. The aperture half-width is measured in units of the standard deviations for the symmetric or antisymmetric $m = 4$ profiles.

The interesting point, however, is that when all the additional energy in the secondary far-field lobes of each pattern is taken into acount the beam quality factors for both the in-phase and alternating-phase arrays turn out to have exactly the same value, namely $M_x^2 = \sqrt{(\pi^2 m^2 - 6)/3} \simeq 7.1$ where $m = 4$ is the number of lobes in the near-field pattern. One can show as a general rule, in fact, that applying suitable binary phase plates to bring all the positive and negative lobes of a purely real but alternating near-field beam profile into phase can never improve the M^2 value for any such profile.

Despite the broad generality of this conclusion, many laser researchers may well consider the symmetric far-field profile in the right-hand column of Fig. 8 to be more useful or of significantly higher quality than the antisymmetric case. An antisymmetric beam with multiple lobes of alternating phase will certainly have a null on axis in the far field (at least for an even number of lobes). Bringing the near-field lobes into phase with a suitable binary optical element will then produce a strong central lobe on axis, although without improving the M_x^2 value; and having a strong central lobe on axis may be important or even essential in certain applications. If the practical requirements of a given application, however, simply require that a large fraction of the total beam power coming from an output aperture of specified size be transmitted into a far-field beam cone of specified solid angle, or into a focal spot of specified spot size, without great concern as to the detailed shape of the beam within these limits, then the M^2 factor will remain a meaningful performance measure for the beam, central lobe or no central lobe. To demonstrate this, Fig. 9 shows the integrated power versus beam angle ("power in the bucket" curves) for the two beams of Fig. 7, as well as for a uniformly illuminated slit of the same width in the near field (which in fact has $M_x^2 = \infty$) and for a single half-sinusoidal lobe of the same total width ($m = 1$ and $M_x^2 \approx 1.136$). It is apparent that for integrated power levels up to 60% or so, the uniform slit, the single half-cycle, and the symmetrized $m = 4$ beams are all

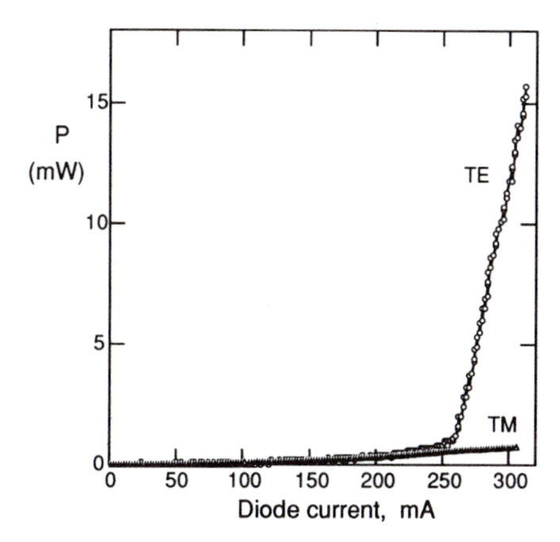

Figure 10. Optical power output in the TE and TM polarizations versus diode current for a wide-stripe ($\approx 60\mu$m wide) GaAlAs diode laser.

nearly equally good, and the antisymmetric $m = 4$ beam is very much worse. At an integrated power of just under 80%, however, the symmetric and antisymmetric cases become equally good (or bad, depending upon the viewpoint), with both being 4 to 5 times worse than the uniform slit of the single half cycle. In fact, if one wishes to capture 90% or more of the integrated power in the beam, then the symmetrized array actually requires almost twice the far-field aperture width as does the antisymmetric array to capture the same total integrated power.

DIODE LASER BEAM QUALITY MEASUREMENTS

As a final example of the utility of beam quality factor measurements, we summarize a number of results from M^2 measurements on a wide-stripe (≈ 60 micron) high-power semiconductor diode laser in which the M_x^2 value in the lateral direction (along the junction plane) and the M_y^2 value in the vertical direction (perpendicular to the junction or the active layer of the diode) are both measured simultaneously as a function of injected current through the diode laser. These measurements are made separately in both the TE and TM polarizations, and over a range of currents extending from well below to well above threshold.

Figure 10 shows the light output versus current for this diode measured separately in the TE (oscillating) and TM (nonoscillating) polarizations. The oscillation threshold region for the TE mode is evidently between 250 and 260 mA input current, while the TM mode never oscillates. Figure 11 then shows the beam quality factors M_x^2 and M_y^2 measured separately in the TE and TM polarizations as a function of input current. Measurements at the lowest currents were taken without a polarizer in front of the ModeMaster and are labelled as TE+TM. For the current range below $\simeq 250$ mA the diode laser output consists entirely of an amplified spontaneous emission (ASE) beam which still provides a sufficiently powerful and collimated enough output to permit direct measurement of the beam quality factor in both transverse directions even far below threshold. In the lateral (or x) direction this ASE output can be viewed as consisting of many transverse modes with very limited spatial coherence, and the measured value of $M_x^2 \simeq 100$ matches reasonably well to the value expected for the aspect ratio of the $\simeq 60\mu m$ wide diode stripe. At threshold, however, the M_x^2 value for the TE mode suddenly but continuously drops to $M_x^2 \simeq 10$ corresponding to a spatially coherent but rather badly filamented mode extending across the full stripe width. (For comparison we have also made measurements on narrow-stripe gain-guided single-mode diode lasers which yield a beam quality factor of $M_x^2 \simeq 1.1$ or better in the lateral direction.)

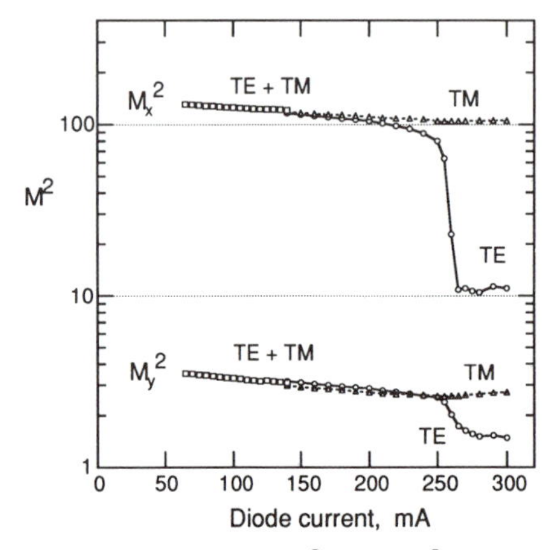

Figure 11. Measured beam quality factors M_x^2 and M_y^2 in the lateral and vertical directions, respectively, versus current for the TE and TM polarizations in the same diode as Fig. 1.

At the same time the M_y^2 value perpendicular to the junction plane of the diode is limited to $M_y^2 \simeq 3$ below threshold, with the TE mode falling to $M_y^2 \simeq 1.5$ in the coherent oscillation region above threshold. In a diode laser of this type the index-guided E field profile perpendicular to the junction is a single lowest-order waveguide mode; and because this mode will have a field profile similar to a double-sided exponential rather than a gaussian it should have a theoretical value of $M_y^2 \simeq 1.4$ to 1.5, in excellent agreement with the measured TE value above threshold. The measured M_y^2 values in Fig. 11 then seem to indicate that for the TE mode below threshold, and also the TM mode at all current levels, a significant amount of a higher-order waveguide mode (perhaps the first odd mode) must also be excited by the ASE, leading to the measured value of $M_y^2 \simeq 3$ below threshold. Both the M_x^2 and M_y^2 values also indicate that the TM mode remains completely below threshold at all of the current values shown.

The ModeMaster instrument used to make these measurements, provides real-time results not only for M^2, but for the other beam parameters W_0 and z_0 as well, measured with respect to a reference plane at the input face of the instrument. The latter two parameters can then be transformed back through the 8 mm focal length collimating lens used to collimate the diode laser's output beam, so as to give the real-beam waist sizes and waist positions of the uncollimated beam emerging from the output face of the diode. Figure 12 shows data for the equivalent waist spot sizes at or just inside the diode laser's output facet. In the vertical or y direction the equivalent spot size is $W_{0y} \simeq 2$ μm below threshold, dropping to $\simeq 0.9$ μm for the oscillating TE mode. The lateral spot size is $W_{0y} \simeq 50$ μm below threshold, indicating that the gain and ASE profile spill over a considerable distance outside the nominal 60 μm stripe width. Again this width drops to $\simeq 20$ μm when the TE mode begins oscillating.

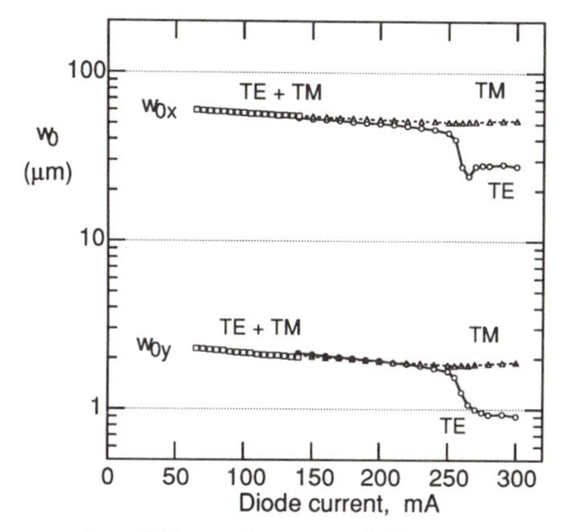

Figure 12. Real-beam spot sizes $W_x \equiv 2\sigma_x$ and $W_y \equiv 2\sigma_y$ for the TE and TM polarizations versus current for the same diode as in Fig. 1.

Figure 13 shows the waist locations in the x and y coordinates, measured as distance behind the 8 mm collimating lens. It is a reasonable assumption that the waist in the rapidly diverging y-directed output will be located at or very close to the laser's output facet. The data indicates that the effective waist location in the lateral direction is then approximately 60 μm deeper into the diode, at least in the ASE regime, corresponding to equivalent ASE sources located at that depth or deeper into the amplifying stripe. The degree of astigmatism in the diode output beam and its variation with diode current is thus clearly indicated. (The unpolarized TE+TM measurements in these figures were taken in a different run from the polarized TE and TM measurements, leading to a small offset in absolute diode positioning, but giving entirely consistent results for diode astigmatism.) The astigmatism increases slightly with increasing current, possibly due to increasing depth of the effective ASE noise sources with increasing gain, as well as increased antiguiding with increased electron density in the stripe. A detailed explanation for the irregular behavior in the waist location for z_{0x} for the TE mode above threshold is less clear, but this could well represent a combination of filamentation, nonlinear gain distortion, and antiguiding in the stripe.

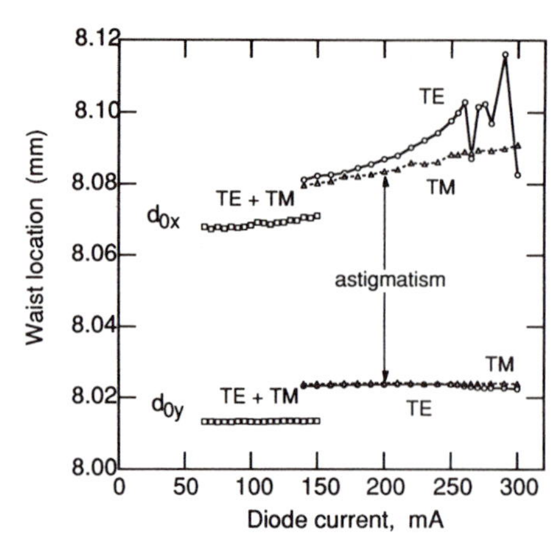

Figure 13. Waist location (measured relative to the entrance face of the collimating lens) for the same diode as in the previous figures.

We believe that these results represent the first accurate measurements of laser beam quality to be taken on a laser of any type below threshold. This data also shows clearly how beam parameter measurements can provide an entirely new way to study diode laser modal properties and oscillation physics both below, above, and continuously through the laser threshold region.

CONCLUSIONS

There exist other alternative beam width and beam quality definitions, as well as other instruments and techniques for measuring laser beam quality parameters besides the particular instrument described in this lecture. The general approach outlined here seems to have many advantages, however, both practical and conceptual, for studying and understanding the modal and spatial properties of laser devices, It can be expected that more careful attention to the measurement and characterization of beam quality for real laser beams will pay significant dividends for improved understanding and improved practical applications of high-power and low-power solid-state laser devices in the future.

ACKNOWLEDGEMENTS

The author appreciates extensive collaboration on the subject of beam quality measurements with Tom Johnston, Jr., and Mike Sasnett of Coherent, Inc.; Tony Kiewitsh, Joseph Ruff, Geoff Fanning, and other students at Stanford University; visitors Manlio Fogliani and Julio Serna; and especially Dr. George Nemes of the Institute of Atomic Physics, Bucharest, Romania. The Stanford research on this topic was supported by the Air Force Office of Scientific Research.

MODE-LOCKED SOLID STATE LASERS

A.I. Ferguson and G.P. A. Malcolm

Department of Physics and Applied Physics
University of Strathclyde
Glasgow G4 ONG
Scotland, UK

ABSTRACT

There has been dramatic improvement in short pulse generation using solid state lasers over the past decade. This has been brought about by improvements in laser gain media, the advent of diode pumping of solid state lasers and the use of nonlinear processes to induce mode-locking. We have seen the reduction in pulse duration of conventional gain media such as *Nd:YAG* by over an order of magnitude, and new tunable materials such as $Ti{:}Al_2O_3$ and *Cr:LiSAF* have been made to produce light pulses with durations well below *100fsec*. I will describe some of the basic ideas and techniques which have been employed in reaching these new levels of pulse duration. These will be illustrated with descriptions of particular systems.

1. INTRODUCTION

Over the past decade a revolution has taken place in the methods of short pulse generation from lasers. Ten years ago the most widely used sub-picosecond light source was the passively mode-locked dye laser. These were relatively inconvenient to use, had restricted tuning range and rather low average output power. Today the mode-locked dye laser has all but disappeared as an advanced source of ultrashort pulses. The demise of the dye laser has been brought about by several factors. Firstly, there have been great advances in solid state lasers. Conventional materials such as *Nd:YAG* and *Nd:YLF* can be conveniently pumped by laser diodes resulting efficient and reliable sources. New materials have also had a tremendous impact. In particular $Ti{:}Al_2O_3$ has revolutionised tunable solid state lasers. The spectral region covered by $Ti{:}Al_2O_3$ could only be covered by the use of several dyes. This broad spectral coverage of $Ti{:}Al_2O_3$ means that it is

Solid State Lasers: New Developments and Applications
Edited by M. Inguscio and R. Wallenstein, Plenum Press, New York, 1993

capable of sustaining extremely short light pulses well below *100fsec*. Other tunable materials are also beginning to have an impact in the technology of short pulse generation. The material *Cr:LiSAF* ($Cr:LiSrAlF_6$) has generated considerable excitement since it not only covers a similar spectral region to that of $Ti:Al_2O_3$ but it is efficient, has a low pump power threshold and can be directly diode laser pumped.

The second ingredient in the revolution has been the development of novel mode-locking techniques. Conventional techniques such as active loss or phase modulation have been extensively used and applied to most materials. Although it was known that many of these materials were capable of sustaining much shorter pulses than could be produced using conventional mode-locking techniques no-one could devise the way to generate these shorter pulses. The breakthrough came when it was realised that if a nonlinearity, caused by the laser pulses themselves, could be used to compress the pulses then the pulses would get shorter, producing even more compression, until the whole bandwidth of the gain medium was utilised. These techniques were at first applied to only specialised soliton laser systems but it was eventually realised that this mode-locking technique was quite general and could be applied to most materials. Materials which have previously only been capable of producing *35psec* pulses *(Nd:YLF)* were then found to be able to produce *1.5psec* pulses of great stability and quality[1]. The combination of these new mode-locking techniques, new materials and new pump sources has totally transformed solid state laser technology and continues to do so. As these new sources are applied to many areas of science and technology we can expect to see many new and exciting developments.

In this article I set the scene of conventional mode-locking techniques and point out their limitations. I then describe the principles of the new nonlinear mode-locking schemes concentrating on additive pulse (or coupled cavity) mode-locking (APM) and Kerr lens mode-locking (KLM). I then illustrate these principles by describing some particular laser schemes including APM and KLM mode-locking of diode pumped *Nd:YLF* and $Ti:Al_2O_3$. Throughout the article I have tried to concentrate on principles and use descriptions of particular systems simply to illustrate important aspects of how these principles are applied. As a consequence this is not a comprehensive review of work in the field and many important and exciting results will not be described. I will also not concentrate on historical scene setting and the work referred to is intended to be illustrative rather than exhaustive.

2. ACTIVE MODE-LOCKING

Active mode-locking techniques are the most widely utilised method for generation of short pulses in solid state lasers. A comprehensive review of active mode-locking has been given by New[2]. A general schematic diagram of an actively mode-locked laser is shown in figure 1. In addition to the gain medium a modulator is placed in the cavity. The modulator which may be an amplitude or phase modulator is driven at a frequency which provides modulation at the cavity mode spacing $c/2\ell$, where ℓ is the length of the cavity. This modulation forces the normally independently oscillating modes to lock together in phase such that the mode amplitudes add in phase at one point in time to form a train of pulses which pass through the modulator at minimum loss, in the case of a loss modulator, or at a phase extremum in the case of a phase modulator. The theory of active mode-locking was developed by Kuizenga and Siegman for the case of a homogeneously broadened gain medium[3]. In this theory a self-consistent solution was sought for a Gaussian pulse passing through a gain medium and then a sinusoidally driven amplitude or phase modulator. The effects of frequency chirping of the pulses, intracavity etalons and dispersion were also included. Among other things this theory showed that the pulse duration was only weakly dependent on the depth of amplitude or phase modulation. The

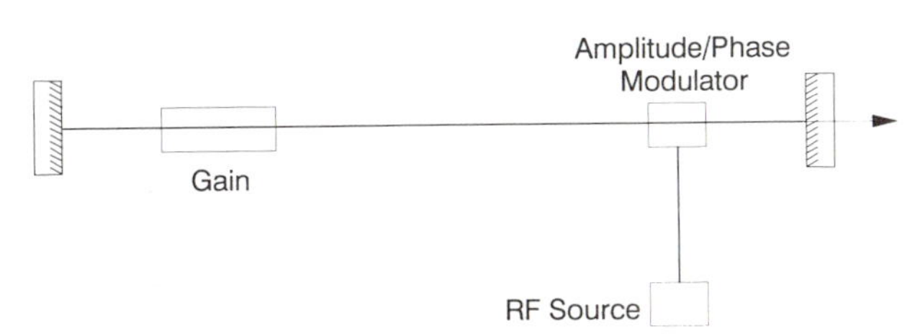

Figure 1. A schematic diagram of an actively mode-locked laser based on amplitude or phase modulation.

reason for this is that the pulse compression at the leading and trailing edges of the pulse is provided by the curvature of the loss/phase modulation. As the pulse gets shorter and shorter the compression becomes weaker and weaker. This sets a limit to the duration of the pulses that can be generated. As a general rule it is unusual for the active mode-locking method to produce pulses of durations less than one thousandth that of the round trip time in the cavity. Typical best figures would be a round trip time of *10nsec* and pulse duration of *10psec*. Over the past five years much of the excitement in mode-locking of solid state lasers has come about because of the development of new nonlinear techniques which have reduced these figures to less than *100fsec* in *10nsec* giving a remarkable 10^5 mark-space ratio, some two orders of magnitude higher than that achieved by the best active mode-locking schemes.

3. ADDITIVE PULSE MODE-LOCKING

A schematic diagram of a typical additive pulse mode-locked laser is shown in figure 3. In this scheme, which is also sometimes called coupled cavity mode-locking, scheme there is no active modulation in the laser cavity. The output from the laser is passed through a medium exhibiting a Kerr nonlinearity in which the refractive index is modified by the light intensity. The light is then reflected off a mirror and returned to the main laser cavity. The path length in the external cavity is adjusted to match the length of the laser cavity or some multiple of this length. Two schemes are illustrated in the figures 2(a) and 2(b) the Fabry Perot scheme and the Michelson scheme.

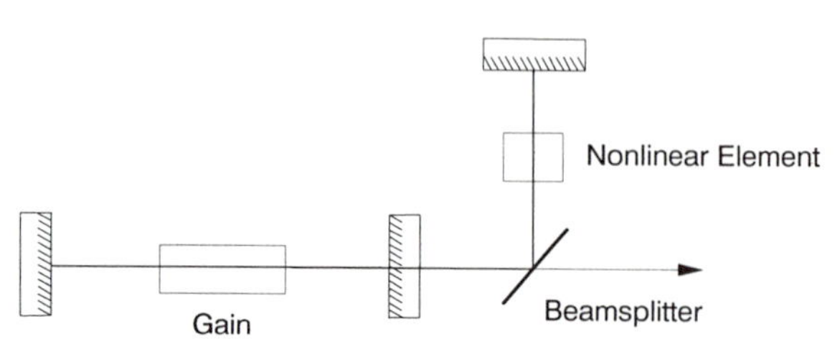

(a) Fabry-Perot type additive pulse mode-locked laser

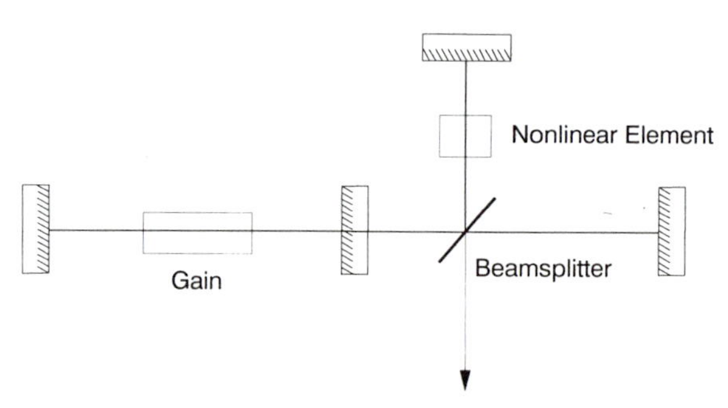

(b) Michelson type additive pulse mode-locked laser

Figure 2

One way to picture what is going on in the mode-locking process is to consider a noise spike or pulse of radiation coming from the laser cavity. It will be double passed through the nonlinear medium. The pulsed nature of this light will mean that it experiences a time-dependent refractive index change. A time varying refractive index gives rise to a frequency chirp. The matching of cavity lengths ensures that the chirped pulse and the laser pulse will interfere at the output coupler. If the phase of the returning pulse from the external cavity is appropriately adjusted it can be arranged that there is constructive interference at the pulse centre and destructive interference at the edges of the pulse. The resulting pulse in the laser cavity will therefore be compressed. This shorter pulse will increase the nonlinear phase shift in passing through the nonlinear medium and the interference will again compress the pulse. This process continues until some bandwidth limitation is reached. The intensity dependent reflectivity of the external cavity can be thought of as a fast saturable absorber.

A theory of additive pulse mode-locking has been developed by Ippen *et al.*[4] A generalised diagram of an additive pulse system is shown in figure 3.

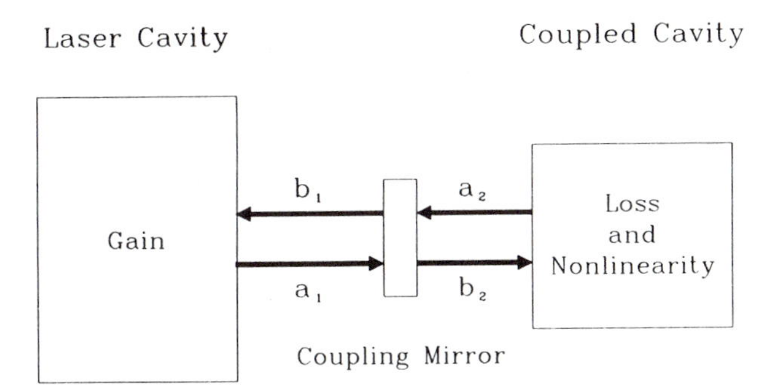

Figure 3. A schematic diagram giving a generalised arrangement for APM systems.

The wave amplitudes at the mirror are given by

$$b(t) = ra_1(t) + (1-r^2)^{1/2} a_2(t) \tag{1}$$

$$b_2(t) = (1-r^2)^{1/2} a_1(t) - ra_2(t) \tag{2}$$

The wave $b_2(t)$ travels through the nonlinear medium, is attenuated by a factor L (< 1) and returns as delayed wave $a_2(t)$. If we ignore dispersion for the time being and adjust the travel distance in the nonlinear arm to match the round trip time in the laser cavity we obtain

$$a_2(t) = Le^{-i(\phi+\Phi)} b_2(t) \tag{3}$$

where $t = 0$ corresponds to the peak of the pulse and Φ is the phase change due to the nonlinearity given by

$$\Phi \equiv \kappa\left[\left|a_2(t)\right|^2 - a_2(0)\right|^2\right] \tag{4}$$

and κ is proportional to the length of fibre and the nonlinear index. The variable ϕ is the phase shift between the two cavities. Note that the peak phase shift $\kappa\left|a_2(0)\right|^2$ has been subtracted from Φ. As a consequence Φ includes a phase bias due to the peak nonlinear phase shift. We thus have

$$b_1 = \frac{1}{(1-r^2)^{1/2}} \left\{ 1 + \frac{r}{L} e^{i(\phi+\Phi)} \right\} a_2 \tag{5}$$

$$a_1 = \frac{1}{(1-r^2)^{1/2}} \left\{ r + \frac{1}{L} e^{i(\phi+\Phi)} \right\} a_2 \tag{6}$$

We can explicitly see the compression mechanism by showing that the reflection coefficient is maximum at the pulse centre and decreases in the wings. The reflection coefficient Γ is given by

$$\Gamma = \frac{b_1}{a_1} = \frac{1 + \frac{r}{L} e^{i(\phi+\Phi)}}{r + \frac{1}{L} e^{i(\phi+\Phi)}} \tag{7}$$

If we assume $L << 1$ then we can expand to obtain

$$\Gamma = r + L(1-r^2) e^{i(\phi+\Phi)} \tag{8}$$

Assuming that Φ is small we have

$$e^{i\Phi} \approx 1 + i\Phi \tag{9}$$

and so

$$\Gamma \approx r + L(1-r^2) e^{i\phi} (1 - i\Phi) \tag{10}$$

In order that $|\Gamma|$ changes maximally with a change of Φ we need $\phi = \pm\ \pi/2$. Recalling that $\Phi = 0$ at the peak and then goes negative in the wings it is clear that $\phi = -\pi/2$ causes the reflection to decrease as Φ goes negative, i.e. at $\phi = -\ \pi/2$ we have

$$|\Gamma| = r + L(1-r^2)\Phi \tag{11}$$

We can thus see that a phase bias of $\phi = -\pi/2$ is needed for pulse compression.

An equation of motion for an APM laser has been developed by Haus *et al*[5]. In this approach the components of the laser break down into four parts. A ring laser is assumed for simplicity. This model is not intended to explain the detailed behaviour of a particular system but rather to show general trends. The model allows for gain, dispersion, self-phase modulation, saturable absorption and phase shifting. Analysis of this system leads to a master equation for the field amplitude $a(t)$ given by

$$\left[-i\psi - (\ell + ix) + g\left(1 + \frac{1}{\Omega_g^2} \frac{d^2}{dt^2} \right) + iD \frac{d^2}{dt^2} + (\gamma - i\delta)|a(t)|^2 \right] a(t) = 0 \tag{12}$$

In developing this equation it has been assumed that all the effects are small and therefore additive and that the system is in steady state. We now give a physical interpretation for each term. The first term *iy* allows for a small carrier frequency shift which manifests itself in a phase shift per pass $y = (Dw_o/c) L_{eff}$ where Dw_o is the frequency shift and L_{eff} is the effective optical length of the resonator. The change in amplitude *Da* due to this phase shift is

$$\Delta a = e^{i\psi}a - a \approx -i\psi a \tag{13}$$

The second term $(\ell + ix)$ arises from the linear loss ℓ and phase shift per pass x and gives rise to an amplitude change of

$$\Delta a = -(\ell + ix)a \tag{14}$$

The third term comes from the gain g which is assumed to be parabolic giving an amplitude change of

$$\Delta a = g\left(1 + \frac{1}{\Omega_g^2}\frac{d^2}{dt^2}\right)a \tag{15}$$

where Ω_g is the bandwidth of the gain line. The term $-iD\frac{d^2}{dt^2}$ comes from the change in amplitude due to group velocity dispersion (GVD) where D is the magnitude of dispersion. The final term allows for both self-phase modulation,

$$\Delta a = -i\delta|a|2a \tag{16}$$

where

$$\delta = \frac{\omega_o}{c}\frac{n_2 d}{A_{eff}} \tag{17}$$

n_2 is the nonlinear index of refraction, d the length of the nonlinear medium and A_{eff} the effective area of the mode, and saturable absorber action described by

$$\Delta a = \gamma|a|^2 a \tag{18}$$

when γ is inversely proportional to the saturation intensity and is positive.

A solution to the master equation (12) is

$$a(t) = A\,\mathrm{sech}(t/\tau)e^{i\beta \ln \mathrm{sech}(t/\tau))} \tag{19}$$

which describes a pulse of duration τ, chirp amplitude β and pulse amplitude A. If this solution is inserted back into the master equation it is possible to find two equations, one which is a constant multiplier of $a(t)$ and the other containing $sech^2$ as a multiplier. These are individually set equal to zero and two equations are obtained

$$-i\psi + g - \ell - ix + \frac{(1+i\beta)^2}{\tau^2}\left(\frac{g}{\Omega_g^2} + iD\right) = 0 \tag{20}$$

$$\frac{1}{\tau^2}\left(\frac{g}{\Omega_g^2} + iD\right)(2 + 3i\beta - \beta^2) = (\gamma - i\delta)A^2 \tag{21}$$

Let us now consider the form of the gain g. For long relaxation times (relaxation time much longer than the round-trip time T_R) the saturation of the gain will be determined by the energy of the pulse and given b

$$g = \frac{g_o}{(1+2A^2\tau / P_sT_R} = \frac{g_o}{1+W / P_sT_R} \tag{22}$$

where $W = 2A^2\tau$ is the energy of the pulse and P_S is the effective saturation power. We can introduce normalised pulsewidth

$$\tau_n = (W\Omega_g^2 / 2g)\tau \tag{23}$$

and normalised dispersion

$$D_n = (\Omega_g^2 / g)D. \tag{24}$$

These can be inserted into equation (21) to give

$$(1/\tau_n)(1+iD_n)(2+3i\beta-\beta^2) = \gamma - i\delta \tag{25}$$

Equating real and imaginary parts we get

$$\frac{3\beta}{2-\beta^2} = \frac{\delta+\gamma D_n}{\delta D_n - \gamma} \equiv \frac{1}{\chi} \tag{26}$$

which gives

$$\beta = -\frac{3}{2}\chi \pm \left[\left(\frac{3}{2}\chi\right)^2 + 2\right]^{1/2} \tag{27}$$

and from equation (25) the pulse width is

$$\tau_n = \frac{2-3\beta D_n - \beta^2}{\gamma} = \frac{-2D_n - 3\beta + D_n\beta^2}{\gamma} \tag{28}$$

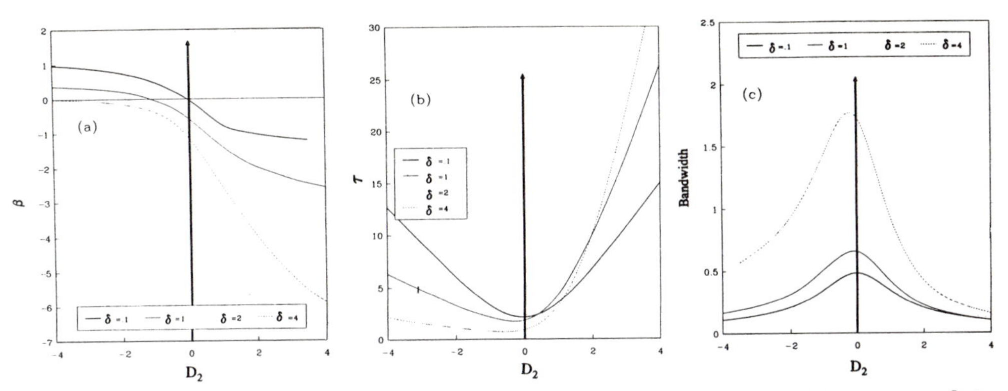

Figure 4. Pulse parameters versus normalized dispersion D_n for $\gamma = 1$ and different SPM parameter δ a) the chirp parameter, b) the inverse normalized pulse with τ and c) the bandwidth $(1+\beta^2)^{1/2}/\tau$.

The normalised bandwidth is given by

$$\omega_n = (1+\beta^2)^{1/2}(1/\tau_n) \tag{29}$$

The normalised pulse width τ_n, chirp parameter β and bandwidth ω_n for saturable absorption parameter $\gamma = 1$ is shown in figure 4 as a function of normalised dispersion D_n. It can be seen that two distinct regimes exist corresponding to positive and negative GVD. For positive GVD the chirp parameter β can be very large. By contrast, for negative GVD short pulse durations are close to being chirp free. In this case the SPM and negative GVD produce a soliton-like pulse compression mechanism. We will see how these ideas are applied to particular laser systems in a later section.

To achieve self-starting additive pulse mode-locking the gain dynamics of the laser medium must allow for the amplification of any intensity fluctuations in the free running cavity, which can then evolve into mode-locked pulses by APM pulse-shaping. The small intensity perturbation caused by beating of the longitudinal modes can be sufficient to self-start additive pulse mode-locking.

An explanation of self-starting of additive pulse mode-locking based on the gain dynamics has been proposed by Ippen *et al*[6]. The general condition for self-starting is given as

$$\frac{\kappa}{g} > \beta\sigma_e\tau_p \tag{30}$$

where, κ is a constant proportional to the losses and nonlinearity in the coupled cavity, g is the single pass saturated gain of the medium, β is a constant close to unity which is dependent on the pulse shape, σ_c is the gain emission cross section and τ_p is the initial pulse duration. As completely passive self-starting of mode-locking will depend on an initial randomly generated pulse duration from noise or mode-beating processes, the importance of the emission cross section σ_e is clear. This explains why Ti:sapphire (σ_e =

$3 \times 10^{-19} cm^2$), Nd:YAG ($\sigma_e = 6.5 \times 10^{-19} cm^2$), Nd:glass ($\sigma_e = 4.3 \times 10^{-20} cm^2$) and Nd:YLF ($\sigma_e = 3.7 \times 10^{-19} cm^2$) will self-start, while colour centres like $KCl:Tl^o(1)$ ($\sigma_e = 3.7 \times 10^{17} cm^2$) require an additional mode-locking mechanism to provide a sufficiently short initial pulse duration before additive pulse mode-locking can then enhance the mode-locking by further pulse shortening.

The actual threshold for self-starting requires that the amplitude of the initial pulse is sufficient to overcome the spoiling effect of spurious intra-cavity reflections. These reflections should be minimised to ease self-starting and also to prevent self-Q-switching due to the long upper state lifetime exhibited by the gain media which favour self-starting.

4. KERR LENS MODE-LOCKING

Recently another novel nonlinear passive technique for ultrashort pulse generation by self-mode-locking has been reported. This self-mode-locking process has been proposed to be due to a Kerr self-focusing effect in an intracavity nonlinear medium, often the gain medium itself, which when combined with an intra-cavity aperture forms an intensity dependent loss mechanism. This has been widely termed Kerr Lens Mode-locking (KLM). The aperture can be a hard aperture inserted into the cavity or an effective soft aperture formed by gain guiding in the laser medium. It has also been suggested that the final pulse duration achieved by femtosecond lasers mode-locked by this process is determined by the counterbalancing between self phase modulation and group velocity dispersion producing a soliton-like pulse-shaping mechanism, similar to the case of APM.

High intensities of laser radiation can cause a modification to the refractive index due to the optical Kerr effect given by

$$n = n_o + n_2 I \tag{31}$$

For a laser beam focused in a Kerr medium the modification of refractive index leads to a lensing in the medium. The focal length of this Kerr lens has been calculated, assuming a parabolic approximation for the index of refraction and an effective focal length much longer than the Kerr medium to be

$$f = \frac{w^2}{4 n_2 I_o L} \tag{32}$$

where w is the spot size at the focus, n_2 the nonlinear index of refraction, I_o the peak intensity and L is the length of the Kerr medium.

By appropriate choice of the Kerr medium in a laser cavity a fast self-focusing can be produced. The position of the Kerr medium in the cavity can also be chosen to achieve a mode-size decrease near the output coupler for increased intensity, as shown in figure 5. Placing a slit aperture at the output coupler therefore creates an intensity-dependent loss mechanism. This results in self-mode-locking in a similar manner to a fast saturable

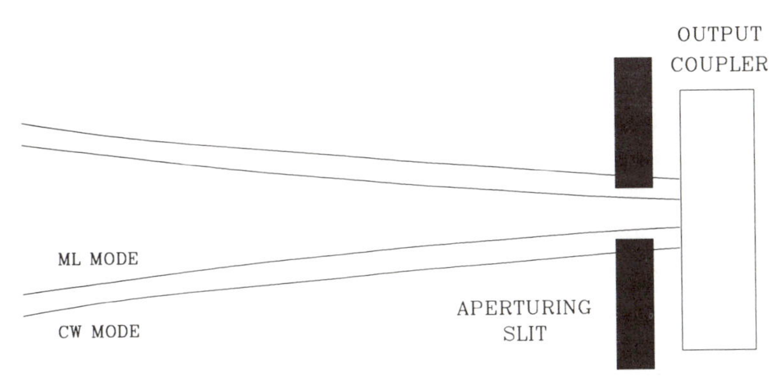

Figure 5. Self-mode-locking by the Kerr lens mechanism, showing decreased mode size for increased intensity produced by lensing in the Kerr medium.

absorber effect. The aperture causes low loss for high peak intensity corresponding to mode-locked operation and large loss for continuous wave operation, thereby favouring mode-locking.

The fast intensity dependent process in the case of self-mode-locking is therefore due to Kerr lensing, while for additive pulse mode-locking the variable phase mirror produced the corresponding effect. This mechanism drives the mode-locking, while other effects including self-phase modulation and group velocity dispersion again play an important rôle in determining the steady-state mode-locked behaviour. The rôle of these other processes can be considered in exactly the same manner as for additive pulse mode-locking.

A high intensity noise burst or initial mode-locked pulse of sufficient intensity is required momentarily to form a Kerr lens of sufficient strength to favour stable mode-locked operation. Self-mode-locking is therefore not a self-starting process. Several processes for initiating self-mode-locking have been investigated. These include a severe mechanical perturbation, such as by tapping one of the mirror mounts, evolution from mode-beating and mode-dragging by moving a cavity mirror or intra-cavity plates. Continuous self-starting can also be achieved by use of an intra-cavity active modulator, a saturable absorber dye, a quantum-well coupled cavity or synchronous pumping of the laser.

5. DIODE PUMPED Nd:YLF

The diode pumped *Nd:YLF* laser system illustrates the rapid developments that have taken place in the mode-locking of solid state laser systems. Figure 6 is a schematic diagram of the development of mode-locking in diode pumped *Nd:YLF*. The basic laser cavity shown in figure 6(a) has become the standard design for mode-locked diode pumped *Nd* systems[7]. The pump laser is either a single *GaAlAs* diode array or a pair of such arrays coupled together using a polarization beamsplitter to combine the two beams. The power of the laser diodes has improved over the years and the latest diodes to be employed in these schemes are *3W* diodes with a *500μm* aperture.

The diode laser radiation is collected by high numerical aperture optics and beam shaped using a combination of spherical and cylindrical lenses. The beam is brought to a focus on the rod with an approximately square shape of about *200μm* dimension. The laser cavity is adjusted such that the mode volume is larger than the pump spot size to encourage TEM_{00} operation. The cavity is an astigmatically compensated three mirror design. In this scheme the change in diameter of the laser beam on entering the Brewster angled rod, with the consequent change in diffraction of the beam, is compensated by the use of an off-axis concave mirror. The radius of curvature of the mirror and the distances between the rod, mirror and output coupler are chosen for pump beam size matching. The rod which is generally *10mm* in length can be made of almost any *Nd* based crystal. We will describe *Nd:YLF* based systems since the combination of properties of *Nd:YLF* is quite attractive for a number of applications. The oscillating wavelength of the *Nd:YLF* laser can be chosen to be *1.047μm* or *1.053μm* by appropriate orientation of the electric field vector of the laser light with respect to the optic axis. The rod is dielectrically coated on one end to be a high reflector at the laser wavelength and highly transmitting at the pump wavelength of about *797nm*. The output coupler is a plane mirror with a wedged rear surface and the output coupling is in the region of *10%* to *15%*. The wedge on the output coupler and the Brewster angle on the crystal ensure the elimination of spurious reflections which upset all mode-locking schemes. This design provides a long collimated arm where modulators can be placed. The cavity length can be adjusted for operation at different repetition rates. Frequencies in the range *20MHz* up to *1GHz* have been operated. For two *3W* pumps the typical output power from this system with nothing in the cavity approaches *2W* at *1.047μm*.

A conventional actively mode-locked system is shown in figure 6(b). In this case a modulator is placed in the cavity, usually near to the end mirror, and driven at a frequency which provides either loss or phase modulation at the cavity mode spacing. These techniques produce pulse durations from *Nd:YLF* which are in the region of *10psec*. This pulse duration is about an order of magnitude larger than *Nd:YLF* is capable of and so a number of schemes have attempted to access shorter pulses. This has involved techniques which use high amplitude modulation, high frequency modulators and high phase retardation phase modulators but the active schemes seem incapable of giving reliable and clean pulses much below *10psec* in *Nd:YLF*.

In an attempt to exploit more of the bandwidth of *Nd:YLF* the additive pulse scheme shown in figure 6(c) was developed[1]. In this scheme there is nothing in the laser cavity but part of the output is coupled into a single mode optical fibre. The light passes through the fibre and is then reflected back into the cavity with the aid of a mirror. The path length is carefully adjusted so that it matches the cavity length or a multiple of that

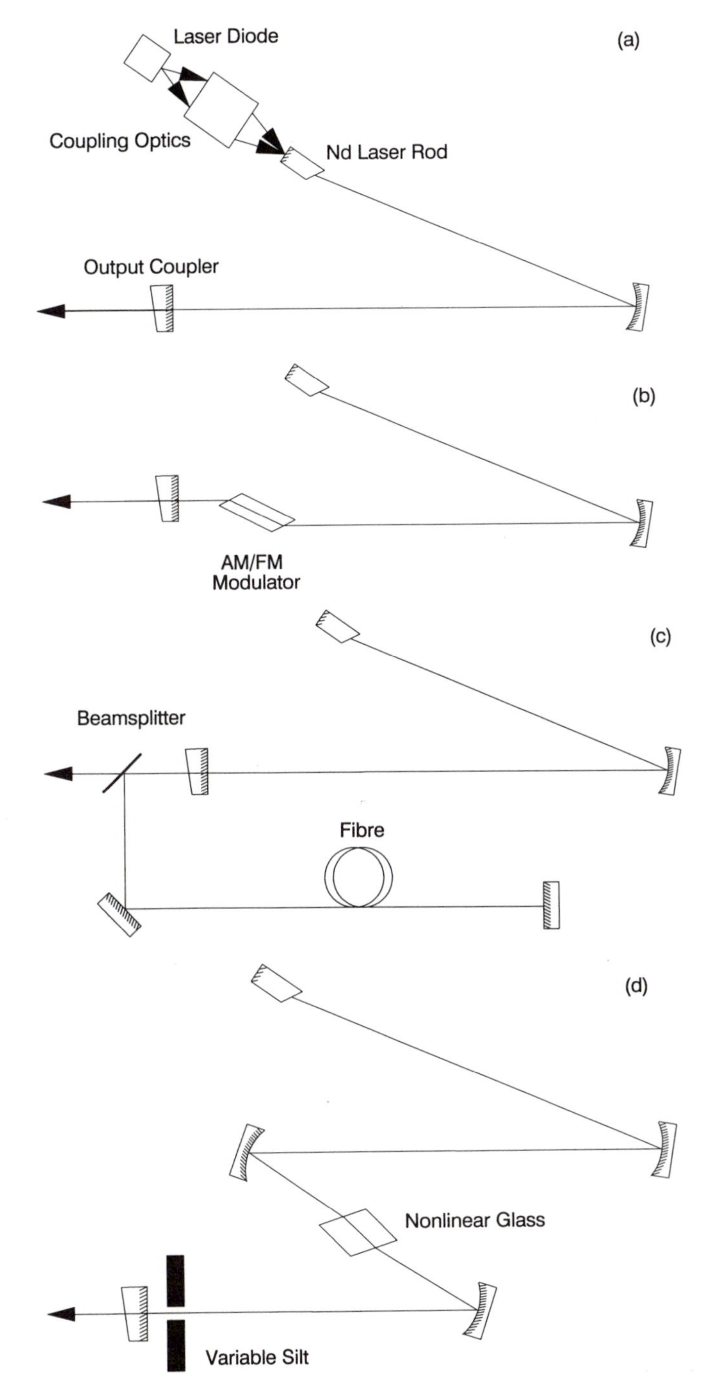

Figure 6 Schematic diagram of diode-pumped lasers. a) Basic astigmatically compensated cavity, b) actively mode-locked system, c) APM system and d) KLM system.

length. Typical figures for the external cavity length are *2m* with a fibre length of about *1m*. For a given laser output power the power into the fibre is carefully adjusted to cause APM mode-locking. In the earliest experiments this power was about *90%* of the laser power leaving only *10%* useful output. However, with two *3W* pumps it is possible to extract about *1W* of usable output power and couple about *200mW* into the fibre. As was indicated in section 3 it is crucial that the phase of the light being returned to the main cavity should be appropriately adjusted. This is accomplished by monitoring the laser power as a function of the position of the piezo mounted mirror at the far end of the fibre. At a certain position the phase will force mode-locking. By comparing the laser power at this position with a reference level it is possible to build a servo-system to keep the system mode-locking. The shortest pulses to date generated using this scheme in *Nd:YLF* have been *1.5psec*[1]. The pulse train is very quiet and well-behaved with the autocorrelation fitting well to $sech^2$ pulses. Although *Nd:YLF* is capable of generating pulses of perhaps half this duration this has not yet been achieved. The time bandwidth product of the pulses was measured to be *0.33* which is close to the bandwidth limit and indicates that the pulses are almost unchirped. Schemes for further compression of these pulses by use of intracavity dispersion compensation have been attempted but with limited success. In materials with much larger bandwidth, e.g. *Nd:glass* these schemes have produced pulses as short as *400fsec*.

Kerr lens mode-locking has also been demonstrated in diode pumped *Nd:YLF*[8]. The schematic diagram for a KLM *Nd:YLF* system is shown in figure 6(d). In this case the cavity has been modified to allow the incorporation of an intracavity focus and a piece of Brewster angled glass. This allows a tight focus to be obtained in the glass without upsetting the mode-matching of the pump and laser beams. A slit at the output coupler ensures that the cavity discriminates against *cw* operation and encourages high intensity mode-locking. The incorporation of a piece of glass in the cavity and a slit drops the output power a little compared to the open cavity, with typically approximately *1W* available for use. The pulse duration obtained with the system was *5psec* with the bandwidth being transform-limited. It is not clear why the pulses are not as short as in the APM system. The additional dispersion of the intracavity glass is a possibility. This aspect is currently under investigation.

6. TUNABLE GAIN MEDIA

Tunable lasers based on transition metal ions have completely changed the face of laser physics over the past few years. The most successful materials have been $Ti{:}Al_2O_3$ and *Cr:LiSAF* with tuning ranges of approximately *670nm* to *1100nm* [9] and *830nm* to *1050nm* respectively[10]. The large gain bandwidths of these materials make them ideal sources for ultrashort pulse generation. Early attempts at mode-locking such very broadband gain media proved to be rather disappointing. Conventional mode-locking techniques were not very satisfactory at taking pulses much below *10psec* despite the fact that the gain medium was capable of generating pulses well below *100fsec*.

The first dramatic progress in mode-locking of $Ti{:}Al_2O_3$ was provided by the *APM* technique[11]. A typical arrangement is shown in figure 7(a). This scheme had already been applied to colour centre lasers both in the soliton sustaining region of optical fibres and in the non-soliton sustaining region[12]. Part of the output of the $Ti{:}Al_2O_3$ laser was directed into a piece of single mode optical fibre. The light from the fibre was redirected back into the $Ti{:}Al_2O_3$ laser such that the cavity lengths were matched in a similar way to that described in section 5 for diode pumped *Nd:based* lasers. Pulses of *200fs* duration

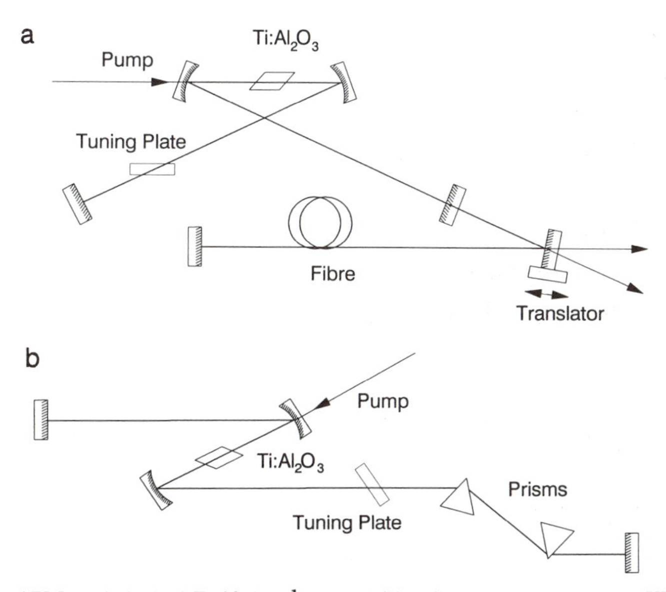

Figure 7. (a) An APM mode-locked $Ti{:}Al_2O_3$ laser and (b) a dispersion compensated KLM mode-locked $Ti{:}Al_2O_3$ laser.

were obtained by this technique using a pair of dispersing prisms in the cavity[11]. It is well known that a pair of prisms can be adjusted such that the effective path length difference corresponds to negative dispersion. This compensates for the normal positive dispersion of components in the cavity. If the prism separation is adjusted such that when the beam passes through the apex of the prisms the overall dispersion in the cavity is negative, it is possible to adjust the dispersion by translating the prisms such that the beam passes through more glass. The optimum group velocity dispersion can then be incorporated to give shorter pulses, minimum chirp and best stability according to the theory set out in section 3.

A remarkable result which has been observed in $Ti:Al_2O_3$ is that by misalignment of the cavity mirrors such that the laser is close to a region of instability and by ensuring multi-mode operation, the laser will self-mode-lock[13]. Pulses in the region of *10psec* were observed with only a birefringent tuning device in the cavity. A version of this laser which has been modified to include dispersion compensation with prisms is shown in figure 7(b). It was further found that by use of a slit at an appropriate point in the cavity it was possible to generate pulses with a TEM_{00} beam profile. Pulses with duration as short as *17fsec* have been generated using this scheme[14]. This approach to the generation of ultrashort pulses has been so successful that it has also led to several commercial systems being made available in a very short time.

The mechanism for self-mode-locking of the $Ti:Al_2O_3$ laser is that of Kerr lens mode-locking. In this case the gain medium itself acts as a Kerr lens[15] which behaves as a fast saturable absorber. Quite recently the KLM technique has been applied to other gain media. One of the most exiting developments has been KLM in *Cr:LiSAF*[16]. Pulses with durations as short as 50fsec have been reported. One of the most attractive features of *Cr:LiSAF* is that it has a relatively low threshold and can be pumped at 670nm where diode laser pumps are available. This opens up the exciting prospect of a directly diode pumped sub-picosecond light source. The compactness, reliability and potentially low cost of such a system will open up many more exciting possibilities and a greater number of application areas where the complexity and expense of conventional lasers are prohibitive.

REFERENCES

1. P F Moulton, J. Opt. Soc. Am. B., **125** (1986).
2. G C H New, Rep. Prog. Phys, **46**, 877, (1983).
3. D J Kuizenga and A E Siegman, IEEE J. Quantum Electron., **QE-6**, 709 (1979).
4 E P Ippen, H A Haus and L Y Liu, J. Opt. Soc. Am. B, **6**, 1736 (1989).
5. H A Haus, J G Fujimoto and E P Ippen, J. Opt. soc. Am. B, **8**, 2063 (1991).
6. E P Ippen, L Y Liu and H A Haus, Opt. Lett, **15**, 183 (1990)
7. G T Maker and A I Ferguson, Appl. Phys. Lett., **53**, 1675 (1988).
8. G P A Malcolm, P F Curley and A I Ferguson, Opt. Lett, **15**, 1303 (1990).
9. G P A Malcolm, A I Ferguson, Opt. Lett, **16**, 1967 (1991).
10. S A Payne, L L Chase, C K Smith, W L Kway and H W Newkirk, J Appl. Phys. **66**, 1051 (1989).
11. J Goodberlet, J Wong, J G Fujimoto, P A Schulz, Opt. Lett. **14**, 1125 (1989).
12. K J Blow and D P Nelson, Opt. Lett, **13**, 1026 (1988).
13. D E Spence, P N Keen and W Sibbett, Opt. Lett. **16**, 42 (1990).
14. Ch Spielmann, P F Curley, T Brabec, E Wintner, A J Schmidt and F Kransz, "Design Considerations for femtosecond Ti:sapphire oscillators", in Digest of Meeting on Ultrafast Phenomena (1992).
15. D K Negus, L Spinelli, N Goldblatt and G Feugnel, Digest of Meeting on Advanced Solid State Lasers (Optical Society of America, Washington DC 1991) PDP4.
16. P Li Kam Wa, B H T Chai, H S Wang and A Miller, Digest of Conference on Lasers & Electro-optics (Optical Society of America, Washington DC) paper CTuCll (1992).

HIGH POWER ROD, SLAB, AND TUBE LASERS

Ulrich Wittrock

Festkörper-Laser-Institut Berlin GmbH
Straße des 17. Juni 135
1000 Berlin 12

INTRODUCTION

This lecture is about high power solid state lasers for materials processing. The fundamental problems of power scaling are discussed. Three approaches to overcome these problems are the multi-rod laser, the slab laser, and the tube laser. Each concept provides a different solution and has its specific advantages and disadvantages. The multitude of materials processing applications require lasers with quite different specifications and there is no single "best" laser. For example, there is need for cw, q-switched, and pulsed (ms-pulses) operation. Q-switching, some materials processing applications, and second harmonic generation (SHG) make polarized beams desirable, while in other applications a randomly polarized beam is advantageous.

Requirements on beam quality (the beam parameter product) are also very different, but three distinct regimes can be identified: A few-times diffraction limited beams are necessary for deep cutting and drilling and for efficient SHG. Most high power solid state lasers are used in conjunction with optical fibers, because the possibilty of beam delivery through an optical fiber is one of the major advantages of solid state lasers when compared to CO_2 lasers. The necessary beam quality is then set by the fiber core diameter and numerical aperture. Commonly used fibers require beam parameter products between 40 and 200 times that of a diffraction limited beam. Finally, there are applications which can be done without fibers and which do not require a small focal spot size and long Rayleigh range. For example, in some cases of welding and hardening. Here, lasers of even poorer beam quality, which can't be coupled to optical fibers, can be used.

All these lasers should have a total efficiency as high as possible because increased efficiency generally results in reduced costs. The processing speed in an application is usually limited by the maximum average power of the laser and also affects the costs in many production systems.

The main obstacles in designing a high power laser with good beam quality are thermal effects in the active medium. Thermal stress limits the maximum output power that

Solid State Lasers: New Developments and Applications
Edited by M. Inguscio and R. Wallenstein, Plenum Press, New York, 1993

can be obtained from a crystal of given dimensions; thermal lensing causes the beam parameter product to depend on the pump power, and thermo-optical aberrations degrade the beam quality. In this article, we first review the important thermal effects and then discuss output power, beam quality, and efficiency of multi-rod, slab, and tube lasers with stable, folded stable, and unstable resonators.

HEAT GENERATION IN THE ACTIVE MEDIUM

The creation of an inversion also creates heat in the active medium because the difference in energy between the pump photons and the laser photons is released to the medium as heat. Ideally, this would be the only heat generating process. It can be minimized by pumping the lowest pump band only, e. g. pumping of Nd:YAG with laser diodes at 808 nm. In real lasers, additional heating is caused because the quantum efficiency of the laser transition is less than unity. There are also non-radiative downward transitions from the pump band to the ground state,[1] and there is absorption of the pump light and the laser beam by impurities, and by stable and pump-light-induced color centers. Some of these processes depend on the spectral composition of the pump radiation. The formation of color centers, that absorb at 1 μm, has recently been identified to be a nonlinear (two-photon) process and, therefore, also depends on the pump light intensity.[2] The nonradiative transitions from the upper laser level are suppressed with increasing laser beam intensity because the proportion of stimulated emission increases. Thus, the heating power also depends on the laser beam intensity.

It is useful to introduce the normalized heating power χ, defined as the ratio of the heating power generated in the active medium, P_{heat}, to the extractable power at the laser wavelength, P_{excit},

$$\chi = \frac{P_{heat}}{P_{excit}} \tag{1}$$

P_{excit} is the product of the excitation efficiency η_{excit} and the electrical pump power

$$P_{excit} = \eta_{excit} \cdot P_{el} \tag{2}$$

and the output power is the product of the extraction efficiency P_{excit} and η_{excit},

$$P_{out} = \eta_{extr} \cdot P_{excit} \tag{3}$$

Since heating of the active medium is the main limitation to high power operation, the inverse value of χ has the character of a figure of merit for solid state laser materials. For Nd:YAG, χ has a value between 0.8 and 2, depending on the type of pumping, the Neodymium concentration and the quality of the crystal. This value is low compared with most other laser materials and together with the good thermo-optical properties and the low lasing threshold it makes Nd:YAG the material of choice for high average power operation.

From Eqs. (1) and (3) we obtain the relation between the heating power and the output power,

$$P_{heat} = \frac{\chi}{\eta_{extr}} \cdot P_{out} \tag{4}$$

The extraction efficiency usually has a value between 0.5 and 0.8. The heating power is thus between 1 and 4 times higher than the output power and can amount to several kW for a single laser crystal.

THERMAL EFFECTS

Temperature Profiles

Assuming a uniform heat source density in the active material, the heat conduction equation has analytical solutions for infinitely long rods, slabs, or tubes. A rod shows a parabolic temperature profile and so does a slab that is being cooled at its two large surfaces only. A tube with equally cooled inside and outside surfaces exhibits a slightly asymmetric temperature profile, which nevertheless is nearly parabolic. Figure 1 shows calculated temperature profiles for the three geometries at equal heating power per unit length. Cross sectional views of the active media are also shown with dimensions that are typical for Nd:YAG crystals used in high power solid state lasers. At these dimensions, the height of the temperature profile in the slab is only 36% of that in the rod, and in the tube it is only 10% of that in the rod.

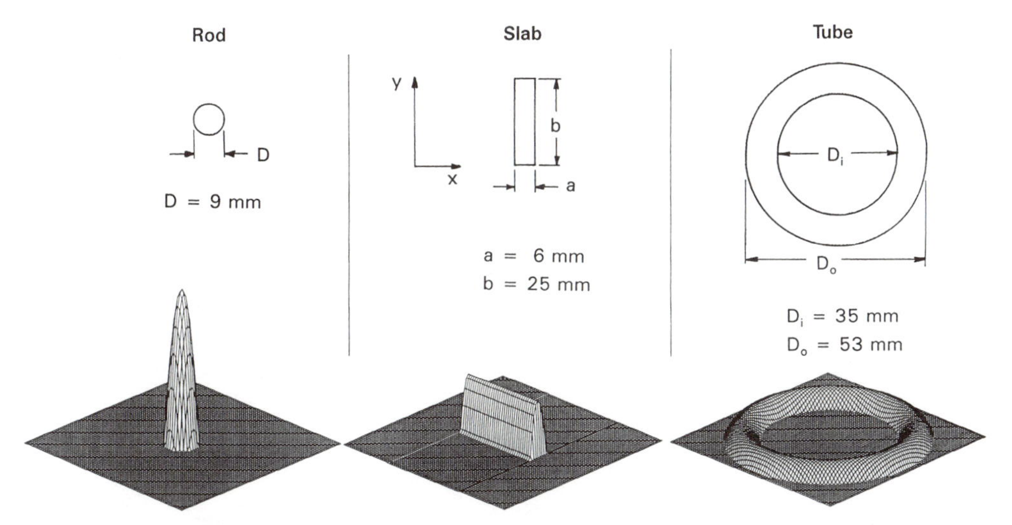

Figure 1. Temperature profiles in a rod, a slab, and a tube at equal heating power per unit length.

Thermal Stress

The temperature profiles lead to stress in the laser material. The components of the thermal stress can be calculated in the plain strain approximation.[3] In rod and tube lasers the thermal stress has circular symmetry, the principal stress components being σ_θ, σ_r, and σ_z in the azimuthal, radial, and longitudinal directions, respectively. The maximum stress occurs at the surface. In the case of a rod it is,

$$\sigma_{s,rod} = \frac{\alpha E}{8 \pi k (1-\nu)} \cdot (P_{heat}/l)_{rod} \tag{5}$$

Here, α is the coefficient of thermal expansion, E is the modulus of elasticity, k is the heat conductivity, ν is the Poisson number, and $(P_{heat}/l)_{rod}$ the heating power per unit length in the rod. The stress does not depend on the rod diameter. Therefore, the only way of power scaling rod lasers is to use longer rods (for Nd:YAG, up to 200 mm are available) or several rods.

In the case of a slab, the stress components are σ_y and σ_z; the third component, σ_x, is zero. See Fig. 1 for the slab coordinate system and dimensions. The stress at the slab surface is inversely proportional to the aspect ratio b/a of the slab,

$$\sigma_{s,slab} = \frac{\alpha E}{12 k (1-\nu)} \cdot \frac{a}{b} \cdot (P_{heat}/l)_{slab} \tag{6}$$

Thus, the thermal fracture limit of slabs can be increased by making them thinner or wider.

The stress in a tube with a wall thickness small compared to its mean diameter can essentially be treated like that in the slab that would be obtained when the tube were cut lengthwise and unrolled. The width b_t and thickness a_t of this slab are

$$\begin{aligned} b_t &= \pi \cdot (R_o+R_i) \\ a_t &= (R_o-R_i) \end{aligned} \tag{7}$$

where R_o and R_i are the outer and inner tube radius, respectively.

Assigning a fracture stress to brittle materials such as YAG is very difficult because it would depend on the flaw size distribution at the surface[4]. Also, the heating power is generally not known exactly. The fracture limit of flashlamp-pumped Nd:YAG rods is experimentally established to allow up to 40 W of output power per centimeter of rod length with stable resonators of high extraction efficiency:

$$(P_{out,max}/l)_{rod} \approx 40 \text{ W/cm} \tag{8}$$

This value can be used to estimate the maximum output power from slabs and tubes without knowing either the absolute value of the normalized heating power χ nor the maximum allowed surface stress. We assume, for simipilicity, that there is a certain maximum safe surface stress $\sigma_{s,max}$ that depends on the material only, although this is not quite true.[xx] If the surface stress in a slab and a rod have the same value $\sigma_s^s = \sigma_s^r = \sigma_{s,max}$ we obtain from (3), (4), and (8) the ratio of the maximum output power per unit length of the slab to that of a rod,

$$\frac{(P_{out,max}/l)_{slab}}{(P_{out,max}/l)_{rod}} = \frac{3 b \eta_{extr,slab} \chi_{rod}}{2 \pi a \eta_{extr,rod} \chi_{slab}} \tag{9}$$

Provided that the normalized heating power χ and the extraction efficiency are about the same in the slab and the rod laser we get:

$$\frac{(P_{out,max}/l)_{slab}}{(P_{out,max}/l)_{rod}} \approx \frac{3}{2\pi} \cdot \frac{b}{a} \tag{10}$$

This means that slabs with an aspect ratio $b/a = 2$ can deliver the same output power as a rod of equal length. A typical Nd:YAG slab of $b = 25$ mm and $a = 6$ mm, and a pumped length of 125 mm can thus deliver up to 1 kW of output power.

Combining Eqs. (3), (4), (6), and (8) we obtain the corresponding ratio of output powers for a tube and a rod,

$$\frac{(P_{out,max}/l)_{tube}}{(P_{out,max}/l)_{rod}} \approx \frac{3}{2} \cdot \frac{R_o+R_i}{R_o-R_i} \tag{11}$$

The advantage of the tube geometry is that much larger aspect ratios b_t/a_t can be realized with tubes than with slabs. The output power from a tube, with an inner diameter of 35 mm and an outer diameter of 53 mm can be 7.3 times higher than that for a rod of equal length. For a tube with the same innner diameter and 41 mm outer diameter, it can even be 19 times higher.

Thermal Deformations

The thermal stress leads to small deformations at the the end surfaces of rods, slabs, and tubes. Koechner has described this effect in more detail for rods,[5] where it is of minor importance. It is quite important in zig-zag slabs, as we will see later, and is also significant in some tube lasers.

ROD LASERS

Thermal Lensing

The parabolic temperature and stress profiles in the rod cause a corresponding variation in the refractive index. A parabolic index profile acts as a spherical lens. The power of the thermal lens (in diopters) is proportional to the electrical pump power. The constant of proportionality is called the *rod sensitivity factor* D_{el}.[5] The sensitivity factor for a typical Nd:YAG rod of 9.5 mm in diameter in a pumping chamber with 5.3% excitation efficiency is approximately $D_{el} = 0.38\ \mathrm{m}^{-1}/\mathrm{kW}$.

Laser resonators are characterized by their *g-parameters*,[6] with $g_{1,2} = 1\text{-}L/R_{1,2}$, where L is the mirror spacing and $R_{1,2}$ the radius of curvature of mirrors 1 and 2. The thermal lens changes the effective g-parameters and with varying pump power the resonator moves on a straight line in the g-diagram [Fig. 2]. The lowest beam parameter product of stable resonators is obtained close to the limits of stability in the g-diagram. It can be shown that the wider the range of pump power over which the resonator is stable, the larger is the maximum beam parameter product which will be encountered when the pump power is varied [Fig. 2]. Nevertheless, if the thermal lens were free of aberrations, resonators could be designed to provide nearly diffraction limited beams at the maximum pump power.

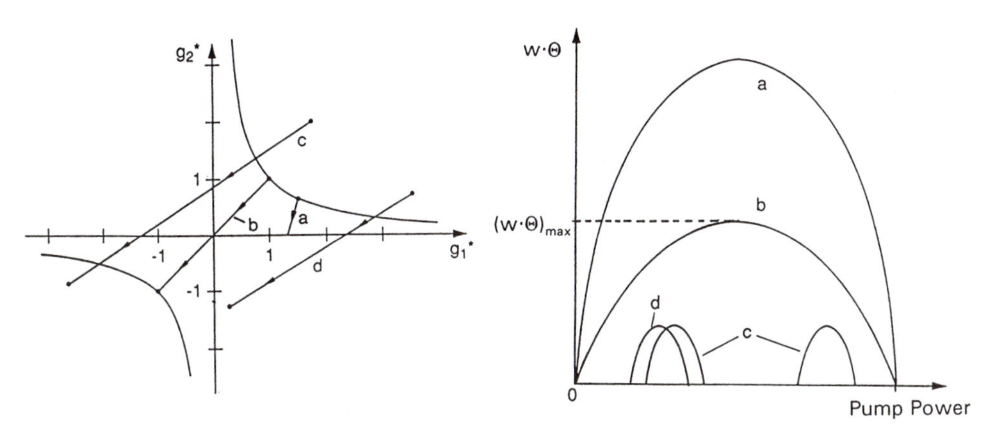

Figure 2. Left: Path of stable resonators through the g-diagram under varying pump power. Right: The corresponding variation of the beam parameter product.

Stress-Induced Birefringence

So far, we have neglected the fact that the lensing contribution of the thermal stress is different for light polarized parallel and perpendicular to the principal axes of the stress tensor. Just like the principal axes of the stress tensor, the principal axes of the resulting indicatrix are oriented radially and azimuthally everywhere in the rod. This would not present a problem if the laser mode were purely radially or azimuthally polarized. However, a field of this type would have a singularity at its center where the radial and azimuthal directions are not defined. The field would have to vanish at the rod axis where the gain is usually the highest. Thus, a mode in either the radial or azimuthal polarization eigenstate does not normally occur in rod lasers. It only occurs with resonators at the limits of stability in the *g*-diagram, when the resonator becomes unstable for one polarization. Such modes have high losses and low extraction efficiency. Predominantly radially polarized modes sometimes also occur in unstable resonators with annular output beams where they have lower losses than azimuthally polarized ones. In general, the modes contain a mixture of both polarizations and for such fields the thermal lens becomes bifocal. The difference in the focusing powers is approximately 18% for Nd:YAG rods.

Thermo-Optical Aberrations

The parabolic solutions of the heat conduction equation in Fig. 1 were based on the assumption that the heat source density in the active material was constant, and that the thermal conductivity was independent of the temperature. Both of these assumptions are not true. Imaging of the flashlamp into the laser rod with a specularly reflecting elliptical pumping chamber produces a nonuniform energy distribution in the rod, with a peak at the center of the rod and some rotational asymmetry due to the position of the lamp. Moreover, at high average power, temperature differences of up to 40 K occur in Nd:YAG rods. The thermal conductivity of YAG varies by about 15% over this temperature range. Taking these two effects into account, the temperature profiles are no longer parabolic. The thermal lens has (positive) spherical aberration and astigmatism. A deviation from the parabolic phase shift by 5 λ at the outer region of a rod of 8 mm diameter has been measured at 10 kW of pump power.

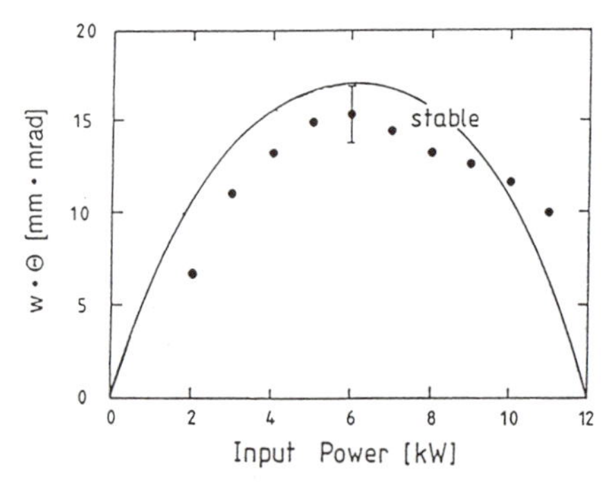

Figure 3. Measured beam parameter product of a planar resonator as a function of the pump power. At high pump power $\theta \cdot w$ remains high due to thermo-optical aberrations.

The bifocal nature of the thermal lens and the aberrations prevent the beam parameter product from returning to the low-power values as the pump power is increased.[7] Figure 3 shows experimental results of a rod laser. Compared to the aberration-free theoretical results from Fig. 2, only a minor decrease of the beam parameter product at high pump powers can be seen.

Resonators for Multi-Rod Lasers

Stable Multi-Rod Resonators. We have already seen that the maximum output power of rod lasers only increases with the rod length. Up to 600 W can be obtained from rods of 15 cm pumped length. A very simple and efficient way of achieving higher power is to put several rods in series into one planar resonator. If the rods are evenly spaced between the mirrors, as depicted in Fig. 4a, their thermal lenses form a periodic lensguide with beam waists in the mid-planes between the rods and on the mirrors. The planar multi-rod resonator with module length L becomes unstable at the same pump power per rod as a planar single rod resonator of length L. The maximum beam parameter products of both resonators are also equal. A multi-rod resonator with m rods, however, travels m times through the g-diagram before it goes unstable.[8] On each transit, it passes through the confocal point at $g_1 = g_2 = 0$. A slightly asymmetrical resonator, however, will pass the confocal point at some distance and go unstable for a small range of pump powers. Figure 4b shows how the path through the unstable region gets longer with each pass through the g-diagram. It is for this reason that multi-rod resonators with more than six rods are not practical. The bifocal thermal lens actually helps a little because, for a small deviation from the $g_1 = g_2$ - diagonal, the laser will remain stable for either the radial or the azimuthal polarization. Maximum output power from our cw-laser with six pumping chambers was 1.95 kW with a beam parameter product of 44 mm·mrad.[8] In our notation the beam parameter product is defined as the product of the beam waist radius in the near field (or half-width, for rectangular beams) that encloses 86% of the energy and the corresponding half angle in the far field.

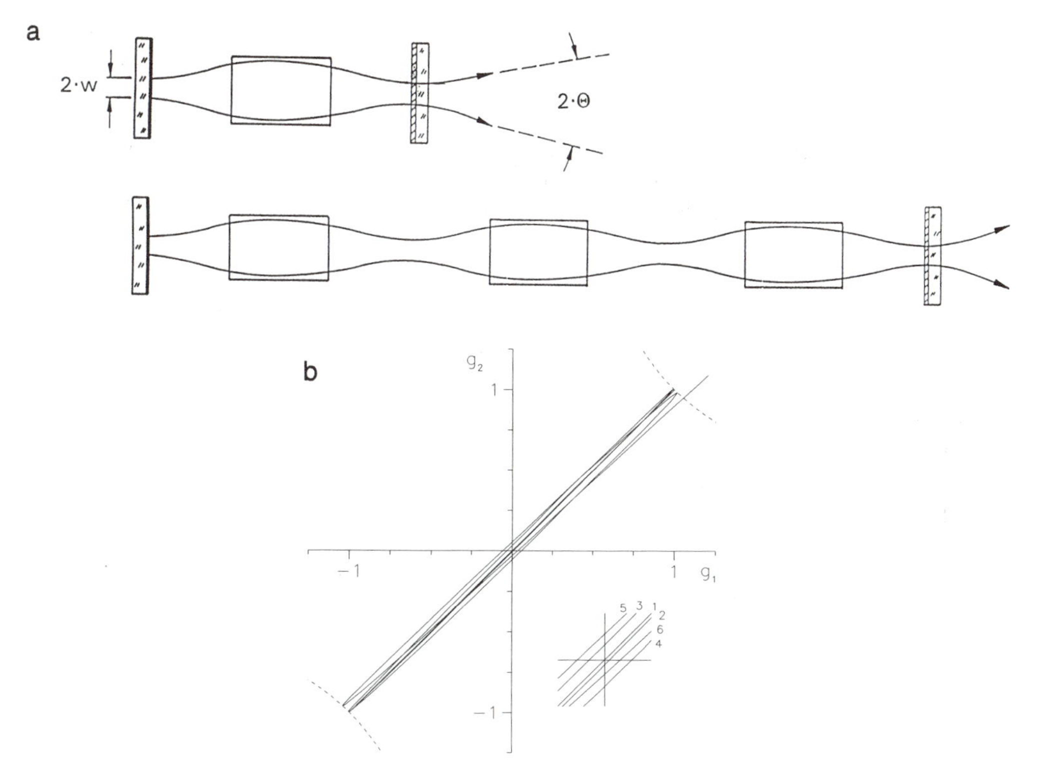

Figure 4. **a:** Setup of a symmetrical stable multi-rod resonator with the thermal lens as a lensguide. **b:** Path of a slightly unsymmetrical multi-rod resonator through the g-diagram.

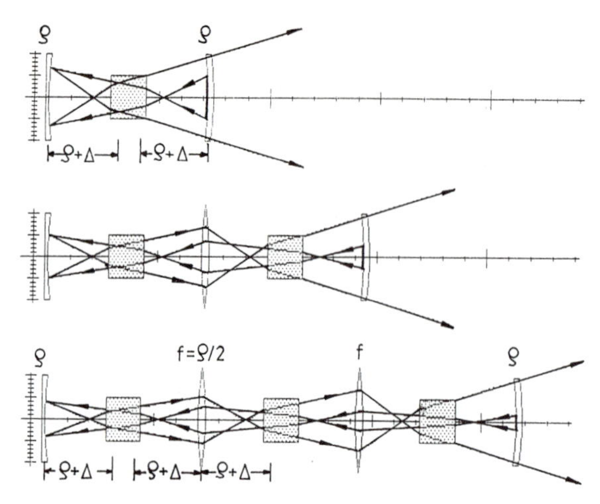

Figure 5. The near-concentric resonator with one, two, and three rods.

Unstable Multi-Rod Resonators. In unstable resonators the mode is magnified on each round trip. Therefore, the lensguide can no longer be periodic with equal mode size in each rod as in stable multi-rod resonators. Moreover, the thermal lens changes the waist locations in the resonator. The resonator has to be carefully designed in order to avoid that at some pump power a beam waist occurs near one of the rod end faces and destroys them. The mode fill factor has to be maximized for optimum efficiency but clipping of the beam at the rod apertures has to be avoided. Unlike in a stable resonator, the mode does not adapt automatically to the rod diameter and losses of the beam at apertures on its last trip through the resonator reduce the output power. All this complicates the application of unstable resonators to multi-rod lasers.

Figure 5 shows a near concentric unstable resonator with two internal focal points that was used with a pulsed multi-rod laser.[9,10] This resonator with a variable reflectivity mirror was chosen because it has a low sensitivity of the magnification to thermal lensing and a good beam quality. Output power and beam parameter product as a function of the pump power per rod are shown in Fig. 6 and Fig. 7. Maximum output power was 680 W with two rods and the addition of the third rod produced only a slightly higher output power of 700 W. The main reason for this is that the fill factor is decreasing with the number of rods due to the non-periodic mode size in this unstable lensguide. The beam parameter product was 7 mm·mrad with two rods and increased to 10 mm·mrad with three rods because of the thermo-optical aberrations.

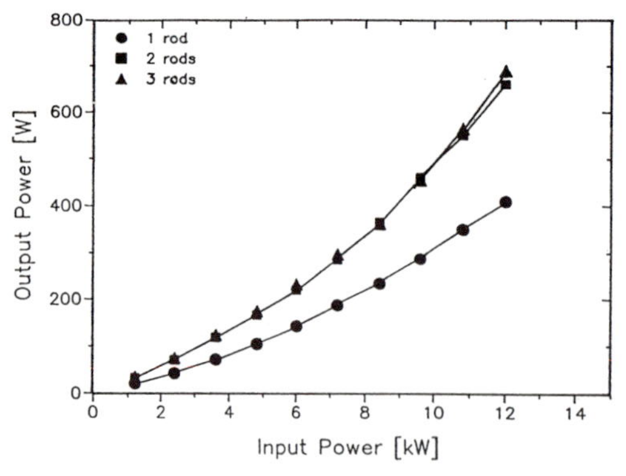

Figure 6. Output power of the near-concentric unstable resonator.

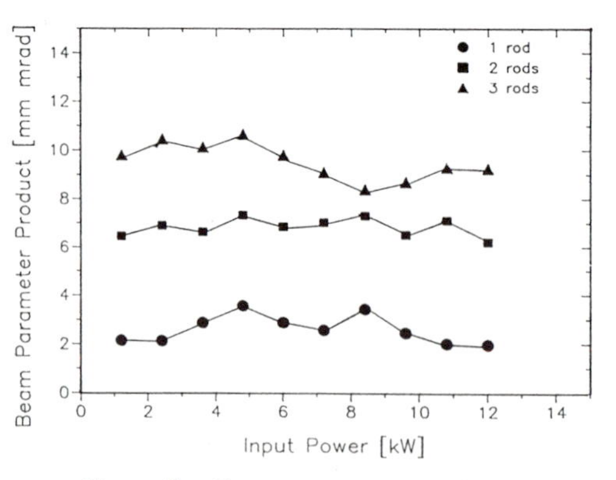

Figure 7. Beam parameter product of the near-concentric unstable resonator.

SLAB LASERS

In contrast to the situation in rods, the direction of the principal axes of the thermal stress in the rectangular slab is constant throughout the cross section of the slab, except for the region near the edges. A laser beam, linearly polarized parallel to one of these axes, does not experience bifocusing or depolarization. In slab lasers with a zig-zag optical path there is also no focusing action of the temperature and stress profiles in the x-direction because all parts of the beam wavefront pass through the same temperature and stress distributions [Fig. 8].[11,12,13] This is the great advantage of the zig-zag slab concept.

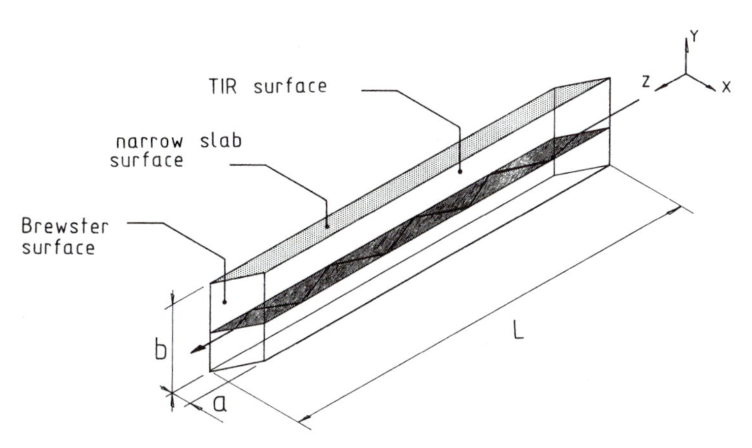

Figure 8. The slab concept. The zig-zag beam experiences no net focusing in x-direction by the temperature profile. Homogeneous heating and cooling across the width of the slab also prevent focusing in y-direction.

Thermo-Optical Aberrations

A slab would only be absolutely free of thermo-optical aberrations if it were infinitely long and wide. A finite slab has significant distortions at the Brewster tips (*end effects*) and at the upper and lower edges (*edge effects*). Numerical Finite Element Methods have to be used to calculate the 3-dimensional temperature and stress distributions that cause these effects.

We have to look in some more detail at the technical aspects of slab lasers in order to understand the end and edge effects, and see how they can be minimized. The zig-zag path of the laser beam through the crystal is maintained by total internal reflection (TIR) at the large polished surfaces. Often, a dielectric coating of 1 to 2 μm SiO_2 is applied to these surfaces because then the water seals can be placed anywhere without destroying the TIR of the laser beam. The problem of contamination of the TIR-surfaces by the cooling water is thus also avoided.

Sealing a slab at its fragile tips is technically very difficult. Therefore, in many designs the slab is not cooled over its full length but only over the central, pumped part, and the Brewster tips extend into air. In this case, there is no temperature control of the slab tips. Consequently, thermo-optical distortions can arise.

Homogeneous heating and cooling of the slab across the width of the TIR-surface are crucial in order to avoid temperature gradients in the y-direction, where no zig-zag beam-averaging occurs. However, even with perfectly homogeneous heating and cooling across the slab width, the finite slab height causes nonuniform stress distributions near the free

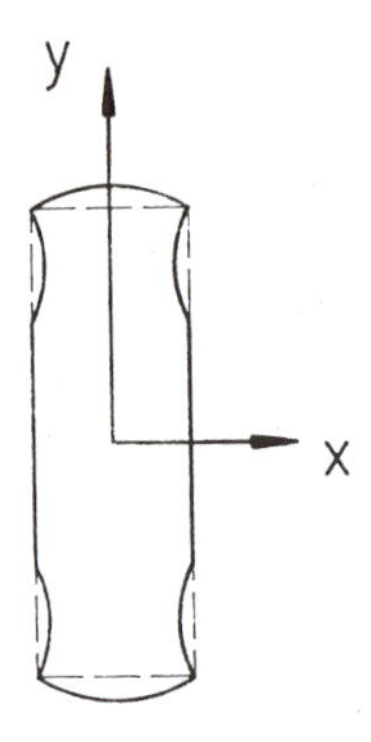

Figure 9. Deformation (strongly exaggerated) of the slab cross section under pumping due to the stress distribution near the free narrow slab surfaces.

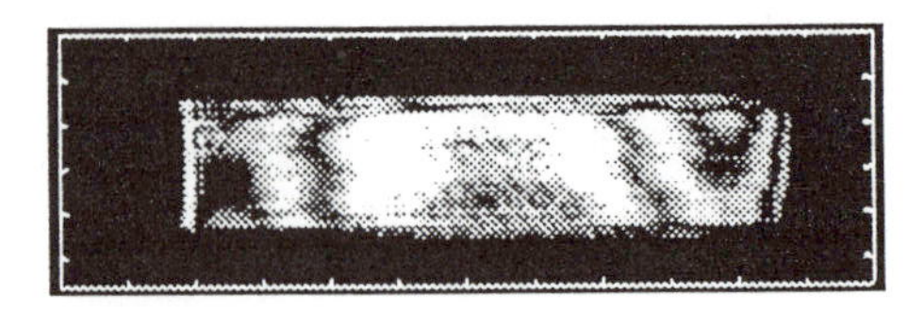

Figure 10. Mach-Zehnder interferogram (λ=632 nm) of a slab at 10 kW pump power. The distortion is approximately 3 fringes.

narrow slab surfaces. The TIR surfaces become slightly bulged at the edges, deflecting the laser beam as it bounces off them [Fig. 9]. The nonuniform stress also depolarizes the beam, causing losses at the Brewster tips. The edge effects can be minimized, to some extent, by active temperature control of the narrow slab surfaces. Water-cooled rails which are in thermal contact with the narrow slab surfaces are used for this purpose.

Figure 10 shows the interferogram of a slab with edge temperature control but uncooled Brewster tips at 10 kW of pump power.

Resonators for Slab Lasers

Stable and Folded Stable Resonators. The difficulty of obtaining low-order modes of large diameter in stable resonators is well known. The highest-order mode that can oscillate is determined by the ratio of the limiting aperture in the resonator to the fundamental mode size. In slab lasers, the critical dimension is the large slab width *b*. The highest mode order is approximately,

$$n_{max} \approx \left(\frac{b}{w_{00}} \right)^2 \tag{12}$$

where w_{00} is the fundamental mode size in the slab. In principle, w_{00} can be maximized by making the resonator very long or by employing near-planar or concave-convex resonators. But these resonators approach the limit of stability in the *g*-diagram and become very sensitive to perturbations such as mirror misalignment or thermo-optical aberrations in the slab. However, the rectangular cross section of the slab provides for an elegant solution to this problem. The beam can be folded several times in the slab, thereby simultaneously reducing the beam width and increasing the resonator length. A particularly simple and efficient setup is obtained by using prisms as depicted in Fig. 11.[14,15,16] As the beam width is reduced by *m*-times folding from b to $b/(m+1)$, the mode order n_{max} decreases as $(m+1)^{-2}$, according to Eq. (12). Again, the resonator becomes sensitive to misalignment when it becomes too long. This is avoided by the oscillator-amplifier configurations in Fig. 11. A partially reflecting dielectric coating is applied to one half of the hyputhenuse surface of a prism and forms one resonator mirror. In the resonator in Fig. 11 c, only the central part of the slab, where the thermo-optical aberrations are lowest, is used for the oscillator. The rest of the slab volume works as a single pass amplifier for the beam. The beam width is reduced but the resonator length is not increased. Figure 12 shows the experimental results that were obtained with the resonators from Fig. 11. The observed reduction of the beam parameter product agrees well with Eq. (12).

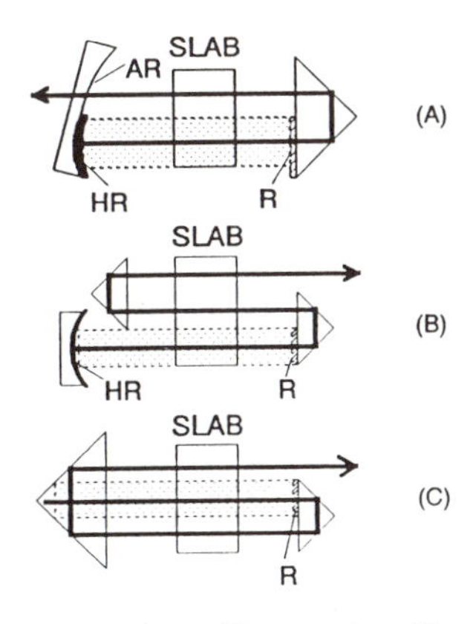

Figure 11. Folded prism resonators in oscillator and oscillator-amplifier configuration.

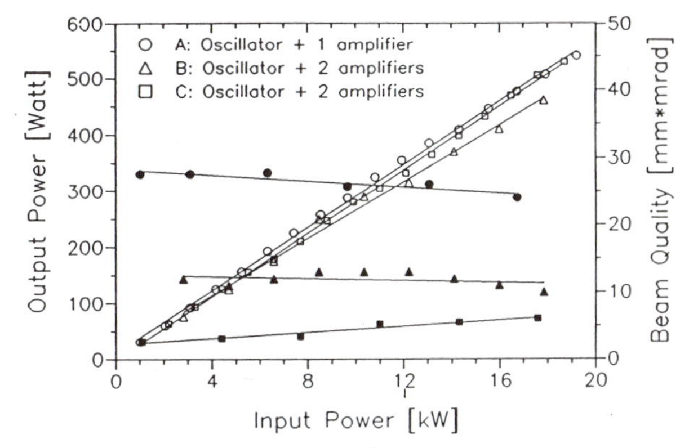

Figure 12. Output power and beam parameter product of the resonators from Figure 11. Resonator *C* provides the best beam quality.

Unstable Resonators. Unstable resonators can provide low beam parameter products from active media of large cross section. Figure 13 shows different unstable resonators we have been using with our slab lasers.[17] An interesting type is the asymmetrical off-axis unstable resonator. It has smaller side lobes in the far field than the on-axis type, with a hard-edged mirror, the far field of which is the Fraunhofer diffraction pattern of a double slit. Another advantage is that the output coupling can be optimized simply by shifting the output mirror. A severe disadvantage, however, is that the resonator has to be realigned with varying pump power because the optical axis lies at the lower edge of the slab, in the region of the greatest optical distortions. The best beam quality was obtained with the on-axis unstable resonator using a variable-reflectivity mirror. The mirror has a rectangular high-reflecting spot of 3.4 mm by 16 mm with a peak reflectivity of 85% and a super-Gaussian profile. The resonator was a negative-branch resonator with one internal focus and g-parameters $g_1 = -0.14$ and $g_2 = 0.43$. Figure 14 shows the resonator and the experimental results. Output power from an Nd:YAG slab of 7 x 26 x 191 mm^3 was 475 W at 18 kW pump power. The maximum beam parameter product was 3 mm·mrad in the x-direction. Lower beam parameter products could not be achieved with this slab because of the aberrations at high pumping powers.

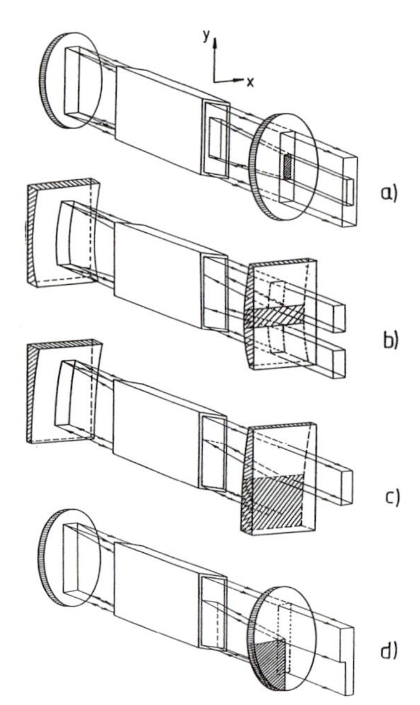

Figure 13. Slab resonators:
a: on-axis unstable,
b: on-axisplane-unstable,
c: off-axis plane-unstable,
d: off-axis unstable

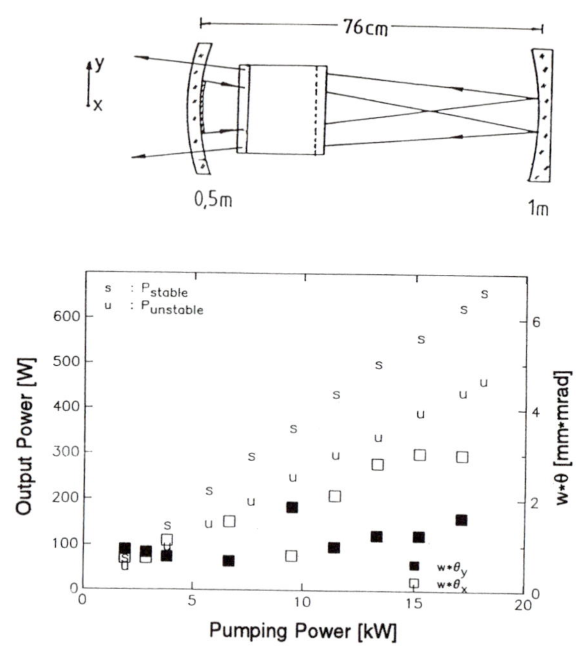

Figure 14. Top: On-axis negative branch unstable resonator with variable reflectivity mirror; bottom: experimental results.

TUBE LASERS

The tube laser is the least known of the three types of lasers we are discussing in this article, and we will therefore give a brief description of this laser concept. We have seen already that the tube geometry is very well suited for delivering high average power from a single laser crystal due to the low thermal stress. Another advantage of tube lasers, which are excited with flashlamps that are placed inside the tube, is the very efficient absorption of the pumping radiation in the active material. This is especially important in low-doped materials such as Nd:YAG.

The first internally-pumped tube laser was built in 1971 with a Nd:Glass tube of 31.8 mm inner diameter, 38 mm outer diameter, and 152 mm length.[18] The maximum output power was 12 W but the efficiency of 1.3% and the beam quality of this laser were not superior to rod lasers and this was attributed to the poor optical quality of the Nd:Glass tube.

The concept was essentially not considered again until 1987, when design studies were published by Saito and coworkers.[19] They built an externally pumped system employing a Nd:YAG tube of 17 mm inner diameter, 30 mm outer diameter, and 154 mm length. Only two flashlamps in a rather simple pumping chamber of rectangular cross section were used to pump the tube. The experimental results were published as an internal report only.[20] A planar resonator was used and the maximum output power was 280 W, the total efficiency was 0.9%, and the beam parameter product was 83 mm·mrad. Recently, a German laser manufacturer has studied externally-pumped systems again and built a laser with a Nd:YAG tube of similar dimensions. 1 kW of output power, 4.1% efficiency and a beam parameter product of 65 mm·mrad at 800 W of output power have been reported.[21]

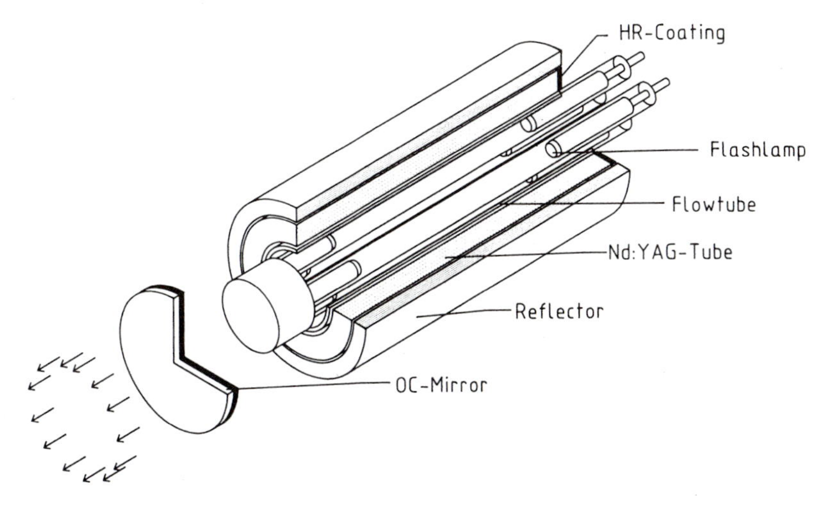

Figure 15. Internally pumped Nd:YAG tube laser. The Nd:YAG crystal of 35 mm inner diameter, 53 mm outer diameter, and 101 mm length is pumped from the inside by four flashlamps.

The first internally-pumped Nd:YAG tube laser was built in 1990 at our institute.[22] A schematic drawing of the system is shown in Fig. 15. We used a Nd:YAG tube of 35 mm inner diameter and 53 mm outer diameter. The boule showed a defect and the tube had to be cut to a length of 101 mm. This resulted in a pumped length of 85 mm, as opposed to the flashlamp arc length of 100 mm. Four standard Krypton flashlamps of 7 mm bore diameter are used, connected in series in pairs, and enclosed by one common flowtube. The high reflecting resonator mirror is formed by a dielectric coating which is directly applied to the planar end surface of the tube. The electrical connections and the water channels for cooling the inner surface of the tube and the lamps all go through the rear end of the tube. There are no obstructions of the annular beam. We have pumped the tube with up to 15 kW of pump power and obtained 1 kW of output power with a planar resonator [Fig. 16]. The slope efficiency with this resonator is 9% and the total efficiency is 7.5% at up to 300 W.

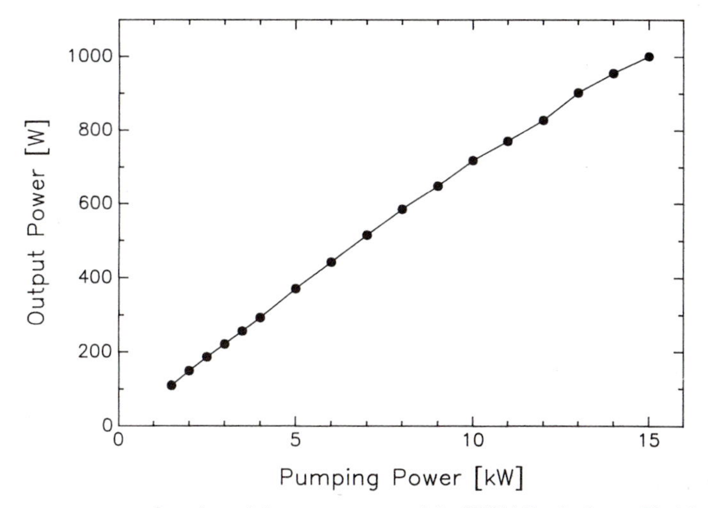

Figure 16. Output power as a function of the pump power of the NdYAG tube laser. Flashlamp puls energy was 988 J in 2 ms-pulses. Slope efficiency is 9% and total efficiency 7.5% at up to 300 W.

Apart from the Nd:YAG tube we have also used a Nd:Glass of 35 mm inner diameter, 41 mm outer diameter, and 132 mm length. The tube was made from Schott LG760 laser glass, doped with 6 wt% Nd. The maximum output power was over 90 W. With Nd:Glass rods of equal length no more than 15 W is possible without risking fracture of the rod.

Thermal Lensing

The temperature and stress distributions in the tube wall have approximately annular parabolic shape. This results in a toric lens, a cylindrical lens 'bent' to an annulus. The flat annular end surfaces also change their shape slightly under pumping and acquire a parabolic curvature (of the kind that the temperature distribution in Fig. 1 represents). This adds to the lensing. For short tubes with large wall thickness, the relative contribution of this effect is more significant than in rods; in our Nd:YAG tube it amounts to approximately 15%.

Stress-Induced Birefringence

The stress-induced indicatrix ellipsoid in the tube wall has the same orientation as in rods. If planar or toric stable resonators are used, the output beam of the tube laser has an annular shape, with the inner and outer radius of the laser crystal. The near field is zero on the optical axis. It is therefore feasible that the radial and azimuthal polarisation eigenmodes can lase with the same loss factor per round trip and the same mode fill factor than any other polarization state. Experiments with the Nd:Glass tube laser have prooved this. The glass is athermal for rod lasers; the strong lensing contribution of the thermal stress is

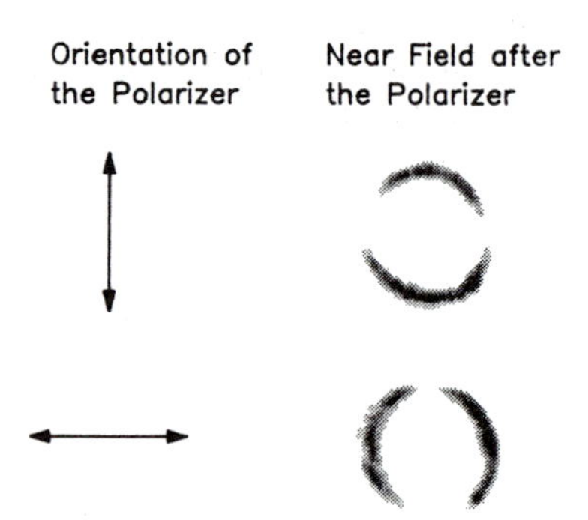

Figure 17. In the Nd:Glass tube, the negative power of the thermal lens for the azimuthal polarization leads to a radially polarized mode.

partly balancing the contribution due to the temperature dependent variation of the refractive index. The tube laser has a negative refractive power for the azimuthal polarization of $D_{el,\Theta}$ = -0.6 m^{-1}/kW and a positive power of $D_{el,r}$ = +0.7 m^{-1}/kW for the radial polarization. Under pumping, the planar resonator becomes stable for the radial polarisation only. Therefore, the mode of the laser is purely radially polarized at all pumping powers. Figure 17 shows the field after transmission through a polarizer. Radially polarized laser beams may have applications in particle beam acceleration.[23]

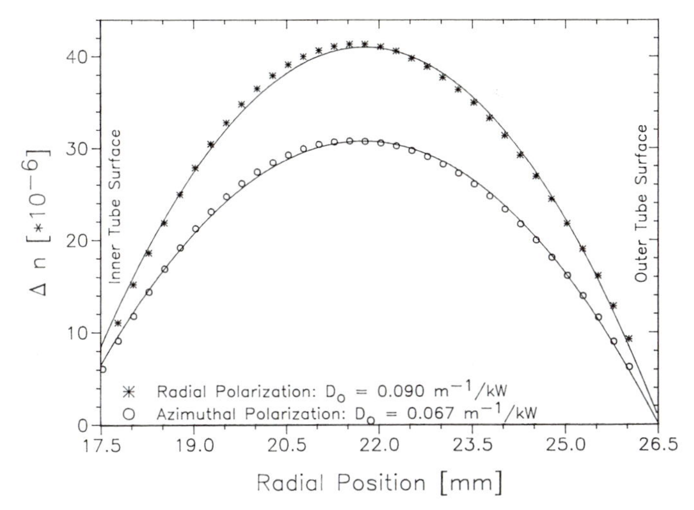

Figure 18. The refractive index profiles in the Nd:YAG tube at 4kW of pump power. The deviation from the parabola-fit is the aberration of the lens.

The Nd:YAG tube laser has a positive refractive power for both polarizations, the values are $D_{el,\Theta} = 0.067\ \mathrm{m^{-1}/kW}$ and $D_{el,r} = 0.09\ \mathrm{m^{-1}/kW}$ in a pumping chamber of 11% excitation efficiency. A planar resonator becomes stable for both polarizations under pumping and, experimentally, no preferred polarization could be observed.

Thermo-Optical Aberrations

Again, as in the case of rods, the toric thermal lens has aberrations. First of all, the temperature and stress distributons have some asymmetry due to the curvature of the tube wall. In addition, the heat source density is inhomogeneous, decreasing from the inner surface of the tube towards the outer surface by about 25%. This causes additional deviation from the parabolic profile in the tube wall. Figure 18 is the result of a measurement of the refractive index profile in the Nd:YAG tube wall. (The effect of end surface curvature can't be seperated from that of the index profile in this measurement.) Also shown in the figure is a parabolic fit to the measured data points. The deviation from the parabola can be seen clearly.

The Nd:YAG tube also has significant aberrations in the unpumped state due to the azimuthal variation of the refractive index in a Czochralski-grown Nd:YAG boule. An interferogram with a HeNe laser at 632 nm reveals 4.7 λ of peak-to-valley wavefront distortion (double pass).

Resonators for Tube Lasers

Planar Resonator. The most straightforward resonator to use with a tube laser is a planar resonator that becomes a toric stable resonator due to the thermal lens in the tube wall. As usual, the field is written as a product of the radial and azimuthal eigenfunctions,

$$E(r,\theta) = E_n(r) \cdot e^{il\theta} \tag{13}$$

The radial eigenfunctions are Gauss-Hermite polynomials divided by $r^{1/2}$ and the azimuthal eigenfunctions are simply cosine-functions,

$$E_n(r) = \frac{V_n\left(r - \frac{R_o - R_i}{2}\right)}{\sqrt{r}} \tag{14}$$

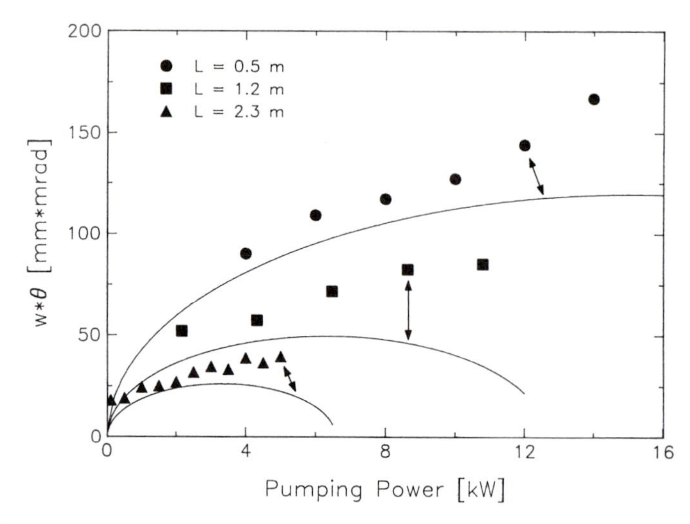

Figure 19. Beam parameter product of the Nd:YAG tube laser with planar resonators of different lengths. The solid curve represents the values derived from the calculated radial mode order.

It can be shown that a toric resonator has very little azimuthal mode discrimination.[24] The radial mode order n, however, is determined by the refractive power of the thermal lens. Since the lens is rather weak compared to rod lasers (approximately 20% at equal pump power), the radial mode order is low. Assuming, for now, that the fundamental azimuthal mode is oscillating, the far field divergence of the beam is given by radial mode order. The beam parameter product that determines the spot size and Rayleigh range of a beam, when focused by an ordinary lens, is the product of the far field divergence angle (half angle) and the beam radius at the waist in the near field. The waist of the annular beam is located at the planar output mirror and its radius is the outer tube radius. This whole-beam waist is larger than the waist of the radial mode in the tube wall by a factor ρ,

$$\rho = \frac{R_o}{(R_o - R_i) / 2} \tag{15}$$

The beam parameter product of an annular beam with radial mode order n is thus also larger than the beam parameter product of the compact beam of a rod of equal mode order by this factor ρ. For example, the beam parameter product of the fundamental radial-azimuthal mode is not λ/π, as usual, but

$$(w \cdot \theta)_{00} = \frac{2\,R_o}{R_o - R_i} \cdot \frac{\lambda}{\pi} \tag{16}$$

For our Nd:YAG tube we have $\rho = 5.9$ and for the Nd:Glass tube $\rho = 13.7$.

The fact that the beam parameter product of an annular beam is larger by the factor ρ than that of a compact beam of equal mode order is a serious drawback of the tube geometry. Measured beam parameter products of the Nd:YAG tube laser at different resonator lengths are shown in Fig. 19. Also shown is the theoretical beam parameter product as determined by the radial mode order only. It can be calculated from the fundamental mode size in the equivalent strip resonator with a cylindrical thermal lens. The equivalent strip resonator would be obtained if the tube were cut lengthwise and unrolled.

In principle, the thermal lens should not affect the azimuthal mode structure. Experiments, however, show that, with increasing pump power, the azimuthal mode order in-

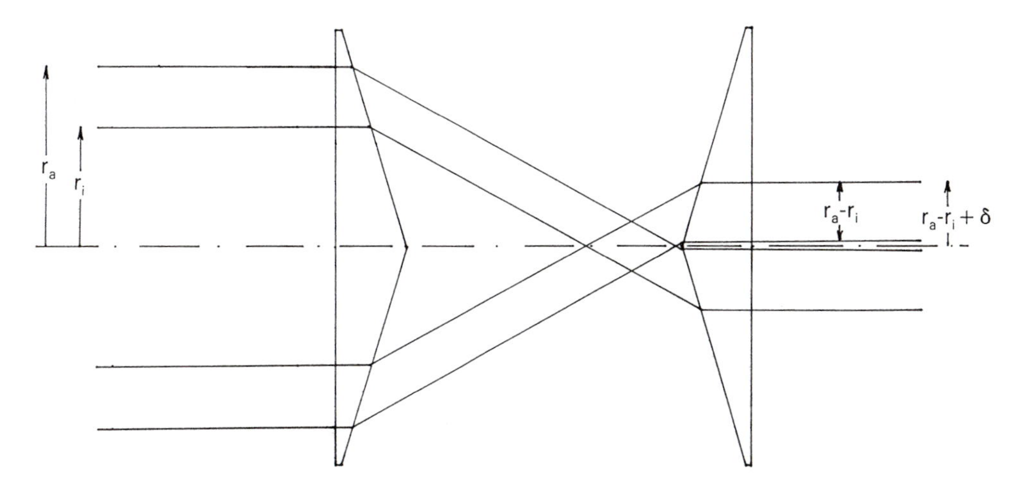

Figure 20. The axicons compact an annular beam. If the divergence doesn't increase, the beam parameter product is reduced by the same factor as the outer beam diameter.

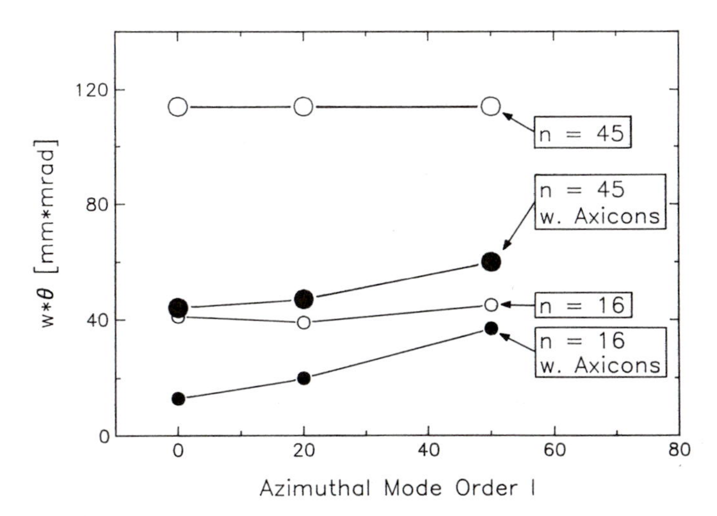

Figure 21. Calculated reduction of the beam parameter product of an annular beam of 35 mm i.d. and 53 mm o.d. by an axicon telescope for different mode orders n and l.

creases as well. The difference in Fig. 19 between the theoretical curves from the radial mode order and the measured beam parameter product is caused by the azimuthal mode structure.

Axicon-Telescope. Figure 20 shows an 'axicon-telescope' that can remove the central hole of an annular beam and compact it without increasing its divergence. The figure also shows how the outer diameter of the beam shrinks. The beam parameter product is reduced by the same ratio:

$$\frac{(w\cdot\theta)'}{(w\cdot\theta)} = \frac{R_o - R_i}{R_o} \tag{17}$$

where the primed beam parameter product is the one after the axicon telescope. The fundamental mode would then have a beam parameter product of

$$(w\cdot\theta)'_{00} = 2\cdot\frac{\lambda}{\pi} \tag{18}$$

However, an axicon telescope does not increase the beam divergence, only if the azimuthal mode order is low enough. Figure 21 shows the calculated beam parameter products with and without the axicons as a function of the azimuthal mode order. The calculations were performed numerically by propagation of modes of radial order n and azimuthal order l through the axicons using the Kirchhoff-Integral. It can be observed from Fig. 21 that, whether or not the beam parameter product of an annular beam can be reduced with an axicon telescope depends both on n and l. For low radial mode orders, it will only be reduced if the azimuthal mode order is also low, while for higher radial mode orders the reduction will also take place at higher azimuthal mode orders.

As mentioned earlier, an important limit for the beam parameter product of solid state lasers is the value of about 40 mm·mrad that is required for coupling into standard quartz fibers. From Fig. 19 it is clear that at high pump powers the beam parameter product of the Nd:YAG tube laser with a stable planar resonator is too large for fiber coupling. The use of an axicon telescope would probably also not suffice. We have to look for other resonators with an improved beam quality.

Many resonators have been developed for CO_2 lasers and chemical lasers with annular gain media.[25,26,27,28] Nevertheless, it is not easy to adapt them to solid state lasers. First of all, the thermal lens has to be taken into account. It can only be compensated by toric surfaces which are difficult to manufacture with the necessary precision for the 1 μm wavelength of the Nd:YAG laser. Some resonator designs for annular lasers require intracavity axicons.[27] These resonators are generally very sophisticated and sensitive to misalignment and may not be suitable for an industrial laser.

Multipass Herriott Cell. A common resonator for annular CO_2 lasers is the Multipass Herriott Cell.[28] Two spherical mirrors can sustain an off-axis beam that traces a circle on each mirror. The beams lie on the surface of a hyperboloid of revolution. This resonator would have the advantage of generating a compact beam of small diameter (equal to the tube wall thickness) with much better beam quality, similar to the case of folded resonators for slab lasers. Unfortunately, the wavefront aberrations in the unpumped Nd:YAG tube sum up over the very long path of the beam through the crystal. The beam expands and suffers losses at the annular tube aperture. For this reason, the efficiency was only 2% when we used this resonator with the Nd:YAG tube.

Azimuthally Unstable Resonator. Recently, an unstable resonator has been proposed for annular gain media that was derived from the Multipass Herriott Cell.[29] One of the two Herriott Cell mirrors has a surface that is formed so as to allow a beam to work its way under multiple bounces, in the azimuthal direction, to the coupling hole. The surface has a non-constant curvature in both the radial and azimuthal direction. The complicated shape of the mirror can be approximated by a tilted planar mirror and the toric lens in the tube wall, as illustrated in Fig. 23. Our experimental results with this resonator are shown in Fig. 24. The beam parameter product of 25 mm·mrad is low enough for fiber coupling and the output power was 500 W at 10 kW pump power.

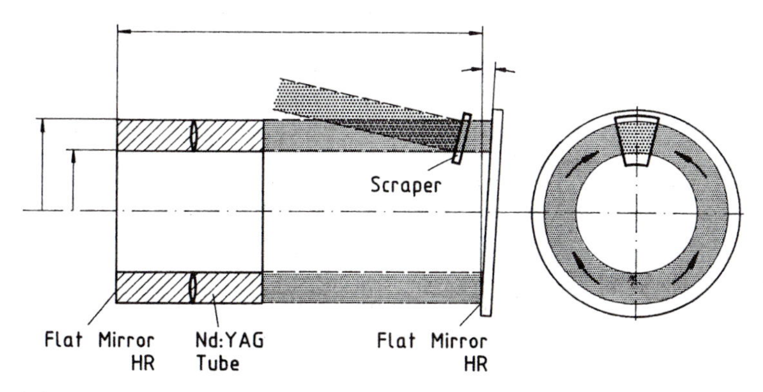

Figure 22. Azimuthally unstable resonator with two planar mirrors, toric thermal lens and scraper mirror.

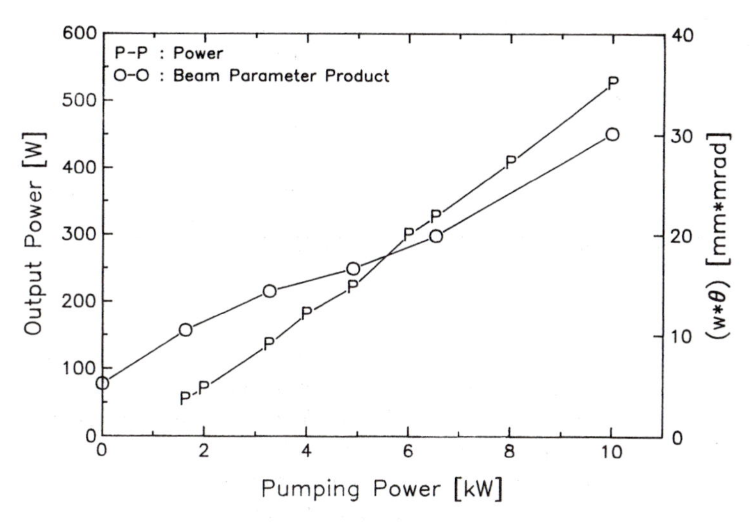

Figure 23. Output power and beam parameter product of the azimuthally unstable toric resonator. The total efficiency is 5% and the beam parameter product is low enough for fiber coupling.

CONCLUSION

It is not possible to give an adequate overview over the current state of high power Nd:YAG laser technology and the research efforts that are under way to improve it in a short article. The wide variety of applications has led to the development of different laser concepts. We have discussed the multi-rod, the slab and the tube laser. Table 1 summarizes the performance of these lasers. Listed are the data of commercially available systems in the rows marked '*c:*', values that have been demonstrated in the lab are marked '*d:*', and values that are possible relying on currently available Nd:YAG crystal size and laser technology are marked '*p:*'. It should be kept in mind, however, that the field is under intense development and further improvements can certainly be expected.

Table 1. Performance of high power Nd:YAG lasers.

		Multi-Rod		**Slab**		**Tube**	
		stable	un-stable	stable	un-stable	stable	un-stable
$P_{out,max}$ [W]	c:	2000		500		-	
at $w\cdot\theta > 40$ mm·mrad	d:	2400	↓	1200	↓	1000	↓
	p:	3000		1400		4000	
$P_{out,max}$ [W]	c:	1800	-	400	-	-	-
at $w\cdot\theta = 5...40$ mm·mrad	d:	2000	750	600	600	400	500
	p:	2500	1000	1000	800	500	?
$P_{out,max}$ [W]	c:	-	-	-	350	-	-
at $w\cdot\theta < 5$ mm·mrad	d:	-	-	-	800	-	-
	p:	?	?	?	?	-	?
Efficiency [%]	c:	5		3		-	-
at $w\cdot\theta > 40$ mm·mrad	d:	5.5		6		7.5	5
	p:	5.5		?		9	?
$P_{out,max}$ [W]	c:	150	-	-	-	-	-
Q-switch	d:	?	-	220	-	-	-
(single rod)	p:	200 (?)	-	1000	?	?	?
$P_{out,max}$ [W]	c:	2000	-	-	-	-	-
cw	d:	2400	-	1200	-	-	-
	p:	3000	-	1200	?	?	-
Scaling to higher power		difficult	-	multi-slab	multi-slab	multi-tube ?	?

c: commercially available, d: demonstrated, p: possible with current technology

Acknowledgments

The research on high power solid state lasers at the Festkörper-Laser-Institut Berlin is funded by the German Ministry of Research and Technology under various contracts.

The research at the institute, which is headed by Horst Weber, is carried out by a number of people to whom I am grateful for allowing me to present their results. Some of them have not been published previously. Malte Kumkar is working on high power multi-rod lasers, Ralf Dommaschk, Shalei Dong, and Qitao Lü have developed the slab lasers at our institute, and Bernd Eppich is working with me on the tube laser. Thomas Haase and Norman Hodgson have designed the unstable resonators for the multi-rod and slab lasers and Georg Bostanjoglo has manufactured the variable reflectivity mirrors.

REFERENCES

1. G. Phillips, to be published.
2. G. Phillips, J. Vater, *1.06 μm absorption due to stable color centers in flashlamp-pumped Nd:YAG laser rods*, accepted for publication in Appl. Opt.
3. S. P. Timoshenko, J. N. Goodier, *Theory of Elasticity*, 3rd ed. (McGraw Hill, 1987).
4. J. E. Marion, *Appropriate use of the strength parameter in solid state slab laser design*, J. Appl. Phys. **62**, 1595 (1987).
5. W. Koechner, *Solid-State Laser Engineering*, 2nd ed. (Springer Verlag, Berlin, 1988).
6. A. Siegman, *Lasers* (University Science Books, 1986).
7. N. Hodgson and H. Weber, *Influence of spherical aberration upon the performance of Nd:YAG lasers*, to be published in IEEE J. Quantum Electr.
8. M. Kumkar, B. Wedel, K. Richter, *Beam quality and efficiency of high-average-power multirod lasers*, Opt. and Laser Tech., **24**, 67 (1992).
9. N. Hodgson, G. Bostanjoglo, and H. Weber, *The near-concentric resonator (NCUR) - an improved resonator design for high power solid state lasers*, to be published in Opt. Comm.
10. N. Hodgson, G. Bostanjoglo, and H. Weber, *Multi-rod unstable resonators for high power solid state lasers*, to be published in Appl. Opt.
11. W. S. Martin and J. P. Chernoch, *Multiple internal reflection face pumped laser*, U. S. Patent 3 633 126 (1972).
12. J. M. Eggleston, T. J. Kane, K. Kuhn, J. Unternahrer, and R. L. Byer, *The slab geometry laser - part I: Theory*, IEEE J. of Quantum Electr. **20**, 289 (1984).
13. T. J. Kane, J. M. Eggleston, and R. L. Byer, *The slab geometry laser - part II: Thermal effects in a finite slab*, IEEE J. of Quantum Electr. **21**, 1195 (1985).
14. Q. Lü and J. Eicher, *Off-axis prism resonator for improved beam quality of slab lasers*, Opt. Let. **15**, 1357 (1990).
15. S. Dong, Q. Lü, and J. Eicher, *Folded prism resonators for slab lasers with high beam quality,* Opt. Comm. **82**, 514 (1991).
16. N. Hodgson and T. Haase, *Improved resonator design for rod lasers and slab lasers*, SPIE Proc. **1277**, 88 (1990)
17. N. Hodgson, T. Haase, *Beam parameters, mode structure, and diffraction losses of slab lasers with unstable resonators*, Optical and Quantum Electr., **24** (1992).
18. D. Milam and H. Schlossberg, *Emission characteristics of a tube-shaped laser oscillator*, J. Appl. Phys. **44**, 2297 (1973).
19. Y. Takada, H. Saito, and T. Fujioka, *New type of solid state laser for several kilowatts*, Proc. Soc. Photo-Opt. Instrum. Eng. **801**, 62 (1987).
20. Y. Takada, H. Saito, T. Fujioka, Internal Report No. OQD-89-12, *Investigation of solid state lasers with high output power and good beam quality* (in Japanese), Laser Laboratory, Industrial Research Institute, Kashiwa-shi, Japan (1989).
21. R. Dinger, *The outside-pumped tube laser* (in German), Laser Magazin **5**, 13 (1991).
22. U. Wittrock, B. Eppich, and H. Weber, *Inside-pumped Nd:YAG tube laser*, Opt. Lett. **16**, 1092 (1991).
23. J. R. Fontana and R. H. Pantell, *A high-energy, laser accelerator for electrons using the inverse Cherenkov effect*, J. Appl. Phys. **54**, 4285 (1983).
24. M. Morin and P.-A. Belanger, *Diffractive analysis of annular resonators*, Appl. Opt. **31**, 1942 (1992).
25. A. H. Paxton and J. H. Erkkila, *Annular convergin wave resonator: new insights*, Opt. Lett. **1**, 1066 (1977).
26. T. R. Ferguson and M. E. Smithers, *Toric unstable resonators*, Appl. Opt. **23**, 2122, (1984).

27. Ch. M. Clayton and C. A. Huguley, Proc. Soc. Photo-Opt. Instrum. Eng. **1224**, 359 (1990).
28. A. Duncan, J. G. Xin, and D. R. Hall, *Herriott cell resonators for large-area gas discharge lasers*, Proc. Soc. Photo-Opt. Instrum. Eng. **1224**, 312 (1990).
29. German Patent Application DE 41 23 024 A1 (1992).

SOLID–STATE LASER MATERIALS:

BASIC PROPERTIES AND NEW DEVELOPMENTS

Günter Huber

Institut für Laser–Physik
Universität Hamburg
Jungiusstr. 11
D–2000 Hamburg 36
Fed. Rep. Germany

INTRODUCTION

In the last decade significant progress has been made in the areas of efficient and broadly tunable solid–state lasers. Important steps leading to the revival of solid–state lasers have been tunable transition metal lasers based on trivalent chromium[1,2] and titanium[3], efficient sensitized Nd–lasers[4], and efficient down conversion lasers[5,6] based on Thulium and Holmium. Also the progress in the area of diode pumped lasers has contributed very much to the renaissance in the solid–state laser field. With diode laser pumping it is possible to obtain higher efficiency and to build all solid–state devices with simpler and more compact design. Besides Nd^{3+} various efficient diode pumped rare earth lasers have been operated with Er^{3+}, Tm^{3+}, Ho^{3+} (see for instance Fan and Byer[7] and Esterowitz[8]), and Yb^{3+} (Lacovara et al.[9], Payne et al.[10]).

The successful operation of Cr^{3+} and Ti^{3+} as tunable room temperature lasers has stimulated further research in transition metal ions like Ni^{2+}, Co^{2+}, V^{2+}, V^{4+}, Mn^{2+}, and Cr^{4+}. Up to now only two of these ions could be operated at room temperature: Co^{2+} in a pulsed mode[11] and Cr^{4+} in pulsed and cw operation[12,13].

This paper reviews the basic properties of laser crystals and describes new developments in the field of rare earth and transition metal solid state lasers.

Solid State Lasers: New Developments and Applications
Edited by M. Inguscio and R. Wallenstein, Plenum Press, New York, 1993

Electric dipole transitions within the optically active 3d–electron shell are, in principle, parity forbidden. However, a static acentric electric crystal field or coupling of asymmetric phonons can force electric dipole transitions by admixures of wavefunctions with opposite parity. If the transition metal ion site has inversion symmetry, the transitions have to be vibrationally induced. The 3d–energy levels are also influenced by vibrational coupling. The electron–phonon coupling can roughly be estimated from the corresponding Tanabe–Sugano–diagram[14] which gives the 3d–energy E versus the crystal field strength Dq for an octahedral coordination. High slopes of E(Dq) yield a strong crystal field sensitivity of the energy levels and broad band transition with a big Stokes shift between fluorescence and absorption spectra (4–level laser scheme).

Transition probabilities within the 3d–shell can be described by the matrix element $(\Psi_f|H|\Psi_i)$ with the wave functions $\Psi_{f,i}$. Using the Born–Oppenheimer approximation, one can separate $\Psi(q,Q)$ into a product

$$\Psi(q,Q) = \varphi(q,Q)\chi(Q).$$

The electronic wave function $\varphi(q,Q)$ depends on the electronic coordinate q and also on the vibrational coordinate of the lattice Q as a parameter. The vibrational wave function $\chi(Q)$ depends only on Q. In the harmonic approximation we have to deal with harmonic oscillator wave functions $\chi(Q)$. In the Franck–Condon approximation, transitions between two states are expressed in terms of Franck–Condon overlap integrals $(\chi_f(Q)|\chi_i(Q))$ of thermalized vibrational states, multiplied by the electronic matrix element $(\varphi_f(q,Q)|e \cdot r|\varphi_i(q,Q))$.

Titanium

The free–ion ^2D state of the $3d^1$ configuration splits in octahedral coordination in the T_{2g} groundstate and the E_g excited state. Due to the simple level scheme Ti^{3+} should not have any problems with excited state absorption (ESA).

Figure 1 shows the configurational coordinate diagram of Ti:Al_2O_3. The diagram is derived from the fluorescence and absorption spectra[15] given in Figure 2. In addition to the simple octahedral T_{2g}–E_g splitting a Jahn–Teller splitting of both, ground state and excited state must be taken into account[16]. The spectral line shapes can be explained within a linear E,Txϵ Jahn–Teller coupling of the excited and ground state, respectively. The Jahn–Teller effect in the excited state (3080 cm^{-1}) is much stronger than in the ground state (119 cm^{-1}). The fluorescence spectrum is not double peaked like the absorption spectrum, but is extremely broad. This is due to the three–dimensional nature of the ground–state energy surface with Txϵ coupling. However Figure 1 simplifies the ground–state energy surfaces and gives the highest ($S\hbar\omega$ = 4410 cm^{-1}) and lowest energy surface ($S\hbar\omega$ = 1990 cm^{-1}). The excited Jahn–Teller state with Exϵ coupling has rotational symmetry in the three–dimensional space (so–called "Mexican Hat").

Due to the absence of excited state absorption in the spectral region of the pump and fluorescence band, Ti^{3+}:Al_2O_3 is a very efficient and broadly tunable laser material. The tuning range covers almost the entire fluorescence band.

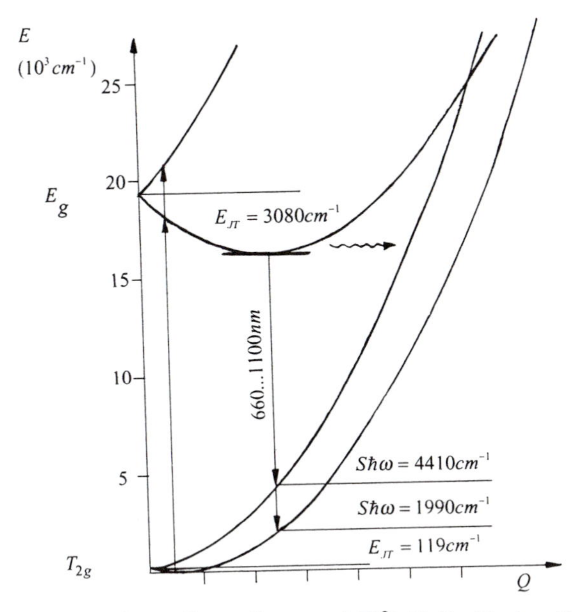

Figure 1. Configurational coordinate diagram of Ti^{3+}:Al_2O_3. Explanations see text.

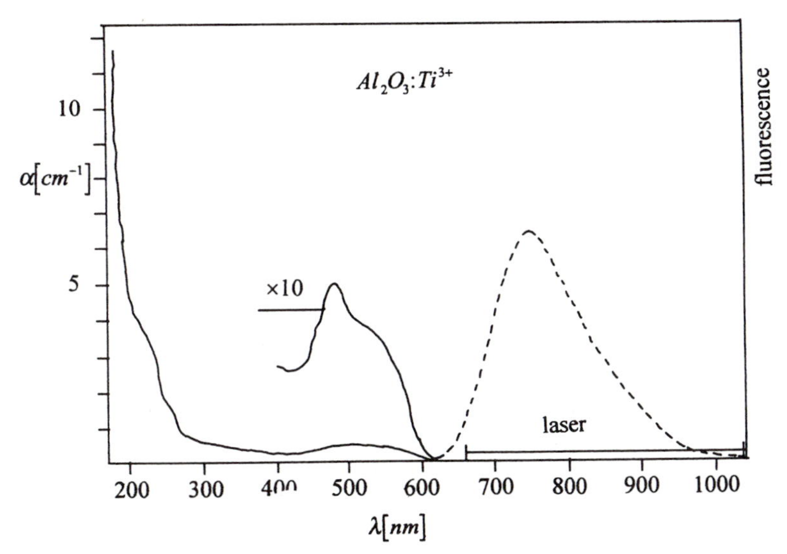

Figure 2. Absorption and fluorescence spectra of Ti^{3+}:Al_2O_3 at 300 K. The tuning range of the laser is indicated.

The room temperature quantum efficiency of $Ti^{3+}:Al_2O_3$ is 80 % as determined from calorimetric measurements and temperature dependent lifetime data[15]. The temperature dependence of the lifetime can be perfectly fitted with the nonradiative quenching model of Struck and Fonger[17]. Figure 3 shows the comparison of the theory and experiment (Albers et al.[15]) with the radiative lifetime of 3.85 μsec, a Huang–Rhys parameter S=11.25, and an effective phonon energy of 392 cm^{-1}. This parameter set corresponds to the tunneling into the upper ground–state potential in Figure 1. The room temperature lifetime is 3.1 μsec.

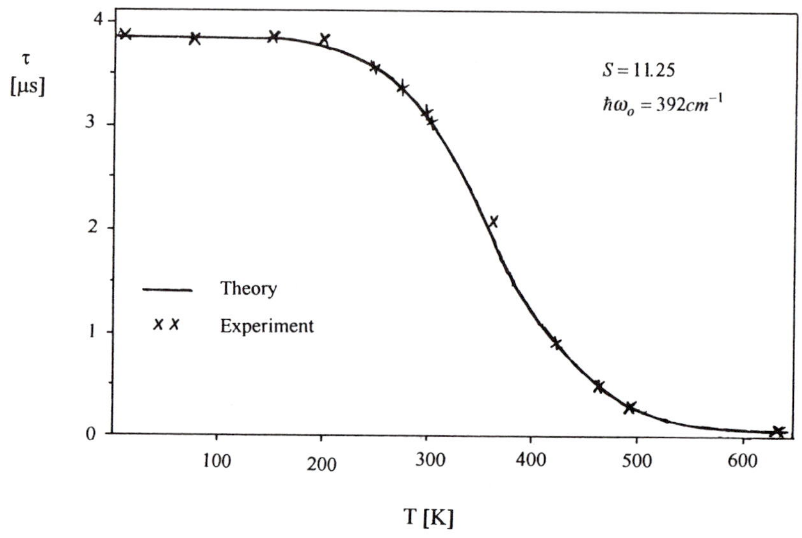

Figure 3. Temperature dependence of the lifetime τ in $Ti^{3+}:Al_2O_3$ and theoretical Struck and Fonger fit with a radiative lifetime of 3.85 μsec.

Today $Ti^{3+}:Al_2O_3$ is already a widely used laser material which can be efficiently pumped by frequency doubled Nd–lasers and cw–Argon lasers. In spite of the short lifetime flashlamp–pumping is also possible with reasonable efficiencies of the order of one percent.

In the meantime Ti^{3+} has been doped in many other host crystals like $Y_3Al_5O_{12}$, $YAlO_3$, $BeAl_2O_4$, $LaMgAl_{11}O_{19}$, $ScBO_3$, and $MgAl_2O_4$. However, there is no crystal reaching the properties of $Ti^{3+}:Al_2O_3$. Most of the crystals do not lase, but from a few crystals laser action with low efficiency has been reported. It was shown by Petermann[18], that in $Ti^{3+}:YAlO_3$ laser action is very unefficient due to a strong excited state absorption overlapping the pump band. This excited state absorption is probably connected with Ti^{3+}–complex centers and might be a problem in most of the Ti^{3+}–doped materials. $Ti^{3+}:Al_2O_3$ has this excited state absorption too, but the transition is located well above the pump band and does not influence the pump process.

Trivalent Chromium

In contrast to ruby, Cr^{3+}–doped materials can also exhibit a broad band 4–level system, if a relatively weak crystal field strength at the Cr–site yields a reduced $^4T_2$–$^4A_2$ splitting (see Figure 4). Due to the thermal population of the $^4T_2$ level and the high transition probability of the spin–allowed $^4T_2$–$^4A_2$ transition, most of the fluorescence can be channelled into the broad band $^4T_2$–$^4A_2$ transition.

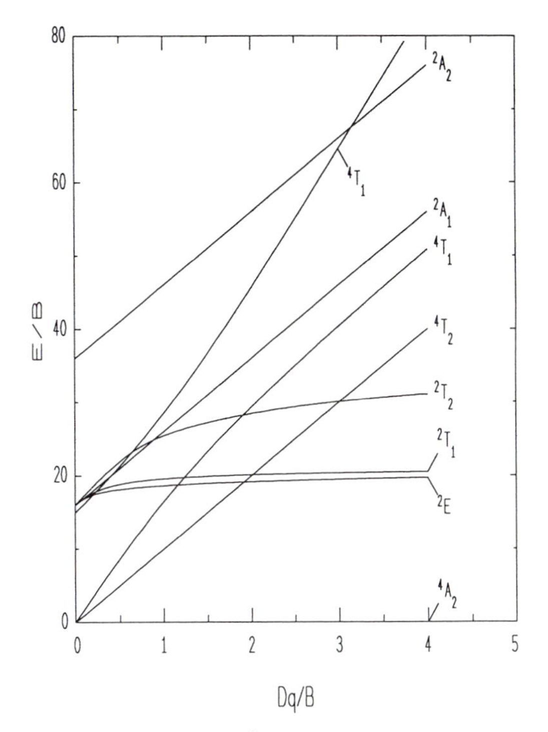

Figure 4. Tanabe–Sugano diagram for a $3d^3$ electronic configuration in octahedral symmetry.

The relative intensity of the ^2E–$^4A_2$ transitions (R–lines) and broad band $^4T_2$–$^4A_2$ transition depends on the enery gap

$$\Delta E = E(^4T_2) - E(^2E)$$

in Figure 4. For instance, ΔE is positive in alexandrite[1] (~ 800 cm^{-1}), nearly zero in GSGG[19] (50 cm^{-1}), and negative in LLGG[19] (–1000 cm^{-1}). As an example, Fig. 5 shows the emission spectra of different garnet crystals. From the top to the bottom, the lattice constant increases with a corresponding decreasing crystal field strength. Thus, the relative intensity of the R–lines near 700 nm decreases and the broad band fluorescence is shifting to longer wavelengths. Following this approach, such behaviour has been realized in many host lattices for Cr^{3+}. If the energy gap ΔE is very small, both levels ^2E and $^4T_2$ are mixed via spin–orbit coupling[19].

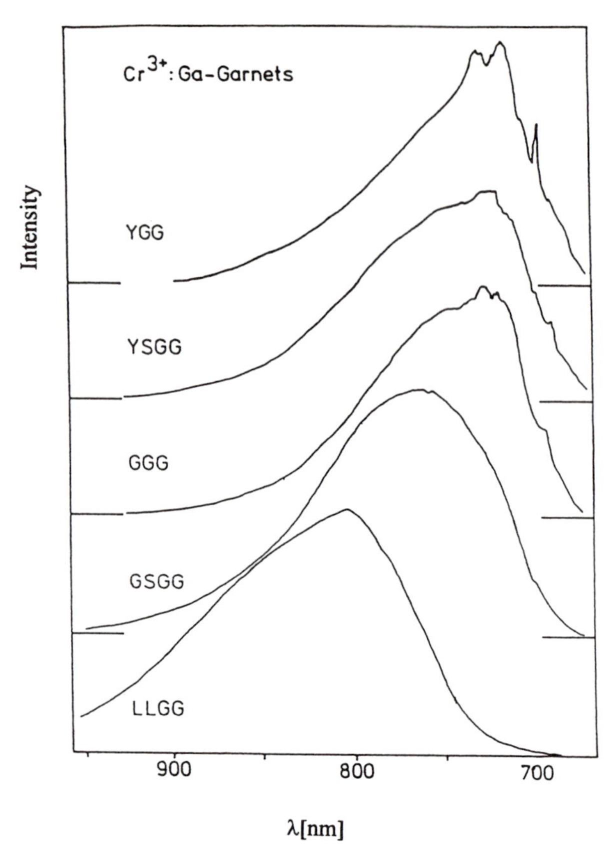

Figure 5. Emission spectra of Cr^{3+} in various Ga–garnets. From the top to the bottom the crystal field strength is decreasing.

The typical laser wavelengths of tunable Cr^{3+} systems are located between 750 nm and 1000 nm depending on the actual crystal field of the host crystal at the Cr–site. The lasers may be pumped by Argon–and Krypton–lasers or by flashlamps.

In contrast to Ti–sapphire, Cr^{3+}–lasers can also be pumped by diode lasers via the $^4T_2$–absorption band in the red spectral region. Recently, diode pumped operation of Cr^{3+}–lasers has been demonstrated in Cr:$LiCaAlF_6$ and Cr:$LiSrAlF_6$ by Scheps[20]. Both colquiriite minerals $LiCaAlF_6$ and $LiSrAlF_6$ are very efficient hosts for the Cr^{3+}–ion[21,22] with slope efficiencies exceeding 50 %.

Cr^{3+}–lasers can also be influenced by ESA processes arising from both ^2E– and $^4T_2$–excited states. In Cr^{3+}:$Gd_3Sc_2Ga_3O_{12}$ the tunability of the laser (0.74 – 0.84 μm) and the slope efficiency (28 %) are limited by ESA at the wings of the fluorescence curve and by pump ESA, respectively. Uniaxial crystals like $LiCaAlF_6$ might have polarized ESA–bands which do not influence special pump and laser polarizations.

Due to the advantage of diode pumping, a revival of Cr^{3+}–lasers is very likely.

Tetravalent Chromium

Room temperature laser action of Cr^{4+} has been achieved in Mg_2SiO_4 (forsterite)[12] and $Y_3Al_5O_{12}$ (YAG)[13]. The absorption and emission spectra (Figures 6,7) of Cr^{4+}:YAG fit very well into the $3d^2$–Tanabe–Sugano diagram (Figure 8) of Cr^{4+} in tetrahedral coordination (equivalent to octahedral $3d^8$–coordination). The broad band fluorescence corresponds to the $^3T_2$–$^3A_2$ transition and shifts to longer wavelength with decreasing crystal field strength. Typically, the energy level scheme is close to the ^1E–$^3T_2$ level crossing point. Thus, the situation is very similar to the low field case of Cr^{3+}. However, the interpretation of the low temperature spectra of Cr^{4+} is very difficult and not yet fully understood[23,24].

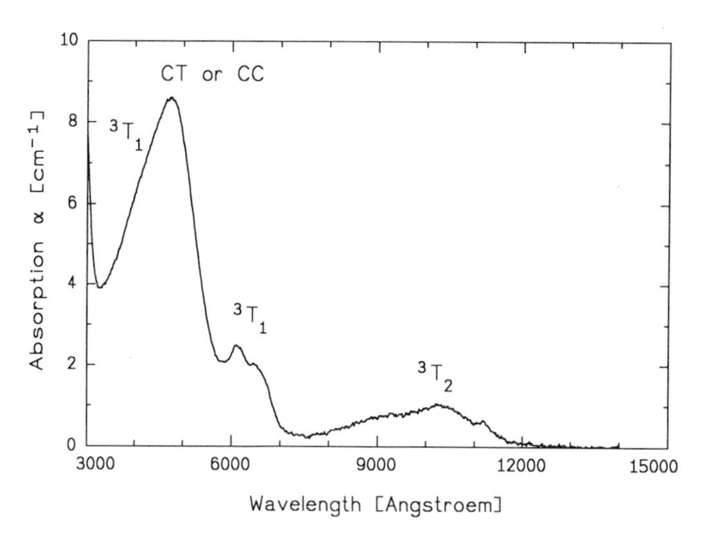

Figure 6. Absorption spectrum of Cr^{4+}:YAG at 300 K.

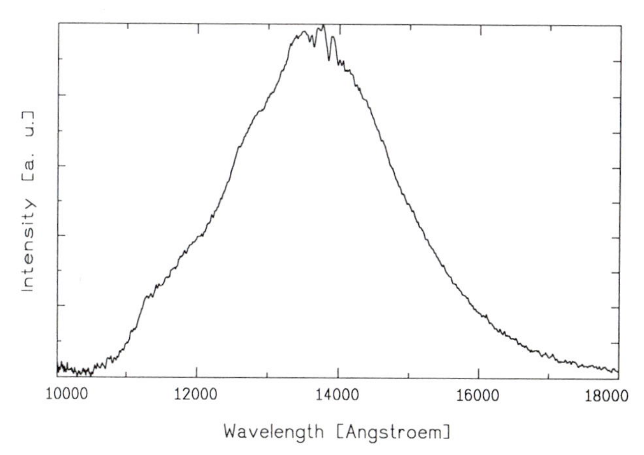

Figure 7. Room temperature emission spectrum of Cr^{4+}:YAG covering the interesting spectral regions near 1.3 μm and 1.52 μm to 1.55 μm.

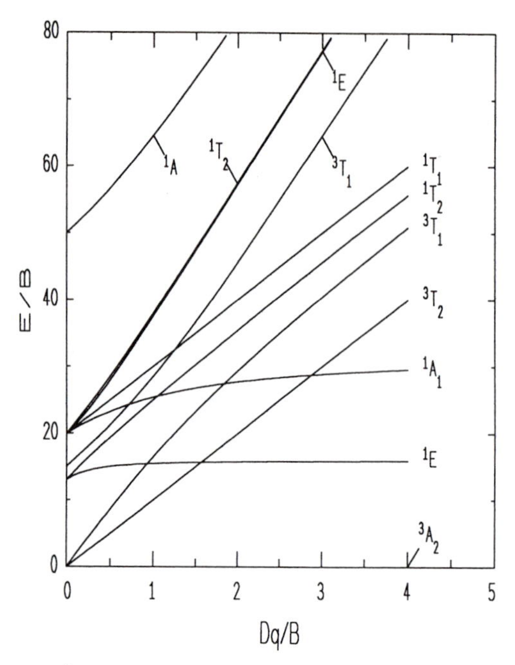

Figure 8. 3d^2–Tanabe–Sugano diagram for tetrahedral symmetry.

The tuning ranges of Cr^{4+}–lasers are located in a very interesting spectral region: 1.17 – 1.35 μm (Cr^{4+}:Mg_2SiO_4) and 1.38 – 1.58 μm (Cr^{4+}:YAG). When pumped with a Nd:YAG laser, slope efficiencies are of the order of 30 % (pulsed operation) and 10 % (cw operation). As in most transition metal ion lasers, the laser performance of Cr^{4+} seems to be limited by ESA. The room temperature lifetimes of the coupled metastable $^3T_2$– and ^1E–levels are 15 μs in Cr^{4+}:Mg_2SiO_4 and 3.6 μs in Cr^{4+}:YAG. In both materials, the lifetime shows a relatively strong temperature dependence, which might be caused by nonradiative processes. The quantum efficiencies of Cr^{4+} laser materials are still not known.

RARE EARTH LASERS

As in the case of 3d ions transitions within the 4f–shell are parity forbidden and admixures of wavefunctions with opposite parity are necessary. However due to the screening of the outer filled $5s^2$– and $5p^6$–orbitals the electron phonon coupling is very weak. At acentric sites one observes zero–phonon–transitions with very weak vibrational sidebands. Therefore the operation of rare earth lasers is restricted to zero–phonon–lines with only very small tunability.

Sensitized Rare Earth Lasers

Cross pumping of Nd^{3+} via Cr^{3+} has been demonstrated to be a very efficient excitation process for broad band pump sources. This is mainly due to an excellent spectral overlap of the broad band $^4T_2$–$^4A_2$ Cr^{3+}–emission and the Nd^{3+}–absorption[4]. Typically, Cr,Nd:$Gd_3Sc_2Ga_3O_{12}$ (Cr,Nd:GSGG) shows improved efficiencies which are 2 to 3 times the efficiency of Nd:YAG. However, GSGG exhibits relative strong thermal lensing which creates problems at higher average output powers above the 50 W–regime.

In addition to Nd^{3+}, also Er^{3+}, Tm^{3+}, and Ho^{3+} can be sensitized by Cr^{3+}. Especially Tm^{3+}– and Ho^{3+}–lasers operate very efficiently near 2 μm wavelength. The pump mechanisms which have been used to invert the Tm^{3+} and Ho^{3+} level systems are shown in Figure 9.

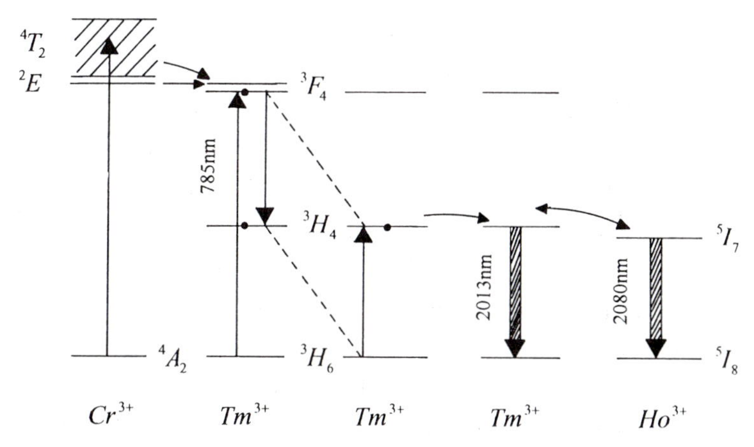

Figure 9. Pumping schemes of Tm– and Ho–lasers. The Cr–pump band is used for flashlamp pumping.

An efficient resonant Cr–Tm energy transfer is present, for instance, in hosts like YAG and YSGG ($Y_3Sc_2Ga_3O_{12}$). At relatively high Tm^{3+} concentrations of the order of $4 \cdot 10^{20}$ cm^{-3}, the Tm^{3+}–ions convert the excitation energy down to the 2 μm–spectral region with a quantum efficiency of nearly 2 by a cross relaxation process. Due to this cross relaxation process the thermal power dissipation in the crystal is drastically decreased.

With flashlamp pumped lasers, one can achieve slope efficiencies of 4 %. Diode pumping at λ = 785 nm yields slope efficiencies above 50 % at room temperature[8].

Laser Pumped Rare Earth Lasers

In addition to Tm^{3+} and Ho^{3+} new rare earth lasers have been operated under cw laser pumping. Table 1 lists interesting crystals which are not based on Nd^{3+}.

Table 1. Examples of recent laser pumped cw rare earth lasers.

Active Ion	Crystal	Wavelength [μm]	Pump laser
Tm^{3+},Ho^{3+}	$Y_3Al_5O_{12}$	2.08	Diode[2,25]
Tm^{3+}	$Y_3Al_5O_{12}$	2.02	Diode[8]
Tm^{3+},Ho^{3+}	$YLiF_4$	2.31	Diode[8]
Yb^{3+}	$Y_3Al_5O_{12}$	1.03	Diode[9]
Yb^{3+}	$Ca_5(PO_4)_3F$	1.04	Ti–Sapphire[10]
Pr^{3+}	$YAlO_3$	0.6139,0.6216 0.6624,0.7195 0.7469,0.7537 0.9960	Argon[26]
Er^{3+}	$Y_3Al_5O_{12}$	1.64	Krypton
Er^{3+}	$Y_3Al_5O_{12}$	2.8	Krypton
Er^{3+}	$YLiF_4$	2.8	Diode[8]
Er^{3+}	$YLiF_4$	1.62 + 2.81	Ti–Sapphire[28]

Ti–sapphire and Krypton laser pumping can be considered as diode laser pump simulations, because powerful diode lasers are avialable in the spectral range of the Ti–sapphire laser and the red Krypton laser lines (650 ... 680 nm).

Figure 10 shows various possibilities for pumping the Er^{3+} laser transitions near 1.6 μm and 3 μm. It is also possible to use deactivators as Ho^{3+} for the lower 3 μm– laser level which has a longer lifetime than the upper laser level. However, it has been shown, that the 3 μm transition is not selfterminating and can be operated in a cw–mode[27] due to strong up–conversion of two Er–ions in the $^4I_{13/2}$–state.

Figure 11 gives the first results of an Er^{3+} cw–cascade–laser[28] operating simultaneously at 2.81 μm and 1.62 μm. In such a situation the 2.81 μm laser transition directly pumps the 1.62 μm laser transition thus increasing the slope efficiency. It should be emphasized, however, that the results of Figure 11 are not yet optimized.

Diode pumped, commercially available Nd–lasers are mainly based on Nd^{3+}:$Y_3Al_5O_{12}$ and Nd^{3+}:$YLiF_4$. But it would be still advantageous to decrease the temperature sensitivity and to increase the pump wavelength tolerance of diode pumped Nd–lasers by larger absorption linewidths. Disordered crystals exhibit broader absorption bands. However, due to the inhomogeneous broadening, the peak cross sections are decreased.

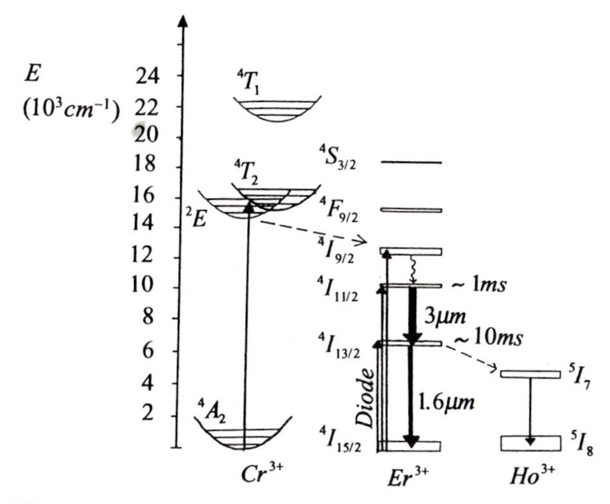

Figure 10. Pumping schemes of 1.6 μm— and 3 μm—Er lasers.

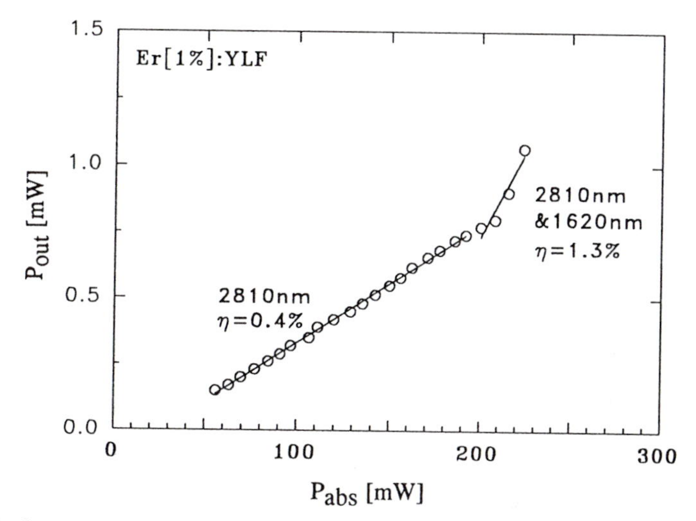

Figure 11. Output power vs. input power of an Er^{3+} cascade laser (1.62 μm + 2.81 μm).

Very recently, diode pumping has been reported for a new laser material Nd:$GdVO_4$, which has high cross sections and homogeneously broadened absorption lines[29, 30]. Figure 12 gives a comparison of the absorption cross sections of Nd:$Y_3Al_5O_{12}$ and Nd:$GdVO_4$. The vanadate crystal with tetragonal symmetry has strongly polarized cross sections with higher peak cross sections and broader linewidths compared to Nd:YAG. At a 1.2 % Nd–doping level the peak absorption coefficient with E || c is 74 cm^{-1}. In addition, Nd:$GdVO_4$ exhibits an accidental $^4F_{3/2}$–degeneracy of the upper Nd–laser level. The emission spectra condense therefore into fewer lines with enhanced cross sections. Figure 13 shows the input–output characteristic of Nd:$GdVO_4$ with a slope efficiency of 54 % at 1.06 μm. The crystal of 1 mm thickness absorbed 87 % of the incident diode pump light.

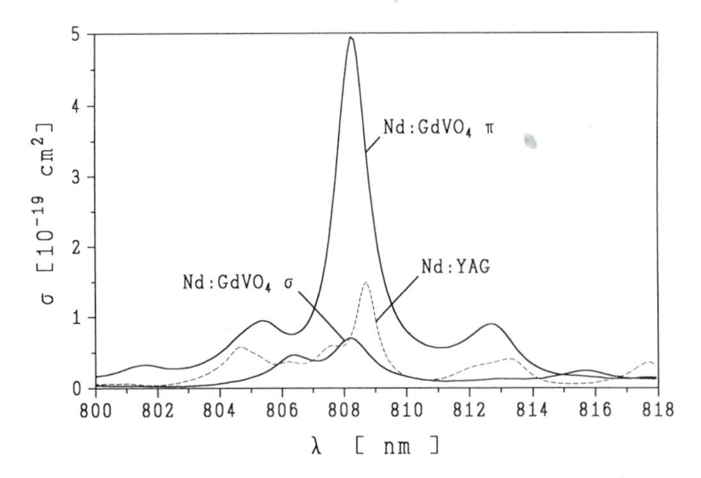

Figure 12. Absorption spectra of Nd:YAG and $Nd:GdVO_4$.

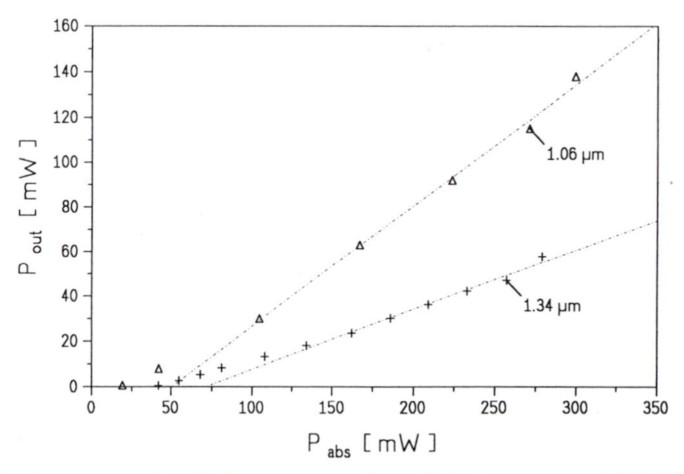

Figure 13. Output power vs. absorbed pump power for a diode laser pumped $Nd:GdVO_4$–crystal.

CONCLUSION

Among the class of vibronic tunable transition metal lasers $Ti^{3+}:Al_2O_3$ and also several Cr^{3+}–lasers have gained practical importance. The renewed interest in Cr^{3+}–lasers is due to the possibility of diode pumping. The laser performance of many transition metal ion doped crystals is decreased by excited state absorption (ESA) processes. Without ESA, transition metal ion laser materials may exhibit slope efficiencies close to the quantum limit as Ti–sapphire and several Cr^{3+}–lasers (emerald, $LiCaAlF_6$:Cr, and alexandrite). Rare earth lasers based on Nd^{3+}, Tm^{3+}, Ho^{3+}, Er^{3+}, and Yb^{3+} are very attractive for diode pumped applications at special wavelengths between 1 and 3 μm. The ions Pr^{3+}, Tm^{3+}, and Er^{3+} have transitions in the visible spectral region.

REFERENCES

1. J.C. Walling, O.G. Peterson, H.P. Jenssen, R.C. Morris, and E.W. O'Dell, Tunable alexandrite lasers, *IEEE J. Quantum Electron.* QE–16:1302 (1980).
2. B. Struve, G. Huber, V.V. Laptev, I.A. Shcherbakov, and E.V. Zharikov, Tunable room–temperature cw–laser action in Cr^{3+}:GdScGa–garnet, *Appl. Phys.* B 30:117 (1983).
3. P. Moulton, Ti–doped sapphire: a tunable solid–state laser, *Opt. News* 8(6):9 (1982).
4. D. Pruss, G. Huber, A. Beimowski, V.V. Laptev, I.A. Shcherbakov, and Y.V. Zharikov, Efficient Cr^{3+} sensitized Nd^{3+}:GdScGa–garnet laser at 1.06 μm, *Appl. Phys.* B 28:355 (1982).
5. B.M. Antipenko, A.S. Glebov, L.I. Krutova, V.M. Solntsev, and L.K. Sukhareva, Active medium of lasers operating in the 2 μm spectral range and utilizing gadolinium scandium gallium garnet crystals, *Sov. J. Quantum Electron.* 16:995 (1986).
6. E.W. Duczynski, G. Huber, V.G. Ostroumov, and I.A. Shcherbakov, CW double cross pumping of the $^5I_7$–$^5I_8$ laser transition in Ho^{3+}–doped garnets, *Appl. Phys. Lett.* 48:1562 (1986).
7. T.Y. Fan and R.L. Byer, Diode–laser pumped solid–state lasers, *IEEE J. Quantum Electron.* QE–24(6):895 (1988).
8. L. Esterowitz, Diode–pumped holmium, thulium, and erbium lasers between 2 and 3 μm operating cw at room temperature, Optical Engineering 29(6):676 (1990).
9. P. Lacovara, H.K. Choi, C.A. Wang, R.L. Aggarwal, and T.Y. Fan, Room–temperature diode–pumped Yb:YAG laser, *Optics Letters* 16(14):1089 (1991).
10. S.A. Payne, W.F. Krupke, L.K. Smith, L.D. DeLoach, and W.L. Kway, Laser properties of Yb in fluoro–apatite and comparison with other Yb–doped gain media, in: "Conference on Lasers and Electro–Optics, 1992", Vol. 12, OSA Technical Digest Series (Optical Society of America, Washington, DC 1992), pp. 540–541.
11. D. Welford and P.F. Moulton, Room–temperature operation of a Co:MgF_2 laser, *Optics Lett.* 13(11):975 (1988).
12. V. Petricevic, S.K. Gayen, and R.R. Alfano, Laser action in chromium–activated forsterite for near–infrared excitation: is Cr^{4+} the lasing ion?, *Appl. Phys. Lett.* 53(26):2590 (1988).
13. G.M. Zverev and A.V. Shestakov, Tunable near–infrared oxid crystal lasers, in: "Tunable Solid State Lasers", Vol. 5 of the OSA Proceding Series, M.L. Shand and H.P. Jenssen, eds. (Optical Society of America, Washington DC, 1989), pp. 66–70.

14. Y. Tanabe and S. Sugano, On the absorption spectra of complex ions II, *J. Phys. Soc. Jpn.* 9:766 (1954).
15. P. Albers, E. Stark, and G. Huber, Continuous–wave laser operation and quantum efficiency of titanium–doped sapphire, *J. Opt. Soc. Am. B* 3(1):134 (1986).
16. R.M. Macfarlane, J.Y. Wong, and M.D. Sturge, Dynamic Jahn–Teller effect in octahedrally coordinated d^1 impurity systems, *Phys. Rev.* 166:250 (1968).
17. C.W. Struck and W.H. Fonger, Unified model of the temperature quenching of narrow line and broad band emission, *J. Luminescence* 10:1 (1975).
18. K. Petermann, The role of excited–state absorption in tunable solid–state lasers, *Opt. Quantum Electron.* 22:199 (1990).
19. B. Struve and G. Huber, The effect of the crystal field strength on the optical spectra of Cr^{3+} in gallium garnet laser crystals, *Appl. Phys.* B 36:195 (1985).
20. R. Scheps, Cr–doped solid state lasers pumped by visible laser diodes, *Optical Materials* 1:1 (1992).
21. S.A. Payne, L.L. Chase, H.W. Newkirk, L.K. Smith, and W.F. Krupke, $LiCaAlF_6:Cr^{3+}$: a promising new solid state laser material, *IEEE J. Quantum Electron.* 24:2443 (1988).
22. S.A. Payne, L.L. Chase, L.K. Smith, W.L. Kway, and H.W. Newkirk, Laser performance of $LiSrAlF_6:Cr^{3+}$, *J. Appl. Phys.* 66:1051 (1989).
23. W. Jia, B.M. Tissue, L. Lu, K.R. Hoffmann, and W.M. Yen, Near–infrared luminescence in Cr,Ca–doped yttrium aluminium garnet, in: "OSA Proceedings on Advanced Solid–State Lasers", George Dubé, Loyd Chase, eds. (Optical Society of America, Washington, DC 1991), Vol.10, pp. 87–91.
24. S. Kück, K. Petermann, and G. Huber, Spectroscopic investigation of the Cr^{4+}–center in YAG, in: "OSA Proceedings on Advanced Solid–State Lasers", George Dubé, Loyd Chase, eds. (Optical Society of America, Washington, DC 1991), Vol. 10, pp. 92–94.
25. T.Y. Fan, G. Huber, R.L. Byer, and P. Mitzscherlich, Spectroscopy and diode laser–pumped operation of Tm,Ho:YAG, *IEEE J. Quantum Electron.* QE–24(6):924 (1988).
26. A. Bleckmann, F. Heine, J.P. Meyn, K. Petermann, and G. Huber, CW–lasing of $Pr:YAlO_3$ at room temperature, in: "Advanced Solid–State Lasers and Compact Blue–Green Lasers Technical Digest, 1993" (Optical Society of America, Washington, D.C., 1993), Vol. 2, pp. 164–166.
27. G. Huber, E.W. Duczynski, and K. Petermann, Laser pumping of Ho–, Tm–, Er–doped garnet lasers at room–temperature, *IEEE J. Quantum Electron.* 24(6):920 (1988).

28. B. Schmaul, G. Huber, R. Clausen, B. Chai, P.LiKamWa, and M. Bass, Er^{3+}:$YLiF_4$ continuous cascade laser operation at 1620 and 2810 nm at room temperature, *Appl. Phys. Lett.* 62(6):541 (1993).
29. T. Jensen, J.–P. Meyn, G. Huber, V.G. Ostroumov, A.I. Zagunmennyi, and I.A. Shcherbakov, Spectrocopic properties and lasing of Nd:$GdVO_4$ pumped by a diode laser, to be published
30. V.G. Ostroumov, I.A. Shcherbakov, A.I. Zagunmennyi, G. Huber, T. Jensen, J.–P. Meyn, Nd:$GdVO_4$ crystal – a new material for diode–pumped solid–state lasers, in: "Advanced Solid–State Lasers and Compact Blue–Green Lasers Technical Digest, 1993" (Optical Society of America, Washington, D.C., 1993), Vol. 2, pp. 52–54.

QUANTUM CONFINED SEMICONDUCTOR LASERS

Roberto Cingolani
Dipartimento di Scienza dei Materiali
Universita' di Lecce
73100, Lecce, Italy

Abstract

We discuss the fundamental properties of solid state lasers based on low-dimensional semiconductor heterostructures. The basic radiative recombination processes and the main technological features are reviewed with special attention to quantum well and quantum wire based devices.

Introduction

In recent years there has been an impressive improvement in the technology of low dimensional heterostructures and in the understanding of the physical properties of quantum confined systems. Among the others, the application of semiconductor quantum wells and quantum wires to high-performances solid state lasers has collected much attention, from both the technological [1] and the fundamental point of view [2]. In this lecture we shortly overview the fundamental properties of the low dimensional systems and their impact on the performances of quantum confined lasers. In the first section we briefly summarize the electronic properties of the quantum confined heterostructures. In the second section the basic aspects of the spontaneous and stimulated emission in semiconductors are addressed. Finally, a few examples of applications are presented in section 3.

1-Electronic states in quantum confined heterostructures

The quantum confinement in semiconductors occurs when the electronic wavefunctions are confined in crystalline regions of extension comparable to the De Broglie wavelength of the electron itself. Such condition can nowadays be achieved by the fabrication of periodic layered structures consisting of a sequence of thin (about 100 Å) small gap semiconductors cladded between thicker wider gap semiconductor layers [3]. This configuration, usually called quantum wells, is shown in Fig.1. In the regime of quantum confinement the electronic eigenstates strongly change due to the additional confinement energy gained by the electron in the potential of the quantum well. In the frame of a simple rectangular particle-in-a-box model with impenetrable barriers and adopting the usual effective mass approximation, the confinement energies of the quantized states with quantum number n are

Solid State Lasers: New Developments and Applications
Edited by M. Inguscio and R. Wallenstein, Plenum Press, New York, 1993

found to scale as [4]:

$$E_{n,e(h)} = \hbar^2/2m_{e(h)}(n\pi/L_z)^2 \tag{1}$$

where e and h indicate electron and hole states, respectively, m is the effective mass and L_z is the thickness of the quantum well.

The true nature of the quantum well/barrier interface can be taken into account in the frame of the so called envelope function approximation, in which suitable boundary conditions for both the electronic wavefunctions and the derivative of the electronic wavefunctions are imposed [4]. Under these conditions the confinement energies can be calculated as the numerical solutions of the following trascendental equations:

$$K_{well}tan(K_{well}L_z/2) = K_{barrier} \tag{2}$$

for even quantum numbers, and

$$K_{well}cotan(K_{well}L_z/2) = -K_{barrier} \tag{3}$$

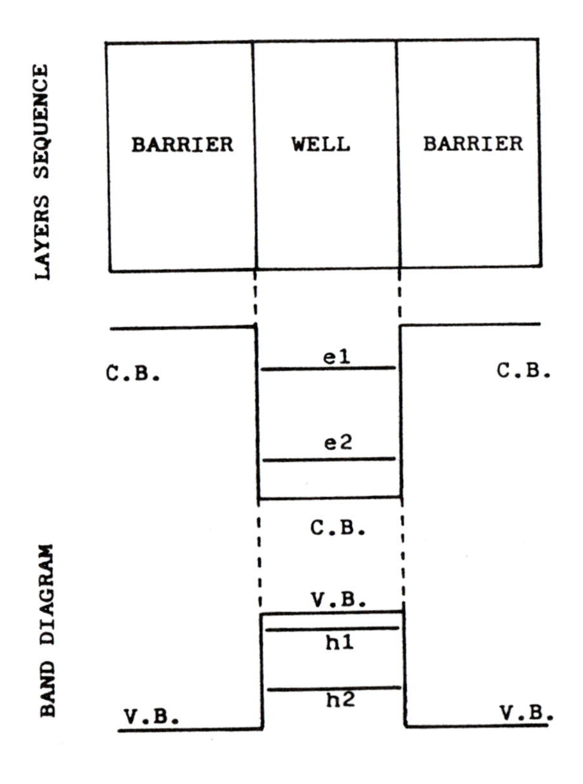

Fig.1 Schematics of a semiconductor quantum well. C.B. and V.B. label the conduction and valence bands, respectively, whereas e_i and h_i indicate the electron and hole confinement energies of the i-th quantum state.

for odd quantum numbers. $K_{well(barrier)}$ is the particle momentum in the well (barrier). As can be deduced from eqs.(1)-(3), there exists a direct dependence of the confinement energy on the physical size of the confining potential region (L_z). This means that the optical transition energies of the quantum confined heterostructures can be tuned over a certain range of energy according to:

$$E_n(L_z) = E_g + E_{n,e}(L_z) + E_{n,h}(L_z) \tag{4}$$

where E_g is the energy gap of the corresponding bulk semiconductor. In Fig.2 we show the variation of the confinement energy of electrons in an infinite GaAs quantum well as a function of the quantum well width (eq.(1)).

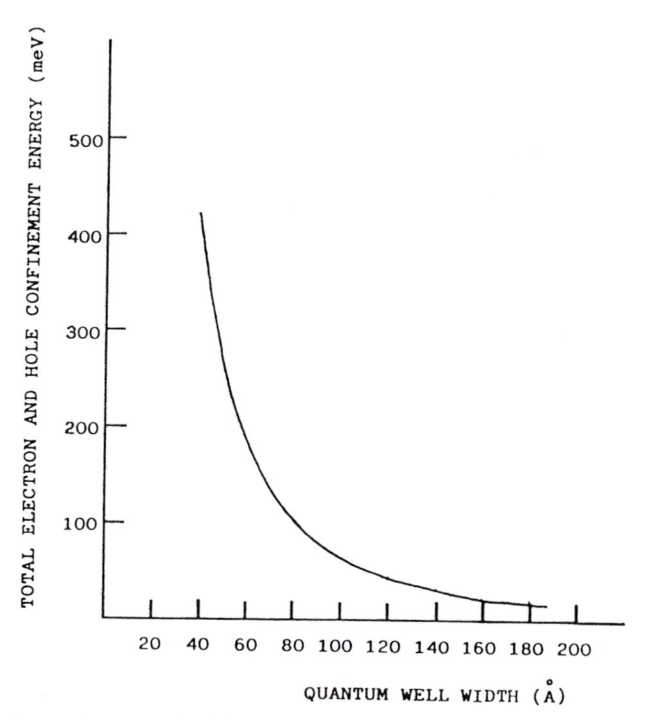

Fig.2 Total confinement energy (eq.(1) for electrons and holes) versus the quantum well width. The material is assumed to be GaAs.

This result is at the basis of the well known concept of band gap engineering, and can be used to artificially taylor the gap of any semiconductor heterostructure by the proper choice of the compositional and structural parameters of the constituent materials. As shown in Fig.3, there are many possible combinations of semiconductors which permit to cover the entire energy spectrum between the infrared and the ultraviolet, as shown in Fig.3.

The most important limitation in the fabrication of different heterostructures is the lattice matching condition, i.e. the requirement that the lattice constants of the layers constituting the heterostructure do not differ by more than about 0.4 %. Although it is nowadays possible to grow strained heterostructures with lattice mismatch as high as 7 % (this is the case of GaAs/InAs quantum wells), most of the technology as been so far developed for the $GaAs/Al_xGa_{1-x}As$ and

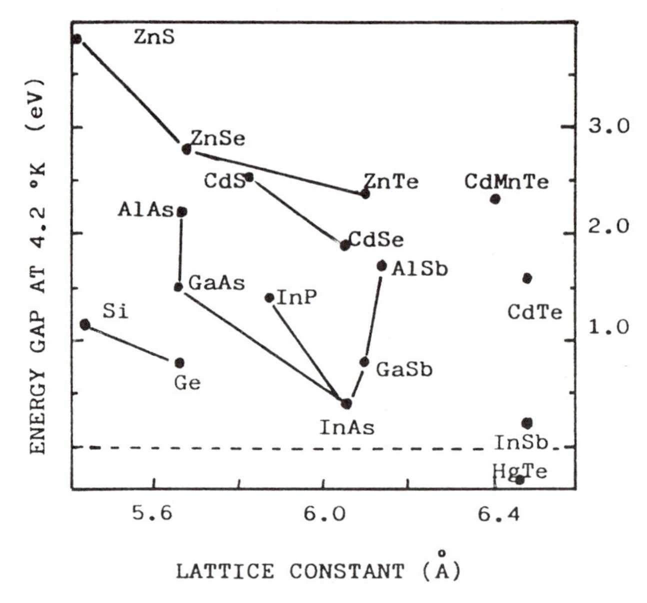

Fig.3 Energy gap versus lattice parameter of different semiconductors. The lines indicate a few possible combinations for heterostructure fabrication.

$Ga_{0.47}In_{0.53}As/Al_{0.48}In_{0.52}As$ material systems having very similar lattice constants. These two material systems cover the spectral ranges 1.4-2 eV and 0.8-1.2 eV respectively, with interesting applications for optoelectronic communications in the low-loss windows of silica fibers. Other important heterostructures are those constitued by II-VI semiconductors, namely ZnS/ZnSe and CdS/CdSe for blue-green emission, and the InAs and GaSb based heterostructures for infrared operation.

Great efforts are today devoted to a further reduction of the dimensionality. In particular people are trying to realize one dimensional heterostructures (so called quantum wires) in which electrons and holes are confined along two directions [5]. These structures are schematized in Fig.4. In this case the electron eigenenergies can be calculated including the confinement energies obtained along the both confinement directions into eq.(4). The ultimate limit of quantum confinement is reached in the zero-dimensional systems (quantum box) where electrons are totally confined along the three directions x, y and z. In addition to the striking advantage of the band gap tayloring, these low dimensional systems exhibit another important property which makes them very interesting for optoelectronic applications. This is the strong change of the density of states (DOS) dispersion depicted in Fig.4. Unlike the case of bulk semiconductors, having a parabolic density of states profile, quantum well systems have a step-like density of states, quantum wires have a $E^{-1/2}$ profile, whereas quantum dots exhibit an ideal delta-like density of states. In all cases, the DOS edges start at the energy of each quantized state. These features strongly modify the optical properties of low dimensional semiconductors and turn out to be very advantageous for the design of coherent emitters based on low dimensional heterostructures.

2-Radiative Recombination Processes in Low Dimensional Semiconductors

The major impact of the above properties is the possibility to get luminescence from low-dimensional systems at any energy depending on the structural parameters of the heterostructure. Further, stimulated emission can also be obtained provided population inversion on a three-level system is established in the crystal. This can be obtained under stationary or transient conditions under intense carrier injection or optical pumping. There exist different recombination mechanisms responsible for the stimulated emission, once population inversion has set in. In large gap semiconductors (like those belonging to the II-VI group) usually excitons play an important role due to the large binding energy which prevents their ionization (either due to thermal effects or to carrier screening). In this case, inelastic excitonic collisions among excitons or with free carriers, provide an efficient means to get a population inversion and subsequent optical amplification of the luminescence. Nevertheless, in most semiconductors of technolgical interest the stimulated emission is caused by free-carrier recombination. This regime, also named electron-hole plasma regime, can be achieved relatively easy with high power laser sources and is characterized by the presence of a high density of carriers distributed in different quantized states of the heterostructures. Electrons and holes in the plasma are described by Fermi-Dirac distribution functions with common carrier temperature (usually higher than the crystal temperature). Assuming momentum conservation in the electron-hole recombination, the spontaneous emission rate reads [6]

$$R_{sp}(E) \simeq \mid M \mid^2 D_{red}(E) f_e(1 - f_h) \tag{5}$$

where $\mid M \mid^2$ is the optical matrix element, $D_{red}(E)$ is the reduced density of states, and $f_{e(h)}$ is the Fermi distribution function of electrons (holes). In this model the quasiparticle broadening due to many body interactions is neglected. A

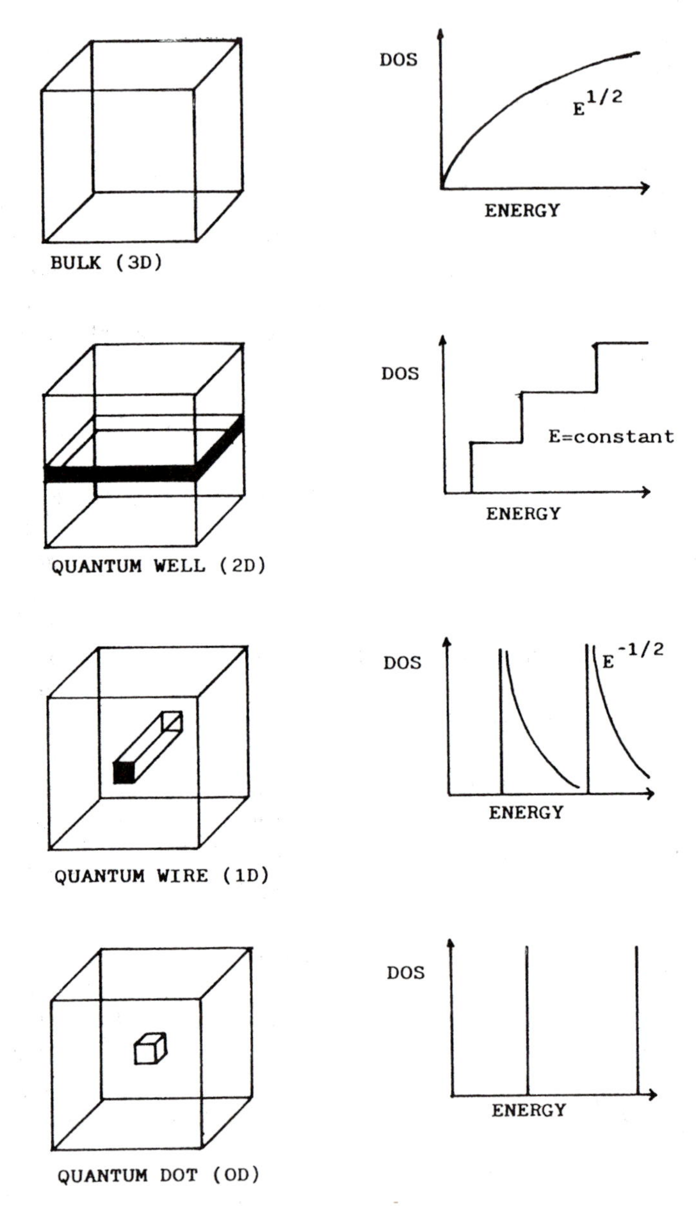

Fig.4 Shape and density of states of low dimensional semiconductors

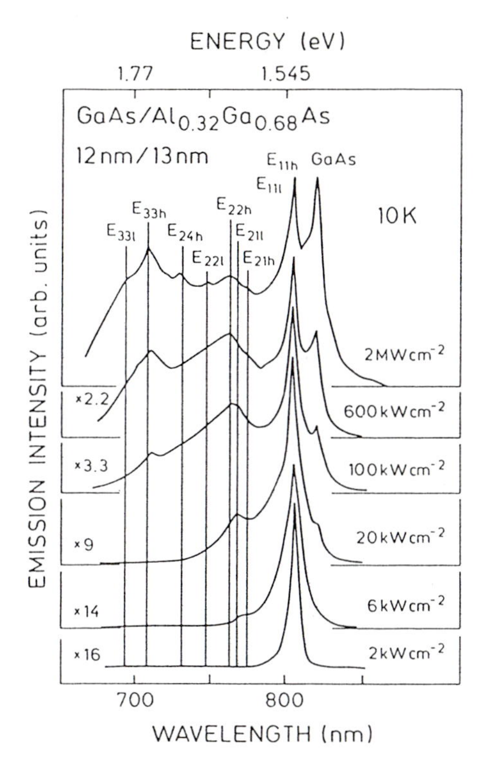

Fig.5 Spontaneous emission spectra of a highly photoexcited GaAs/$Al_{0.30}Ga_{0.70}As$ multiple quantum well structure at 10 K and different excitation intensities. The spectra have been magnified by the indicated factors to be compared.

phenomenological inclusion can be made by means of a convolution of the single particle states with Lorentzian functions $L_{e,h}$. Depending on the carrier density the quasi-Fermi energy rises reaching higher energy quantized states, whereas the band gap edge is found to decrease due to the reduction of the electron-hole pair energy caused by the exchange and correlation interaction in the plasma (band gap renormalization). As a result one expect that, increasing carrier density, light can be emitted from higher energy states (so called band filling). This is shown in the photoluminescence spectra of GaAs/AlGaAs quantum wells in Fig.5, where the progressive filling of higher energy states is directly observed under high photogeneration rates.

The radiative recombination processes of a dense electron-hole plasma in semiconductors are schematically sketched in Fig.6. The chemical potential of the plasma and the renormalized band gap edge are plotted as a function of the total carrier density present in the sample. Qualitatively, the emission spectrum of the crystal at a given carrier density is expected to fall within the two curves as shown in the right hand side of Fig.6. The carrier density labeled n_m corresponds to the carrier density at which the band gap renormalization equals the exciton binding energy (which does not change energy due to its charge neutrality). Under these conditions excitons are screened and the free carrier regime is established in the crystal.

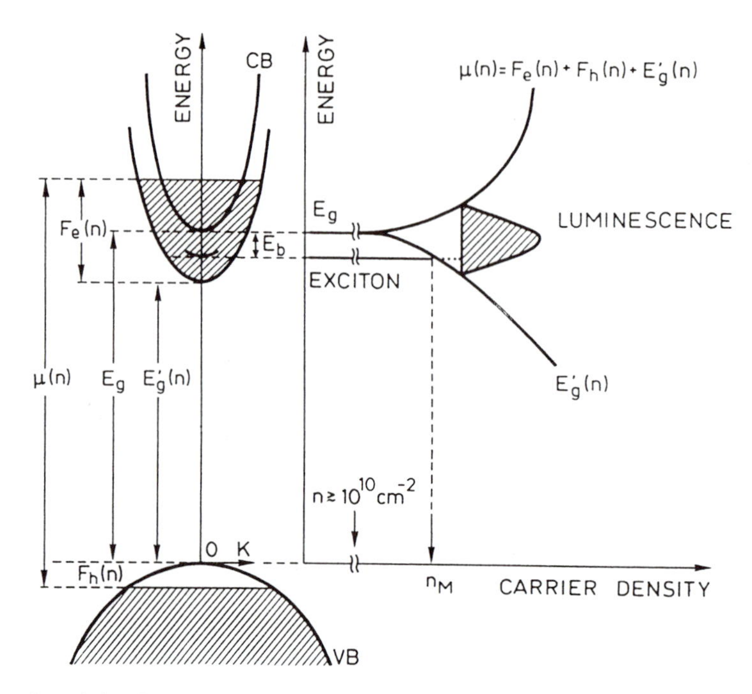

Fig.6 Illustration of the changes of the electronic states induced by the dense electron-hole plasma in semiconductors. E_g is the unperturbed gap, whereas $E_g(n)$ is the carrier density dependent renormalized gap. E_b is the exciton binding energy, $F_{e,h}(n)$ is the density dependent quasi Fermi level of electrons and holes and $\mu(n)$ is the total chemical potential. The carrier density dependence of $E_g(n)$ and $\mu(n)$ are depicted on the right-hand side. The splitting of the curves defines the bandwidth of the luminescence. The scale of the carrier density starts at $10^{10}cm^{-2}$ electron-hole pairs. The density n_M indicate the critical density at which the renormalized band gap merges the exciton state.

Above the stimulated emission threshold the emission line shape changes. This results from the drastic decrease of the carrier lifetime corresponding to the preferential recombination of carriers at a well defined photon energy where optical losses are minimized [7]. The stimulated emission rate reads [6]:

$$R_{st}(E) \simeq | M |^2 D_{red}(E)(f_e - f_h) \tag{6}$$

Under these conditions, the well known sharp stimulated emission spike appears in the optical spectra. The spontaneous and stimulated emission rates are related to each other by:

$$R_{st}(E) = R_{sp}(E)[1 - exp((E - \Delta F)/KT)] \tag{7}$$

where ΔF is the difference in the quasi-Fermi levels of electrons and holes and KT is the carrier thermal energy. The numbers of photons emitted by stimulated emission in a single mode of the radiation field of energy E is given by:

$$Q(E) = \frac{R_{sp}(E)}{N(E)/\tau - R_{st}(E)} \tag{8}$$

where N(E) is the number of modes per unit energy interval and unit volume, and τ represents the photon losses due to scattering, self-absorption and diffraction of the emitted light. It is clear that approaching the condition

$$N(E)/\tau \simeq R_{st}(E) \tag{9}$$

the Q(E) factor diverges, resulting in the sharp stimulated emission peak observed in solid state lasers. The relation (9) can easily be fulfilled by those modes having low losses (i .e. large τ values), in which even a relatively low $R_{st}(E)$ rate is enough to vanish the denominator of eq.(8). At the stimulated emission threshold, inserting eq.(7) into eq.(9), we obtain that the spontaneous emission saturates, and all emitted photons recombine radiatively at the energy of the stimulated emission line (lowest loss mode). Under these conditions, increasing the carrier density does not result in a change of the quasi-Fermi levels but only in a shortening of the carrier recombination time. This situation is shown for the case of a GaAs/AlGaAs quantum well in Fig.7 [8].

Below stimulated emission threshold the spontaneous emission spectrum of the multiple quantum well clearly exhibits the photoluminescence of carriers belong-

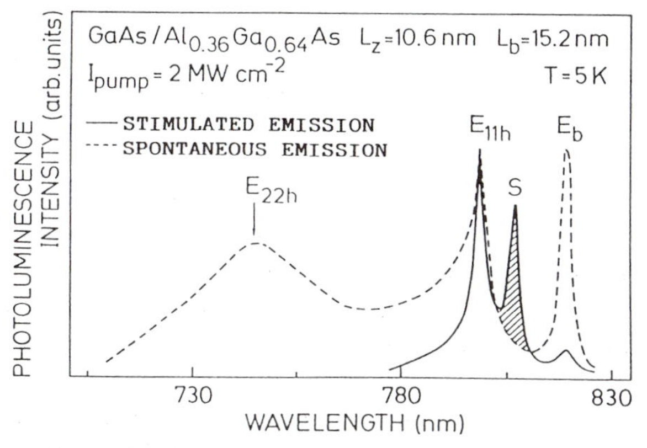

Fig.7 Spontaneous and stimulated emission spectra of highly photoexcited GaAs/AlGaAs quantum well at 10 K. Note the band filling of the n=2 states in the spontaneous spectrum and the saturation of the spontaneous emission when stimulated emission sets in. The curves have been normalized to be compared.

ing to the first and the second conduction and valence quantized states of the quantum well. Above stimulated emission threshold, the spontaneous emission saturates and a sharp stimulated emission peak appears in the low energy side of the spectrum.

3-Fundamental Aspects of Quantum Confined Solid State Lasers

Following the above discussion we will briefly outline the basic features of solid states lasers based on low-dimensional heterostructures. In a semiconductor laser the population inversion occurs between the quasi-Fermi levels of the electrons and holes and the stimulated emission arises around the edge of the renormalized band gap where optical losses are strongly reduced. The active material can be either a semiconductor heterostructure (eg GaAs/AlGaAs) with thicknesses far beyond the quantum size regime (namely about 1 μm) or a low-dimensional heterostructure consisting of a periodic sequence of quantum wells, wires or boxes. The former case, which is nothing but a normal bulk semiconductor laser, is well established and commercialy diffused. Conversely, the latter configuration is the subject of intense technological study. Despite the actracting possibility of exploiting the band gap engineering to tune the laser emission, the quantum confined lasers have another important advantage connected with the modifications of the density of states in low dimensional semiconductors. As schematically represented in Fig.8 there is a smaller density of states to be inverted to reach the lasing threshold, with important benefits in terms of low-injection-rate operation of the quantum well lasers. Infact, at a fixed quasi-Fermi level the electron density $n=\int f_e D_e$ decreases with decreasing the crystal dimensionality, due to the reduced density of states in the considered energy range (shaded areas in Fig.8). At present, lasing thresholds as low as 50 mA/cm^2 have been reached in GaAs quantum well lasers. Another important advantage is the possibility of tayloring the profile of the refractive index in the laser cavity, in order to get good waveguiding characteristics for the device. We should also mention that the optical gain of quantum well lasers remains finite also at low band filling or at low temperatures, i.e. when only a few states close to the band gap edge are populated, due to the constant density of states. The optical gain otherwise goes to zero in bulk lasers due to the vanishing density of states at the bandgap edge (see Fig.8).

On the other hand several physical limitations and technological constraints should be mentioned. First of all, the reduced coupling of the stimulating electromagnetic field with the photon field in the thin quantum well laser cavity. This is depicted in Fig.9, where the small portion of electromagnetic field interacting in the cavity is represented by the shaded area.

In order to increase the coupling in the heterostructure, the special potential profiles shown in Fig.9 are purposely realized. These are realized by exploiting the strong dependence of the energy gap of some III-V ternary alloys (like for the the $Al_xGa_{1-x}As$ shown in Fig.10) which allows us to obtain a graded gap profile of the heterostructure by changing the aluminum content (x) . In this

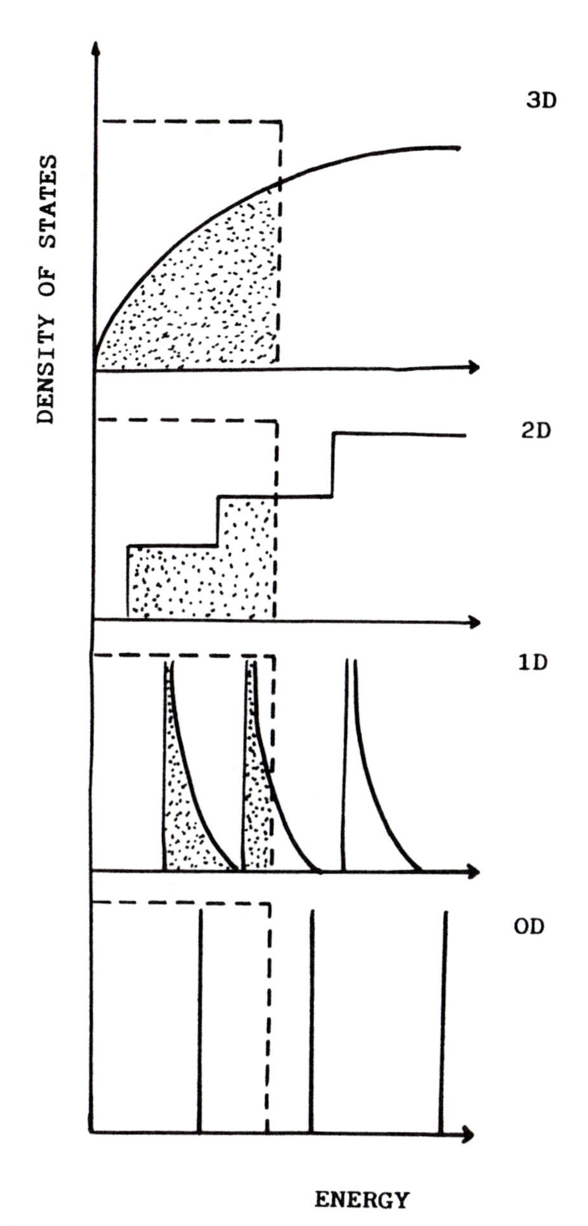

Fig.8 Carrier density in low dimensional semiconductors (shaded areas). The dashed lines indicate the zero temperature Fermi-Dirac distribution function.

case parabolic as well as rectangular confinement regions can be obtained. These potential structures (Separate Confinement Heterostructures and Graded Index Heterostructure lasers) enhance the confinement factor and the trapping efficiency of the injected carriers at the lowest energy emitting state in the quantum well. Further, the refractive index profile guarantees efficient waveguiding of the light in the active layers, with important advantages in the stimulated emission performances (Fig.10). Besides the fabrication complexity of these devices, we should also recall that in quantum well lasers the optical gain tends to saturate at large band filling, due to the constant density of states, untill the population of the higher energy quantum states is achieved.

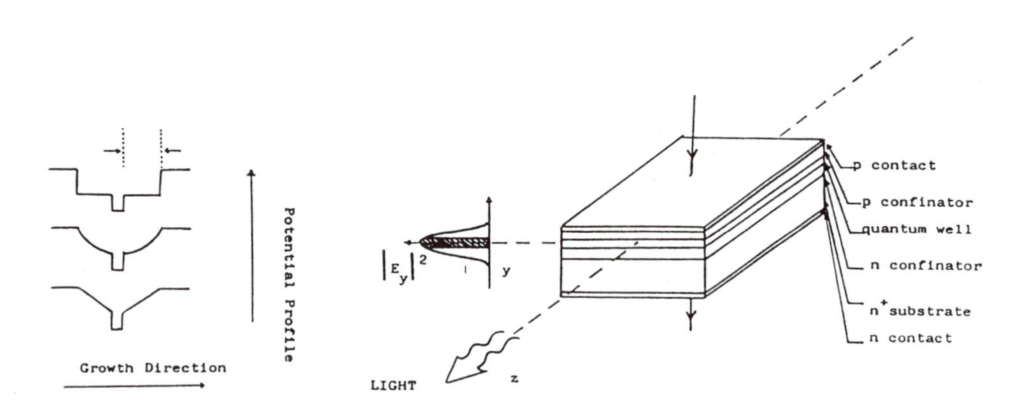

Fig.9 Schematics of a semiconductor laser. The individual layers are identified in the figure. The shaded curve on the left represents the fraction of electromagnetic field interacting with the cavity. Different potential profiles designed to increase the coupling in the cavity (confinement factor) are depicted on the left hand side of the figure.

At present, the most advanced frontier of low-dimensional quantum-confined lasers is related to a further reduction of the dimensionality of the active heterostructure. In particular a reduction of the lasing threshold and a high room temperature performances are expected for quantum wire and quantum box lasers, as a result of the narrowing and reduction of the density of states (see Fig.8). So far only very few results on the radiative recombination of the electron-hole plasma in quantum wires were reported [9], and evidence of laser action in quantum wires [10] and quantum boxes [11] were reported in pioneering works by few groups. In Fig.11 we show the calculated optical gain spectra of bulk, quantum well, quantum wire, and quantum boxe laser systems of comparable sizes. From the theoretical

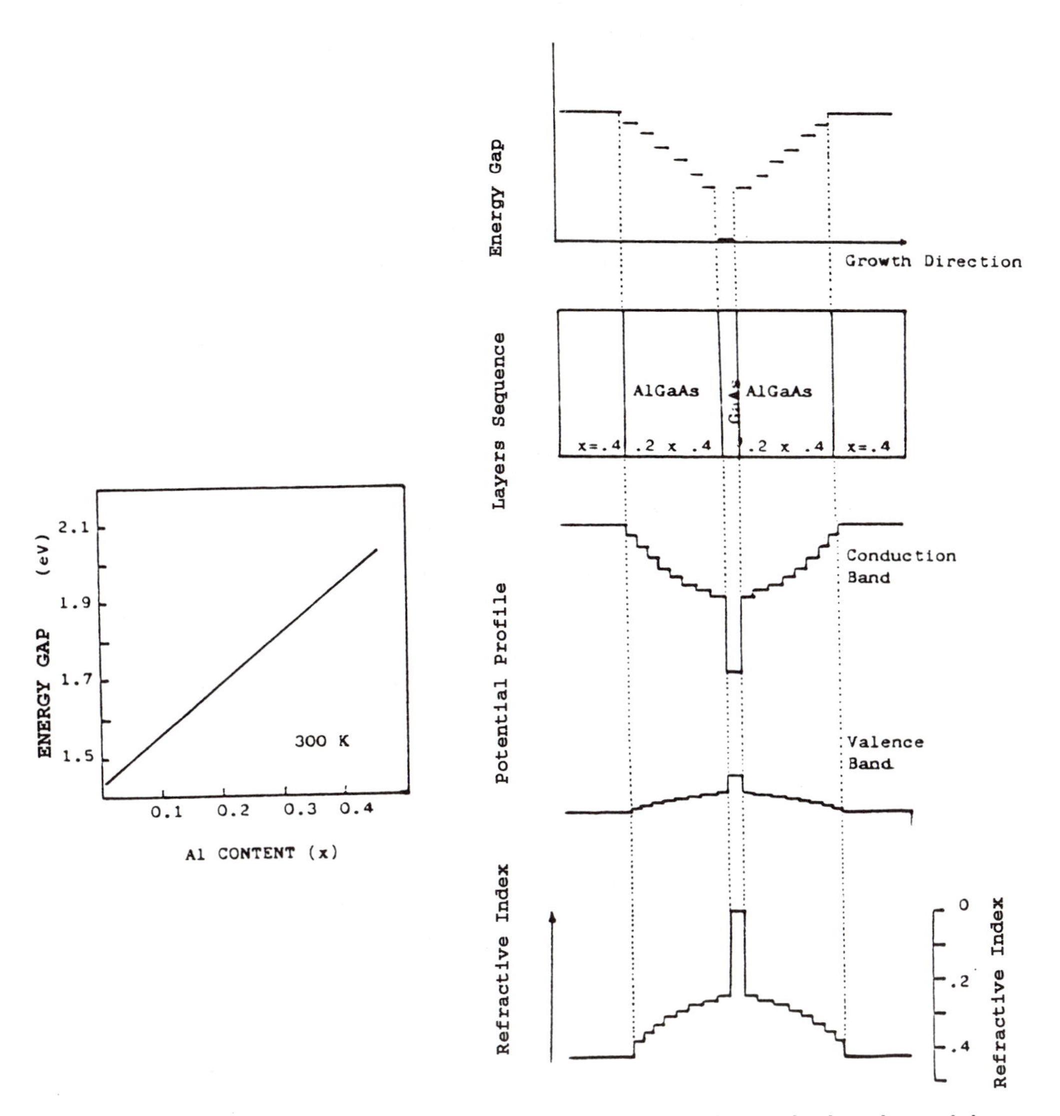

Fig.10 Schematics of a graded index heterostructure laser. Exploiting the dependence of the energy gap of the $Al_xGa_{1-x}As$ on the Al content (left panel), the gap profile can be taylored along the growth direction (top). The layer sequence consists of a GaAs quantum well embebbed into two thick layers of AlGaAs with increasing Al content. This results in the conduction and valence band profiles shown in the figure. The refractive index (bottom) is convenient for light waveguiding.

point of view it is quite clear that low dimensional systems are very much advantageous. Unfortunately, the real technology has to face with a number of problems related to the extreme complexity in the fabrication procedures and in the physics of crystals down to the quantum limit, which drastically reduce the potential performances of these devices. It is a general and reasonable guess that in the near future part of these problems will be removed, allowing the fabrication of true zero-dimensional lasers.

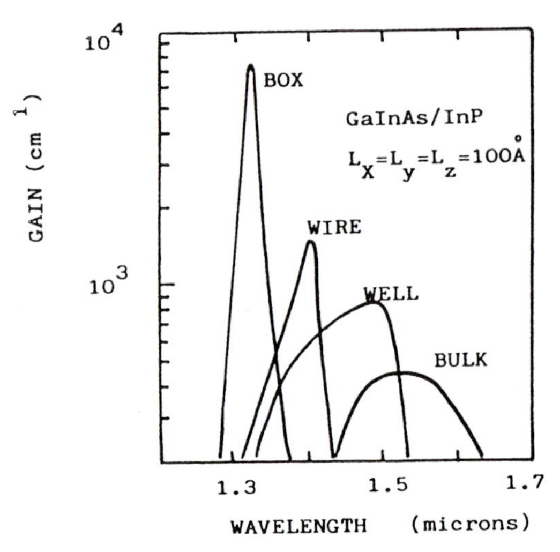

Fig.11 Calculated optical gain spectra of bulk, quantum well, quantum wire and quantum box lasers of comparable size (written on the figure).

References

[1] J.Nagle and J.C.Wiesbuch , Phys. Scr. (1987)

[2] S.Schmitt-Rink, D.S.Chemla and D.A.B.Miller, Adv. Phys. **38**, 89 (1989) R.Cingolani and K.Ploog, Adv. Phys. **40**, 535 (1991)

[3] K.Ploog, in *The Technology and Physics of Molecular BEam Epitaxy*, edited by E.H.C.Parker, Plenum Publ. Corporation, pg.647-679 (1985)

[4] G.Bastard, *Wave Mechanics Applied to Semiconductor Heterostructures* , ed. Les Edition de Physique , Les Ulis Cedex, France (1988)

[5] K.Kash, J. Luminesc. **46**, 69 (1990)

[6] G.Lasher and F.Stern , Phys. Rev. **133**, A553 (1964)

[7] E.O.Goebel , R.Hoeger , J.Kuhl , H.J.Polland , and K.Ploog , Appl. Phys. Lett. **47** , 781 (1985)

[8] R.Cingolani, K.Ploog, A.Cingolani, C.Moro, and M.Ferrara, Phys. Rev. **B42**, 2893 (1990)

[9] R.Cingolani, E.Lage, D.Heitmann, H.Kalt, K.Ploog, Phys. Rev. Lett. **67**, 891 (1991)

[10] E.Kapon, D.M.Hwang, and R.Bhat, Phys. Rev. Lett. **63**, 430 (1989)

[11] Y.Miyamoto, M.Cao, Y.Shingai, K.Furuya, Y.Suematsu, K.G. Ravikumar, and S.Arai, Jpn. J. Appl. Phys. **26**, 1225 (1987)

DIODE PUMPED SOLID STATE LASERS

Robert L. Byer

Center for Nonlinear Optical Materials
Edward L. Ginzton Laboratory
Stanford University
Stanford, CA 94305

INTRODUCTION

In the summer of 1960 T. Maiman[1] reported in Nature the first operation of an optical maser, later to be called a laser. Only one year later this solid state laser was frequency doubled into the ultraviolet in the first nonlinear optical experiment by P. Franken and his group at the University of Michigan.[2] The following year, 1962, saw three independent groups demonstrate almost simultaneously the operation of the injection diode laser. After thirty years, the combination of these three areas of quantum electronics, solid state lasers, diode lasers and nonlinear optics has brought us to the treshold of a major advance in laser science and technology.

This paper on diode pumped solid state lasers will treat three areas; first, the coherence of the diode pumped solid state laser; second, the nonlinear frequency extension of the solid state laser source emphasizing the recent advances in efficient harmonic generation and parametric oscillation of continuous wave, cw, sources; and third, the recent advances and promise of diode pumped high average power lasers . The goal is to bring students in touch with current advances in each of these areas with references to appropriate reviews and to background materials.

The advantages of diode pumped solid state lasers were recognized by Newman[3] shortly after the demonstration of the first diode laser. One year later, in 1964 Keyes and Quist[4] at Lincoln Laboratories demonstrated the first diode pumped solid state laser in $U^{3+}:CaF_2$ pumped by five GaAs diode lasers operating at 4K. Diode pumping was extended to Nd:YAG in 1971 by Ostermayer[5] with a promise for higher output power demonstrated by Conant and Reno in 1974.[6] It was not until the demonstration of the linear multistripe diode array by D. R. Scifres, R. D. Burnham and W. Streifer[6] in 1978 that the promise of efficient, high average power, diode pumped solid state lasers could begin to be realized.

The rapid advances in diode laser pumped solid state lasers has been the subject of several reviews including a general overview of the field by Byer,[7] a more technical review

Solid State Lasers: New Developments and Applications
Edited by M. Inguscio and R. Wallenstein, Plenum Press, New York, 1993

by Fan and Byer,[8] and a recent review article summarizing the various diode pumped solid state laser sources and their properties by Hughes and Barr.[9] In addition, for those looking for engineering technical details, there is a good overview of the design of diode pumped solid state lasers by Kane.[10]

The early research on diode pumped solid state lasers was motivated by the need for a coherent, efficient, long lived laser as the transmitter for global wind sensing from a satellite. The transmitter requirements, as understood in 1980 when the work began, included a 100W average power laser with a diffraction limited spatial beam operating in a single frequency with less than 10kHz linewidth. The laser should also have 10% overall electrical efficiency and operate in an eyesafe spectral region. At that time the state of the art for solid state lasers was well below these requirements in all categories. The research program when initiated, met Edwin Land's challenge "Don't undertake a project unless it is manifestly important and nearly impossible." The early work demonstrated the potential for diode pumping[11] and soon thereafter led to the invention of nonplanar ring oscillator (NPRO)[12] as the local oscillator for coherent laser radar.[13] With the successful demonstration of a highly coherent laser, work began on extending the laser wavelength by diode pumping new ions in crystals and by nonlinear optical methods. This work, and the recent work on high average power diode pumped lasers is reviewed in the second and third sections of this paper.

Historically, every advance in solid state laser technology was preceeded by an advance in the pump source. The diode laser pump source for solid state lasers brings the advantages of high efficiency, long operational life and high spectral brightness for selectively pumping the absorption bands of the ion doped crystals and glasses. The potential for single mode, diffraction limited laser output at high average power and good electrical efficiency has opened new applications to diode laser pumped solid state laser systems. Among these applications are quantum limited interferometric measurements such as gravity wave interferometry[14] and tunable precision optical oscillators and perhaps optical frequency synthesis for high resolution spectroscopy of atoms and molecules.

THE DIODE PUMPED Nd:YAG LASER OSCILLATOR

The coherence of solid state lasers has improved continuously from the early lasers with multiple gigahertz linewidths shortly after the invention of Nd:YAG in 1964 to the single axial mode lamp pumped Nd:YAG lasers with linewidths of 0.2MHz in 1982[15]. It was clear almost immediately that diode pumping would lead to a dramatic reduction in laser linewidth by a reduction in pump power fluctuations compared to lamp pumping and by the elimination of water cooling. Indeed, the first diode laser pumped Nd:YAG standing wave laser oscillator operated with a linewidth of 10GHz[11] but suffered from optical feedback induced instabilities and was limited to relatively low powers to maintain single axial mode operation[16]. The invention of the nonplanar ring oscillator overcame these limitations.

The Nonplanar Ring Oscillator

The nonplanar ring oscillator combined the stability of the monolithic solid state laser with the elements of an optical diode to induce unidirectional oscillation of the ring resonator.

The three elements of an optical diode are the magnetic field induced faraday rotation, the polarizer and the optical wave plate. In the nonplanar ring oscillator the non-normal internal reflectance at the dielectric coated output face of the ring is a partial polarizer; the breaking of the planar symmetry yields the waveplate; and the application of the external magnetic field yields the faraday rotation. In practice, the application of only a few hundred gauss of magnetic field causes the monolithic nonplanar ring oscillator to operate in a single direction, eliminating spatial hole burning, and leading to single frequency operation. The detailed theory of the nonplanar ring oscillator based on symmetry arguments is presented by Nilsson, Gustafson and Byer [17].

Linewidth studies of the diode laser pumped nonplanar ring oscillator were initiated shortly after the initial demonstration. The monolithic structure of the laser coupled with the reduced noise of diode laser pumping led to the expectation that the laser linewidth would be substantially less than that of lamp pumped lasers. Within six years the relative linewidth two monolithic diode pumped lasers was reduced from 10kHz to less than 3Hz[18] and then to below one Hertz.[19]

This work has been extended to stabilization studies of two lasers locked to high finesse Fabry Perot reference cavities. The two interferometers, enclosed in separate vacuum chambers, operated with a finesse of 200,000 and a linewidth at 1064nm of 5kHz. Two diode pumped Nd:YAG lasers were locked independently to these cavities using Pound-Drever locking. The result shown in Figure 1 was a measured Allan variance[20] between the two separate laser systems of less than 3×10^{-14} for delay times between 0.1 and 1sec.[21] For shorter times the noise is nearly white while for longer interval times drift leads to a nearly 1/f variance. The long term drift can be overcome by using an atomic or molecular transition as a reference. The use of the iodine molecule at 532nm as an absolute frequency reference is discussed below.

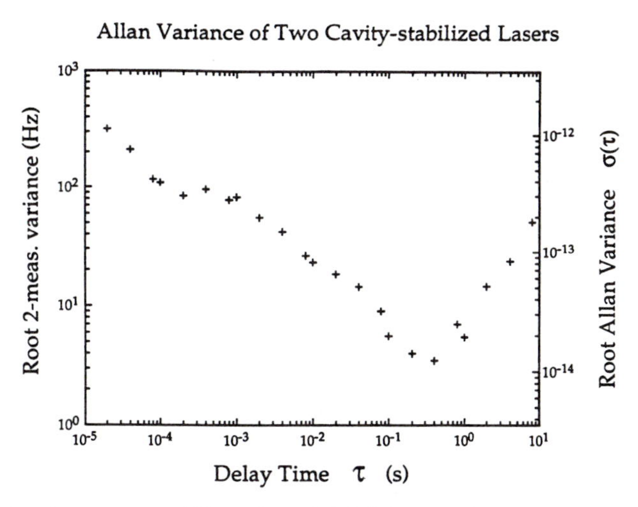

Figure 1. Relative frequency stability of two NPRO Nd:YAG lasers locked to two high finesse Fabry Perot cavities. The minimum value of the root Allan variance is 6Hz for intervals of 0.1 to 1sec.

Injection Locking and Phase Locking

Although the output power of the nonplanar ring oscillator has increased from 3mW in 1986 to greater than 300mW in 1992, another approach is required to maintain the coherence of the local oscillator at ever higher powers in the future. One approach is to use the coherence of a low power master oscillator to control the frequency of a higher power slave oscillator. This idea, called injection locking, was first suggested in 1946 by Robert Adler who applied it to the frequency control of vacuum tube oscillators.[22] Injection locking was extended to solid state lasers by Nabors et. al. in 1989.[23]

In their experiment, Nabors et al. used a 40mW nonplanar ring oscillator with a free running spectral linewidth of less than 20kHz to control the spectrum of a cw, lamp pumped, 13W Nd:YAG oscillator. The power level of the slave oscillator was increased to 18W of cw TEM_{00} mode output in later work. This 20kHz linewidth, single frequency Nd:YAG laser has been applied to efficient harmonic generation and to optical parametric oscillator studies. The success of the early injection locking studies opens the possibility of meeting the high power single frequency laser requirements for applications to remote sensing and gravity wave interferometry.

Optical phaselocking of independent lasers is essential for applications of lasers to coherent communications and for phased array operation. In 1989 Kane and Cheng demonstrated the first fast frequency tuning and optical phaselocked operation of the diode laser pumped nonplanar ring oscillator.[24] A year later, Kazovksy and Atlas demonstrated phaselocked loop operation of a 1320nm Nd:YAG laser.[25] In 1990, Day, Farinas and Byer[26] demonstrated a low bandwidth, less than 10kHz, 1064nm optical phaselocked loop coherent homodyne detection system with a closed loop phase noise of less than 12mrad (0.69degrees of phase angle). This demonstration opened the path for coherent laser oscillators to be used for deep space communication as well as for coherent laser radar applications including global wind sensing.

Phaselocking of lasers was extended by Cheng et. al.[27] who demonstrated serial coherent addition of three Nd:YAG laser oscillators to generate 900mW of cw output power in a single frequency beam. An additional benefit of the injection chaining of three oscillators is longer fail-safe operation for applications in space. In the future two dimensional phased arrays of laser oscillators may be feasible using the diode pumped approach because the phase noise of the diode pumped Nd:YAG laser is two orders of magnitude less than that of the diode laser sources.

Absolute Frequency Stabilization

The Nd:YAG nonplanar ring oscillator has inherent short term frequency stability due to its small size and monolithic construction. However, the frequency drift of the oscillator is -3.1GHz per degree which leads to long term frequency drifts of the order of 10 megahertz per hour. To be useful as an optical frequency standard for deep space communication or for precision interferometric measurements, absolute frequency stability to less than a megahertz per hour must be achieved. This can be accomplished by locking the Nd:YAG laser output or its harmonic to an atomic or molecular transition.

We have locked a frequency doubled Nd:YAG laser operating at 532nm to the hyperfine transitions of molecular iodine using Doppler-free spectroscopy. The resulting two sample Allan variance of the beat frequency between two lasers locked to two independent iodine molecular reference cells is less than 650Hz or 2.3 x 10^{-12} of the laser frequency for time intervals between 24 and 80 seconds.[28]

The Nd:YAG nonplanar ring oscillator is frequency doubled in a monolithic ring doubler of $MgO:LiNbO_3$ to generate between 5 and 25mW of output at the 532nm. One beam of the output is phasemodulated using a electrooptic modulator (EOM) and counterpropagated in the iodine cell relative to a pump beam. The partial saturation of the iodine transition leads to a differential absorption of the carrier and sidebands of the probe beam which is easily detected, demodulated and used to drive a feedback servo loop to control the frequency of the Nd:YAG laser. The relative frequency of the two independently stabilized Nd:YAG lasers is first offset by 80MHz using an acousto-optic modulator (AOM) and detected for frequency stability measurements. Figure 2 shows a schematic of the iodine frequency stabilized Nd:YAG laser experiment.

The FM saturation absorption spectrum of the 18788.3 cm^{-1} transition of iodine is shown in Figure 3. The hyperfine line position frequencies marked at the bottom of the figure were taken by beat note spectroscopy by locking the second laser system to successive hypefine components and averaging for 4 seconds using a HP 5371 time interval analyzer. The high signal to noise level of the Doppler free measurement leads to the improved absolute stability of the Nd:YAG laser relative to earlier helium neon laser experiments.

Figure 4 shows the variation in the beat note frequency over a thirty minute time interval. The free running laser beat note is shown to vary over 10MHz in this time interval due to the temperature drift of the laser crystal. This slow drift is removed by locking to the iodine transition as shown by the straight line of the beat note for the locked laser systems. In this early experiment the relative drift was measured to be less than 10kHz for the 30min time interval. In later work the beat note drift was reduced to less than 2kHz over one minute.

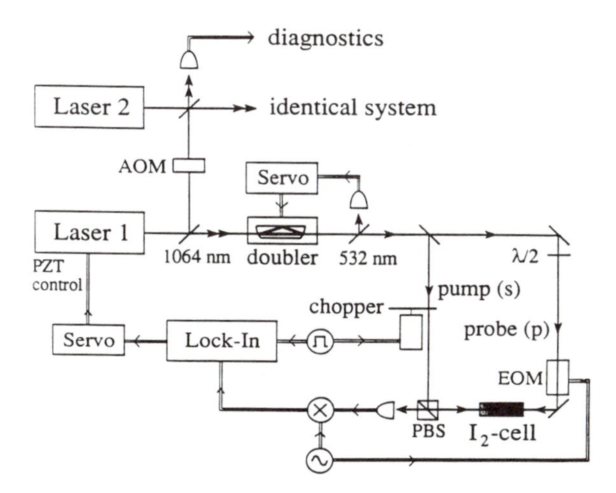

Figure 2. Schematic of the experiment for absolute frequency stabilization of Nd:YAG by Doppler free locking to the hyperfine components of $^{127}I_2$ absorption lines at 532nm.

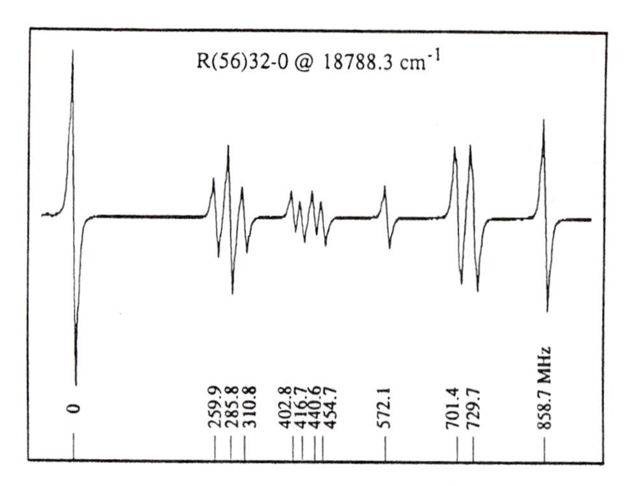

Figure 3. FM saturation absorption of the 18788.3 cm^{-1} line of $^{127}I_2$. Each hyperfine component transtion is 5.2MHz wide at room temperature.

Absolute frequency stabilization to iodine of the doubled Nd:YAG laser is not an ultimate frequency standard but is an intermediate and useful engineering standard that can be placed in any laboratory at reasonable cost. In the future, absolute laser frequency stability undoubtedly will be improved by locking to laser cooled atoms or to ions stored in traps. However, the combination of the iodine stablized Nd:YAG laser oscillator with increased power by injection locking and with frequency extension by nonlinear optical devices may make the nonplanar ring oscillator the 'quartz crystal' of the optical frequency region.

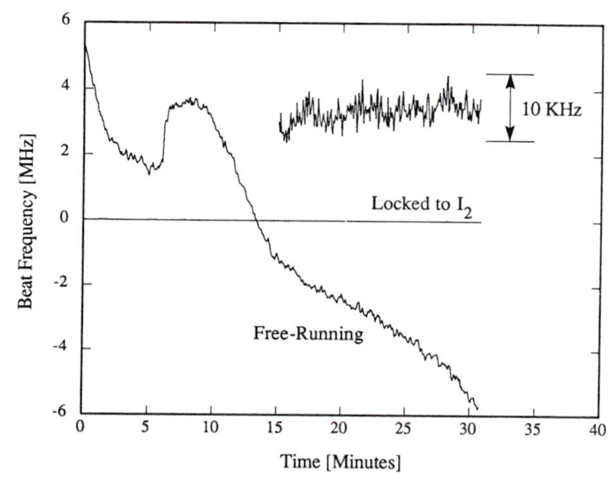

Figure 4. Variation in the free running and the iodine locked laser beat note over a thirty minute time interval. The inset shows an expanded view of the locked beat note over a 15 minute interval.

NONLINEAR FREQUENCY EXTENSION

The availability of single frequency lasers coupled with the recent improvements in nonlinear optical crystals has led to improvements in the conversion of optical radiation by nonlinear techniques. Following the initial demonstration of second harmonic generation in nonphasematched crystal quartz by Franken et. al. [2], nonlinear optical studies were almost exclusively conducted with high peak power lasers. It was not until low loss nonlinear crystals became available that efficient internal laser cavity second harmonic generation or external resonant cavity second harmonic generation became feasible. Resonant enhancement approaches were successful because stable, single frequency Nd:YAG laser oscillators were available under diode laser pumping.

This section of the paper addresses the recent advances in second harmonic generation and optical parametric oscillation using cw, highly coherent, diode laser pumped Nd:YAG lasers. The demonstration of efficient harmonic generation with milliwatts of laser power is testimony to the frequency stability of the laser source and to the low loss of the nonlinear material and optical cavity.

Second Harmonic Generation

Internal Second Harmonic Generation. Efficient nonlinear frequency conversion requires high powers or long interaction lengths in the nonlinear crystal. The obvious approach to power enhancement is to place the nonlinear crystal inside the laser resonator cavity. This approach was first demonstrated for a diode pumped laser by Fan et. al. in 1986 who doubled a Nd:YLF laser using in internal nonlinear crystal of $MgO:LiNbO_3$.[29] That early experiment generated 145μW of green output power for 30mW of diode pump power into the Nd:YLF laser system. The simplicity of intracavity doubling brought with it two complexities: the insertion of elements into the laser cavity introduced loss which decreased the laser efficiency, and the internal cavity nonlinear crystal coupled axial modes of the laser to introduce modulation of the optical power. The latter problem was observed and studied by Baer[30] and was overcome in an experiment by Oka and Kubota[31] who demonstrated internal second harmonic of diode pumped Nd:YAG using a KTP type II doubling crystal with a quarter wave plate set to avoid coupling of axial modes.

External Second Harmonic Generation. An alternate approach is to use an external cavity to resonately enhance the fundamental wave field. This approach was first suggested and experimentally demonstrated by Ashkin et al. in 1966.[32] The early work suffered from multiple axial mode low power lasers and from the high loss of the nonlinear crystals. External resonant second harmonic generation has been explored by Kozlovsky et al.[33] who used a monolithic $MgO:LiNbO_3$ ring frequency doubler to convert 53mW of cw 1064nm to 30mW of single mode green output for a conversion efficiency of 56%.

Figure 5a shows a schematic of the external resonant doubling experiment used by Kozlovsky et al.[33] The incident single axial mode Nd:YAG nonplanar ring oscillator is mode matched and coupled into the lithium niobate ring resonator which has optical coatings applied to its curved surfaces. The incident power is increased from 53mW to approximately

3W of which 1% is doubled to generate 30mW of cw green output. A feedback system is used to keep the incident laser frequency locked to the external resonant cavity.

A study of this approach to frequency doubling discovered that low loss is essential to high efficiency. For example, a round trip loss of 4% limits the doubling efficiency to 20%. A round trip loss of 1% yields 60% doubing efficiency and a round trip loss of 0.5% gives greater than 80% doubling efficiency. To lower the loss the nonlinear crystal length should be decreased, which is counter to the expected length scaling for normal nonlinear interactions. By reducing the lithium niobate crystal length from 25mm to 12.5mm, Kozlovsky et al.[33] improved the doubling efficiency as shown in Figure 5b.

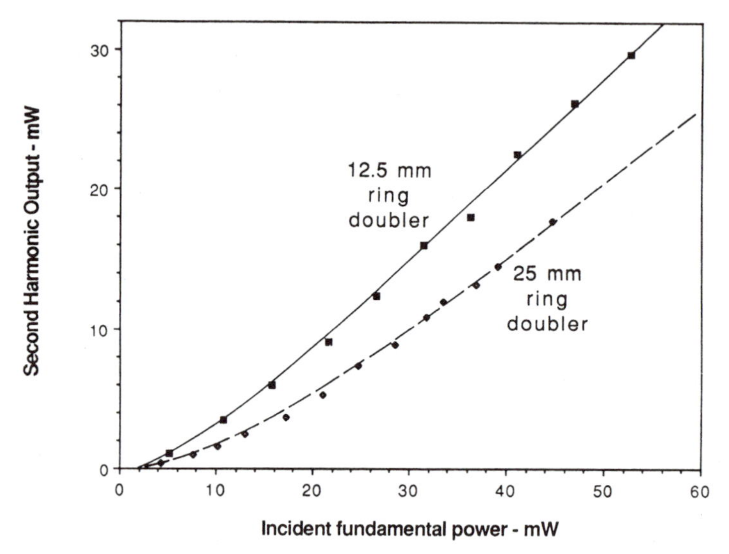

Figure 5 (**a**) Schematic of external resonant second harmonic generation in a lithium niobate monolithic frequency doubler. See Kozlovksy et al.[33] for details.

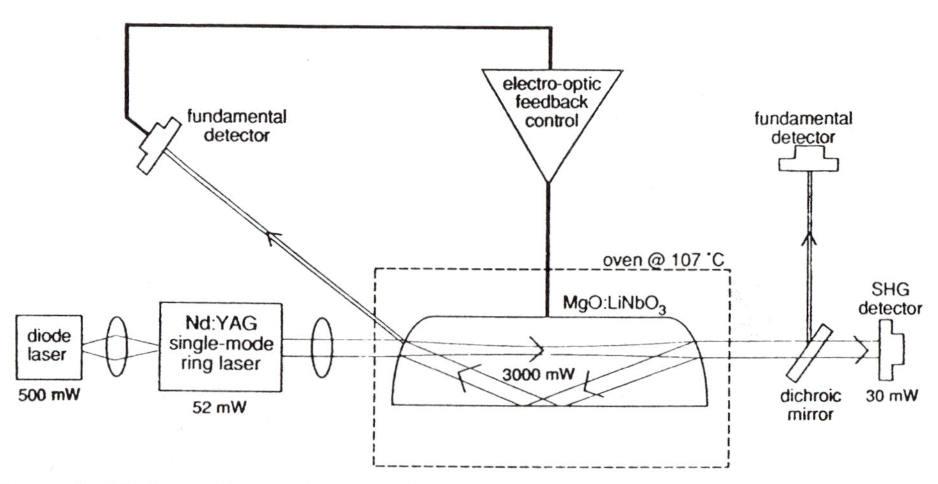

Figure 5 (**b**) Second harmonic output from a 25mm and 12.5mm ring resonator doubler of $MgO:LiNbO_3$. After Kozlovsky et al. [33]

External resonant harmonic generation has been extended to direct doubling of a 100mW cw diode laser by Kozlovsky et al.[36] to generate 41mW of cw blue output. Recently Risk and Kozlovsky[37] have demonstrated external resonant sum frequency generation of a Nd:YAG laser and diode laser in KTP to generate 2mW of 462nm blue radiation which is applicable to reading and writing information on magneto optical storage disks.

The efficiency of external resonant harmonic generation has been improved by recent work of Ady Arie who demonstrated 62% conversion efficiency in an 8mm lithium niobate ring resonant doubler to convert 300mW of 1064nm to 115mW of 532nm. Gersternberger et at.[34] used a three bounce internal reflection ring of lithium niobate 5mm in diameter to convert 300mW of 1064nm to 200mW of 532nm for an external conversion efficiency of 65%. The record for second harmonic generation conversion was set by Ou et al.[35] who used a KTP crystal to double a cw 700mW, 1.08μm Nd:$YAlO_3$ laser to generate 560mW of 0.54μm output at 85% conversion efficiency.

High Average Power External Second Harmonic Generation. External resonant second harmonic generation has been extended to higher average powers by Yang et. al. [38] In this case the external resonant cavity was a bow tie design comprised of discrete optical elements. Figure 6a shows a schematic of the resonator used for the doubling experiments. The laser source was the injection locked, lamp pumped 18W cw Nd:YAG laser source described above. The 18W of fundamental power was modulated in an electrooptic modulator to generate sidebands for FM locking to the external doubling cavity. The beam was mode matched and impedance matched to the external cavity to maximize the coupled power. The external cavity had a 30μm beam waist at the second harmonic nonlinear crystal which was a 6mm long, antireflection coated, crystal of LiB_3O_6 (LBO). The use of the ring resonator configuration eliminated back reflections to the pump laser and permitted the generation of the second harmonic in a single direction. In this experiment, the conversion efficiency reached 35% or 6.5W of second harmonic output for 18W of incident 1064nm power as shown in Figure 6b. The output was TEM_{00} mode and single axial mode with a frequency width equal to that of the master oscillator used to injection lock the 18W Nd:YAG laser.

The use of a discrete component external resonator proved to be convenient for evaluating nonlinear crystals under high power conditions. For example, three types of lithium niobate were studied using this external resonant cavity. The build-up of circulating power at 1064nm in the external cavity allowed the cw damage threshold of congruent lithium niobate to be measured. For antireflection coated crystals the cw surface damage was found to be $10MWcm^{-2}$. For uncoated lithium niobate crystals with clean polished surfaces the maximum cw damage threshold reached $30MWcm^{-2}$. These damage limits allow a $1mm^2$ lithium niobate crystal to handle in excess of 1kW of average power if the phasematching condition could be maintained by proper temperature control.

In a second experiment, D. H. Jundt et al.[39] investigated the performance of lithium-rich lithium niobate created by lithium diffusion into lithium niobate using vapor transport. The vapor transport equilibrated (VTE) lithium niobate nonlinear crystal phasematched for noncritical second harmonic generation of 1064nm at 233.7C. For a 10mm length crystal

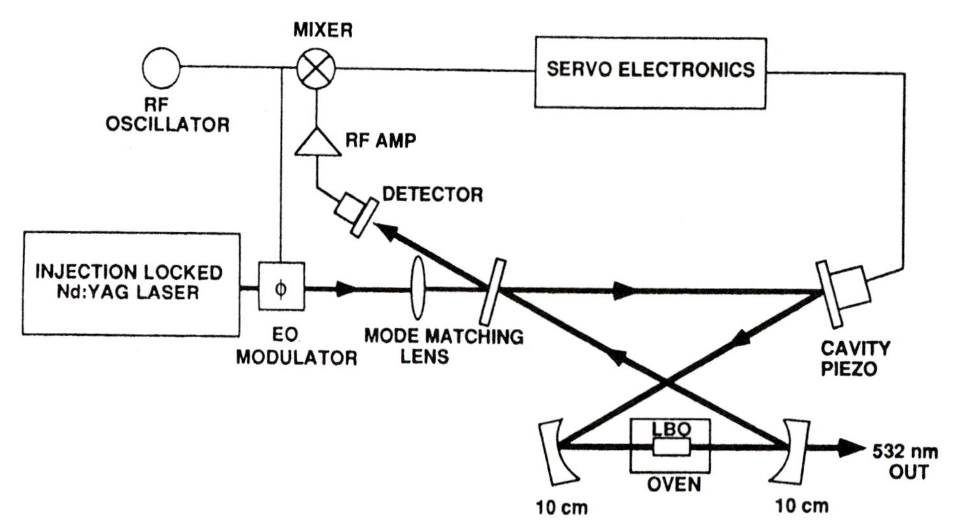

Figure 6a. Schematic of the external resonant second harmonic experiment using a bow tie cavity with discrete optical elements. This resonator allows an easy exchange of nonlinear crystals. See Yang et al.[38] for details.

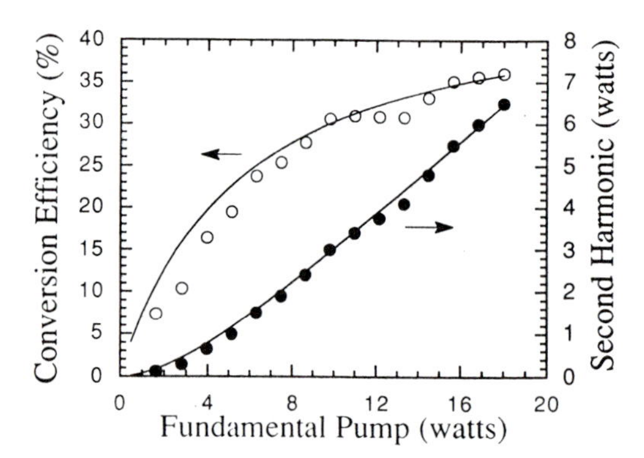

Figure 6b. Conversion efficiency and output second harmonic power vs fundamental pump power for high power external resonant second harmonic generation in LBO.

1.6W of cw green was generated at 69% conversion efficiency showing both the low loss and the lack of photorefractive damage in VTE lithium niobate.[40] Of interest, was the onset of thermally induced phase mismatch as observed by the asymmetry in the phasematch peak caused by absorption of the fundamental beam.

In a third experiment using the external resonant cavity, Jundt et al.[41] generated 1.7W of cw green output for 4.2W of 1064nm input power in a 1.24mm long periodically poled $LiNbO_3$ crystal. Using the laser heated pedestal growth method, three 0.88 diameter rods of lithium niobate were grown with ferroelectric domain reversals every 2.29, 3.47 and 6.31μm for first order quasiphasematched second harmonic generation of blue, green and red radiation. This experiment demonstrated the versatility of quasiphasematching and the high optical quality of the first order periodically poled lithium niobate crystals. Further, following the earlier work by Magel et al.[41] Jundt confirmed that periodically poled lithium niobate does not suffer from optically induced photorefractive index variations even to power levels of greater than $10MWcm^{-2}$.

Optical Parametric Oscillation

Optical parametric oscillators (OPO) have long been studied[42] for their potential to generate continuously tunable radiation over a wide spectral region. With the availability of improved nonlinear materials and stable, single frequency pump lasers, optical parametric oscillators are increasingly being studied. Parametric oscillators operate as doubly resonant devices (DRO) where both the signal and the idler wave are resonated and the threshold is proportional to the product of the losses in each cavity, and as singly resonant devices (SRO) where only one of the waves is resonant and the threshold is proportional to the loss of the resonant wave. In the later case, the singly resonant OPO has a threshold that is of the order of 100 times higher than the doubly resonant OPO. However, the SRO is capable of tuning over a broad spectral range without axial mode jumps and without tuning off the reflectance range of the dielectric mirrors that form the cavity. Because of the higher threshold power for SRO operation, high peak power pumping by Q-switched lasers has been required.[43] The low threshold of the DRO allows cw operation but brings difficulties in continuous tuning because of the need for maintaining the doubly resonant condition for both tunable waves. The tuning characteristics of the DRO have been studied in detail by Eckardt et. al. [44] who showed that continuous tuning of the DRO is possible if two variables, the pump frequency and an applied electric field or temperature, for example, are tuned simultaneoulsy to maintain the doubly resonant condition.

Doubly Resonant Optical Parametric Oscillation. The advent of the diode pumped Nd:YAG laser with stable single frequency operation brought rapid progress to OPO performance.[45] In 1989 Nabors et al. [46] demonstrated cw operation of a $LiNbO_3$ DRO with a measured threshold power of only 10mW. For 20mW of 532nm pumping the OPO operated cw with a conversion efficiency of 85%. This monolithic ring resonator OPO tuned with temperature from 1000nm to 1120nm. Further, it tuned with electric field over a 5THz frequency region near degeneracy for an applied voltage of 924V. The OPO operated in a single axial mode but would hop axial modes when tuned over a wide spectral region.

The monolithic DRO proved to be stable and was shown by Nabors et. al.[47] to reproduce the linewidth of the pump laser of less than 13kHz. At degeneracy the DRO jumped to a phaselocked signal and idler frequency operation at the exact subharmonic of the pump frequency thus forming the first divide-by-two optical frequency divider.

Singly Resonant Optical Parametric Oscillation. The promise of SRO operation with the advantage of continuous tuning without mode hops or cluster mode formation led to continued efforts toward low threshold operation or even cw operation of the SRO. Kozlovsky et al.[48] demonstrated that a monolithic lithium niobate SRO would operate at a high conversion efficiency with a 35W peak power threshold for both a standing wave and a ring resonator configuration. The $MgO:LiNbO_3$ SRO was temperature tuned from 835nm to 1.47μm at a tuning rate of 0.8Thz per degree. For pump pulse lengths of 500nsec derived from an amplified single axial mode laser source, the SRO operated in a single axial mode approximately 20% of the time without any frequency selective elements in the resonator. Operation of the SRO under cw pumping conditions was not possible due to the high theshold of this monolithic device.

The study of the SRO operation under cw pumping was extended by S. T. Yang.[49] Recently, Yang has achieved cw operation of a KTP SRO pumped at 532nm by the green generated from the injection locked, frequency doubled, cw Nd:YAG laser source described above. This is the first demonstration of a cw SRO. The threshold for the SRO is 1.85W with a loss in the resonant signal cavity of 0.85%. Since the threshold is directly proportional to the loss in the resonant wave, it is expected that the threshold can be further reduced with lower resonator loss. Singly resonant operation was assured by operating the SRO with a lithium niobate brewster angle prism within the resonant cavity. The prism ejected the nonresonant idler wave which could be directly measured. For 3.3W of pump power the SRO generated 1.1W of cw idler output for a conversion efficiency of 33% . Furthermore, the SRO operated in a single axial mode under all pumping conditions. Tuning and bandwidth studies of this cw SRO are continuing. However, preliminary results are encouraging because of the high conversion efficiency with single axial mode operation without the need for frequency selective elements.

Advanced Nonlinear Optical Materials

Nonlinear frequency conversion of diode laser pumped solid state lasers requires improved nonlinear materials that can be tailored to meet the requirements of the specific interaction. For bulk crystals the material figure of merit is the ratio of the effective nonlinear coefficient to the loss coefficient or d_{eff}/α . Recent progress in second harmonic generation and optical parametric oscillation has resulted from the reduction in crystal losses not an increase in the nonlinearity. Indeed, a recent careful remeasurement of the nonlinear coefficients of a number of nonlinear crystals has led to a significant reduction in the value of the nonlinear coefficients for KTP, $LiIO_3$, $LiNbO_3$ and a slight increase for BBO.[50]

Quasiphasematching. Phasematching is the most demanding requirement of an optical nonlinear material. The use of birefringence to offset the dispersion of the index of refraction to achieve phase velocity synchronism between the fundamental and the second harmonic waves is well known and widely practiced. However, few crystals have the requisite birefringence to allow phasematched interactions across the entire transparency range of the crystal. An alternative phasematching approach is to use quasiphasematching; that is to alternate the sign of the nonlinear coefficient every coherence length to maintain phase synchronism for the nonlinear interaction. Quasiphasematching was first proposed by Armstrong et al.[51] in 1962 as a method of achieving phasematching. However, a practical means of alternating the sign of the nonlinear coefficient by alternating ferroelectric domains in $LiNbO_3$ was not demonstrated until late 1988 almost simultaneously for bulk $LiNbO_3$ by Magel et al.[41] and for guided wave surface interactions by Lim et al.[52]

The use of guided wave geometries for nonlinear interactions allows the propagation of a confined wave over long distances without the spreading and reduction in intensity caused by diffraction. Lim et. al. [53] applied quasiphasematching in $LiNbO_3$ with a channel waveguide to generate blue radiation at 410nm. This approach has been extended to $LiTaO_3$ by Matsumoto et al.[54]. Recent work in guided wave interactions that use quasi-phasematching has been summarized by Fejer. [55]

Quasiphasematching allows the nonlinear interaction properties to be tailored by controlling the phasematching parameters. Indeed, the synthesis of phasematched interactions by Fourier composition of quasiphasmatched 'chips' has been treated in depth in an article by Fejer et al. [56]. For the first time, quasiphasematching allows the nonlinear crystal to be designed to meet the goals of the interaction independent of the dispersion of the nonlinear crystal. This has led to a rapidly growing emphasis in nonlinear conversion using quasiphasematched interactions for efficient generation of blue light by second harmonic generation of diode laser sources.

Tailored Nonlinear Materials. Recent advances in material synthesis using molecular deposition techniques allows the design and synthesis of artifical nonlinear materials atomic layer by atomic layer. The first experimental demonstration of such design and synthesis was the observation of a very large quadratic nonlinear susceptibility in quantum well AlGaAs by M. M Fejer et al. [57]. The measured nonlinear response at 10.6μm was 28,000pmV^{-1} or 73 times larger than the nonlinear response of bulk GaAs. This work was followed by the design of an asymmetric quantum well structure with enhanced nonlinear response of 320 times that of bulk GaAs.[58] It is now possible to design the nonlinear response of a crystalline medium and to synthesize that design.

When the phasematching is controlled by quasiphasematching, and the interaction region is controlled by guided wave or by special resonator designs[59], and the nonlinear response is designed into the material on an atomic scale, then all aspects of the nonlinear interaction are under the control of the experimenter. We have only recently entered this era of advanced nonlinear materials [60] some thirty years after the first demonstration of a nonlinear optical process in crystal quartz. With the ability to tailor the nonlinear medium we can expect rapid progress in the application of nonlinear frequency conversion to practical devices from the generation of blue light for information storage to the generation of widely tunable infrared radiation for advanced sensing applications.

HIGH AVERAGE POWER DIODE-PUMPED SOLID STATE LASERS

Introduction

The traditional rod shaped solid state laser is pumped from the side by flashlamps for pulsed operation or by arc lamps for continuous wave operation. Joseph Chernoch[61] of General Electric Corporation suggested in 1972 that a zig-zag slab geometry laser may offer advantages in scaling to high average power operation while avoiding the limitations of thermal induced focusing and thermal induced birefringence present in the rod laser. The slab idea was pursued by the Stanford group both experimentally[62] and theoretically[63,64] in the early the 1980s. The early work which used lamp pumping of slab crystalline [65] and slab glass lasers [66,67] set the stage for diode pumping that was to follow.

The early work showed that thermal fracture set the ultimate average power of a solid state laser.[68] The rod geometry laser average power level is independent of the rod diameter and scales linearly with the length. For the slab geometry, the zigzag optical path and the rectilinear geometry allows operation without thermal induced focusing. Depolarization is minimized because the linear polarized light field maps onto the rectilinear heat conducting geometry. Further, the average power level of a slab geometry laser scales as the area of the slab. As an example, a 30cm x 30cm slab glass laser is capable of generating more than one kilowatt of average power before reaching the thermal fracture limit. Beyond these general guidelines, laser design approaches exhibit considerable variation as engineering is an art supported by science. Innovative designs have proliferated in diode pumped solid state lasers as evidenced by the number of articles publised in this field. [69] There is not space in this overview to summarize all of the progress that has been made. Therefore selected articles are cited that provide a sampling of the engineering approaches that have been taken in the design of high average power diode pumped solid state lasers.

Progress in diode laser pump sources

Diode laser are the most efficient source of optical frequency radiation. Further, the power spectral brightness of the diode lasers make them ideal for pumping solid state lasers. The rapid progress in diode lasers can be illustrated by noting that the average power output of a 1cm linear bar diode laser array has increased from 1W in 1978[6], to 12.5W in early 1988,[70] to 38W in mid-1988[71] to 76W in mid-1989[72] and to 120W by early 1992.[73] Although diode lasers are not energy storage optical devices but cw devices, there are advantages in thermal cooling to be gained by operating in a pulsed mode. Progress in one and two dimensional pulsed diode laser arrays has been as rapid as for the cw array devices as summarized and reviewed by Endriz et al.[74] and Beach et al. [75]. The average power levels of two dimensional arrays is limited ultimately by heat removal techniques which include diamond substrates and microchannel coolers [76]. It now appears that average power densities of greater than $1000 Wcm^{-2}$ are feasible for two dimensional diode laser arrays when appropriate cooling and brightness enhancement approaches are taken.

The power densities now available from diode arrays are of the order of the saturation fluence of transitions of ions doped into crystals and glasses. Thus, as first pointed out by Fan and Byer [77,78] it is possible to pump three level solid state lasers at room temperature using diode sources. This in turn opens the potential for pumping new laser transitions at 2μm and other wavelengths that were heretofore not possible to pump by lamps.[79] Recent progress in diode laser pumping of three level laser transitions is discussed by Kubo and Kane [80] in the special issue of the Journal of Quantum Electronics mentioned above.[69]

End Pumped and Side Pumped Solid State Lasers

End pumped laser oscillators offer the optimum geometry for efficiently coupling the diode pump radiation into the gain medium. Thus for low power laser oscillators end pumping is the geometry of choice. As the average power levels increase, thermal focussing becomes a factor in limiting the end pumped laser. [81] At higher average power pumping levels thermal stress fracture sets the ultimate pumping limit of the end pumped laser. [82] The average power limit for end pumped Nd:YAG appears to be less than 100W.

There are novel end pumping arrangements such as end pumping through an optical fiber or end pumping with pump wavelength resonance enhancement. [83] End pumping can also involve pumping a series of separate regions [84] or special optical imaging arrangements to focus the diode laser array into the solid state laser medium [85].

Side pumping of solid state lasers opens new possibilities including side pumping of rod and slab geometry lasers. Since the average power of the rod laser scales with the length of the rod, side pumping offers higher average power possibilities than end pumping. This has been confirmed by Kasinski et al.[86] who demonstrated a 10W average power Q-switched laser with a scaling potential to greater than 1kW of average power. Side pumping of slab geometry lasers offers scaling to high average power levels because the power scales as the area of the slab. Although early results on diode array pumping of slab Nd:YAG by the Livermore group reported only 70W of average power [87] recent results have been reported at the 1100W average power level. This is the highest average power reported from a diode laser array pumped solid state laser.

The Stanford 10W and 100W diode laser pumped slab Nd:YAG lasers. Many applications require high average power lasers that operate in a single transverse mode and a single frequency. Further, some applications are more demanding on the laser source and require long term operation with the possibility of repairing the laser while it operates. Applications that place these stringent requirements on the laser include the laser transmitter for remote sensing of wind on a global scale and the laser source for driving the interferometer for gravity wave detection.

To meet these requirements, the Stanford group has taken the approach of using optical fibers to deliver the diode laser power to the laser gain medium. Further, to allow for scaling to high average power levels, side pumping of slab geometry lasers has been selected. Fiber coupling allows the diode pumping sources and the power supplies to be remotely located from the solid state laser. Fiber coupling also allows the diodes that fail to be replaced without turning off the remaining operating diode lasers. Finally, fiber coupling allows the upgrading of the pump diode lasers to new sources without forcing the redesign of the solid state laser source.

The cw 10W, TEM_{00} mode Stanford Nd:YAG mini-slab laser has been designed to test the fiber coupling approach to diode laser pumping. The laser is pumped by 56 Sony Corporation 1W cw diode lasers coupled to optical fibers. The 300μm core optical fibers are bundled into a single line 24mm in length to deliver the power to a miniature 1.5mm x 1.5mm x 24mm slab of Nd:YAG. The slab is mounted against a gold coated reflector and is cooled from the sides. Figure 7a shows a schematic of the Stanford 10W cw Nd:YAG slab laser. This laser has operated recently with a threshold pump power of 2.8W and an output power of 7.7W for 30W of pumping at the end of the optical fiber as shown in Figure 7b. The measured slope efficiency was 24.3% for this side pumped slab geometry laser when

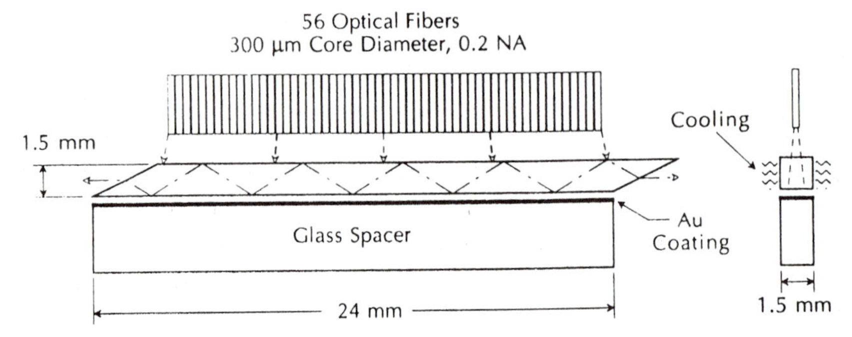

Figure 7a. Schematic of the Stanford 10W cw TEM_{00} mode Nd:YAG minislab laser side-pumped by 56 diode lasers coupled through optical fibers.

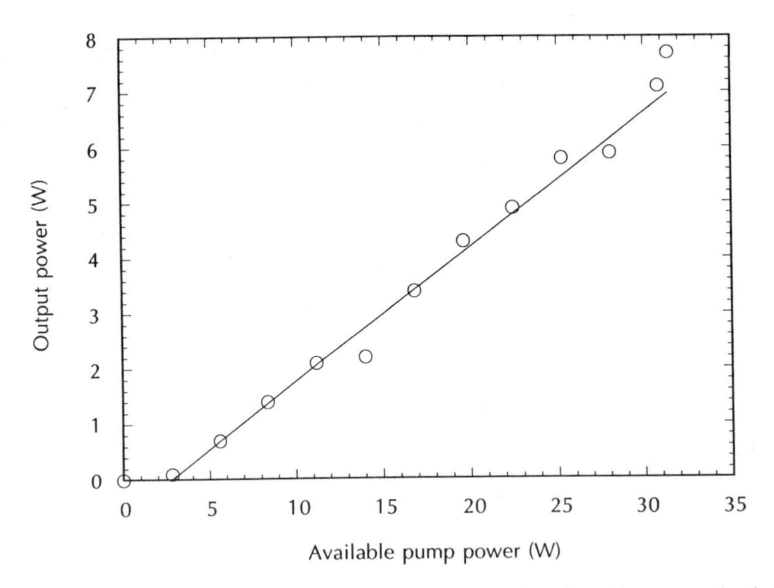

Figure 7b. Output power vs intput diode pump power for the side pumped minislab Nd:YAG laser.

operating in a TEM_{00} spatial mode. Work is underway to operate the laser in a ring cavity and to use injection locking to achieve single frequency operation.

The 10W cw minislab laser is a first step toward the operation of a 100W cw slab Nd:YAG laser pumped by 35 diode arrays that emit 10W of power at the end of the 400μm fiber . This side pumped slab laser uses uniform pumping and cooling through the 6mm wide x 24mm long face of the 1.5mm thick Nd:YAG slab. This uniform pumping and cooling approach should achieve uniform thermal boundary conditions. At the time of this writing, the fiber coupled diode lasers were being delivered and tested. The three dimensional computer analysis of the thermal properties of this Nd:YAG slab laser indicate that near diffraction limited spatial mode operation should be acheived using an unstable-stable resonator configuration with a gaussian variable output coupler. First operation of the 100W laser is expected in mid-1993.

CONCLUSION

Diode pumped solid state lasers offer unprecendented potential for improvements in power and efficiency at decreasing costs in the future. As the market volume for diode pumped solid state lasers increases the price of the diode pump sources decreases. Looking back over the past five years the performance per device doubles each year while the cost per Watt is reduced by a factor of two. The AlGaAs diode laser obeys the same cost vs production volume trends as semiconductor devices in volume production. Thus, at current market growth rates, it is reasonable to predict that diode laser array costs will decline from the current value of $500 per Watt of average power to less than $50 per Watt by 1995. Today it is cost effective to design and build 1W Nd:YAG laser using diode pumping in place of lamp pumping. Past trends of delivered diode pumped Nd:YAG products show that the average output power increases by a factor of 10 every two years at constant cost. If this trend continues, we can expect to see diode pumped Nd:YAG laser products with 10W of average power by 1994 and with 100W of output power by 1996. These lasers will operate with 10% overall electrical efficiency and mean time to failure of geater than 20,000 hours.

Laser sources that operate at 10% efficiency can do real work. The potential applications for such lasers includes uses in medicine, remote sensing, semiconductor materials process and repair, and eventually material cutting, welding and annealing. Diode pumped kilowatt average power Nd:YAG lasers offer a promising alternative to the carbon dioxide lasers because of compact size and optical fiber delivery of the high power radiation.

It is an interesting historical note that widespread application of new sources electromagnetic radiation, from the radio to the laser, takes about thirty years to grow to a market scale that impacts the general economy. It is also interesting to note that in each case the largest market has been entertainment related. Thus the diode laser has become the most widespread source by virtue of its application in the compact disk. Perhaps the diode pumped solid state laser, coupled with nonlinear frequency conversion, will find widespread use in the projection of high definition color images in theaters and homes of the future.

Acknowledgements: This work was partially supported by the United States Government through sponsored research supported under DARPA and the Center for Nonlinear Optical Materials.

REFERENCES

1. T. H. Maiman, Stimulated optical radiation in ruby masers, *Nature* 187:493-494 (1960)
2. P. A. Franken, A. E. Hill, C. W. Peters, G. Weinrich, *Phys. Rev. Lett.* 7: 118-119 (1961)
3. R. Newman, Excitation of the Nd^{3+} fluorescence in $CaWO_4$ by recombination radiation in GaAs, *J. Appl. Phys.* 34:437 (1963)
4. R. J. Keyes and T. M. Quist, Injection luminescent pumping of CaF_2:U^{3+} with GaAs diode lasers, *Appl. Phys. Lett.* 4:50-52 (1964)
5. F. W. Ostermayer, Jr. GaAsP diode pumped YAG:Nd lasers, *Appl. Phys. Lett.* 18:93-96 (1971)
6. D. R. Scrifres, R. D. Burnham, W. Streifer, Phase-Locked semiconductor laser array, *Appl. Phys. Lett.* 33: 6 (1978)
7. R. L. Byer, Diode Laser-Pumped Solid State Lasers, *Science* 239: 742-747 (1988)
8. Tso Yee Fan and R. L. Byer, Diode Laser-Pumped Solid State Lasers, *IEEE Journ. Quant. Electr.* 24: 895-912 (1988)
9. D. W. Hughes and J. R. M. Barr, Laser diode pumped solid-state lasers, *J. Phys. D: Appl. Phys.* 25: 563-586 (1992)
10. Thomas J. Kane, Laser Diode-Pumped Solid-State Lasers, Lightwave Electronics Corporation, 1991
11. Binkun Zhou, Thomas J. Kane, George J. Dixon, and Robert L. Byer, Efficient, frequency-stable laser-diode-pumped Nd:YAG laser, *Optics Letts.* 10:62-64 (1985)
12. Thomas J. Kane and R. L. Byer, Monolithic, unidirectional single-mode Nd:YAG ring laser, *Optics Letts.* 10:65-67 (1985)
13. T. J. Kane, W. J. Kozlovsky, and R. L. Byer, Coherent laser radar at 1.06μm using Nd:YAG lasers, *Optics Letts.* 12:239-241 (1987); T. J. Kane, Binkun Zhou, R. L. Byer, Potential for coherent Doppler wind velocity lidar using neodymium lasers, *Applied Optics* 23: 2477-2481 (1984)
14. A. Abramovici et. al. LIGO: The Laser Interferometer Gravitational-Wave Observatory, *Science* 256:325-333 (1992)
15. Y. L. Sun and R. L. Byer, Submegahertz frequency-stabilized Nd:YAG oscillator, *Opt. Letts.* 7:408-410 (1982)
16. Alan C. Nillson, Thomas J. Kane, and Robert L. Byer, Monolithic nonplanar ring lasers: resistance to optical feedback, *SPIE* 912:13-17 (1988)
17. Alan C. Nillson, Eric K. Gustafson, and Robert L. Byer, Eigenpolarization Theory of Monolithic Nonplanar Ring Oscillators, *IEEE Journ. Quant. Electr.* 25:767-790 (1989)
18. T. Day, E. K. Gustafson, and R. L. Byer, Active frequency stabilization of a 1.062μm, Nd:GGG diode-laser-pumped nonplanar ring oscillator to less than 3Hz of linewidth, *Opt. Letts.* 15: 221-223 (1990)
19. T. Day, E. K. Gustafson and R. L. Byer, Sub-Hertz Relative Frequency Stabilization of Two-Diode Laser-Pumped Nd:YAG Lasers Locked to a Fabry-Perot Inteferometer, *IEEE Journ. Quant. Electr.* 28:1106 (1992)
20. D. W. Allan, Statistics of atomic frequency standards, *Proc IEEE* 54:221, (1966)

21. N. M. Sampas, R. Liu, E. K. Gustafson and R. L. Byer, Frequency stability measurements of two non-planar ring oscillators independently stabilized to two Fabry-Perot interferometers, presented at the SPIE conference 1837, November 16-18, 1992, OPTCON, Boston, MA

22. Robert Adler, A Study of Locking Phenomena in Oscillators, *Proc. IEEE*, 51:1380 (1973)

23. C. D. Nabors, A. D. Farinas, T. Day, S. T. Yang, E. K. Gustafson, and R. L. Byer, Injection locking of a 13W cw Nd:YAG ring laser, *Opt. Lett.* 14:1189-1191 (1989)

24. T. J. Kane and E. A. P. Cheng, Fast frequency tuning and phase locking of diode-pumped Nd:YAG ring lasers, *Opt. Lett.* 13:970-972 (1988)

25. L. G. Kazovsky and D. A. Atlas, A 1320nm experimental optical phase-locked loop, *Photon. Technol. Lett.* 1:395-397 (1989)

26. T. Day, A. D. Farinas and R. L. Byer, Demonstration of a Low Bandwidth 1.06μm Optical Phase-Locked Loop for Coherent Homodyne Communication, *IEEE Photon. Technol. Lett.* 2:294-296 (1990)

27. E. A. P. Cheng, T. J. Kane, R. W. Wallace, D. M Cornwell, Jr. Injection chaining of diode-pumped single-frequency ring lasers for free-space communication, Lightwave Electronics Corp. and NASA Goddard Space Flight Ctr. 1990

28. Ady Arie, Stephan Schiller, E. K. Gustafson and R. L. Byer, Absolute frequency stabilization of diode-laser-pumped Nd:YAG lasers to hyperfine transitions in molecular iodine, *Opt. Lett.* 17:1204-1206 (1992)

29. T. Y. Fan, G. J. Dixon, and Robert L. Byer, Efficient GaAlAs diode-laser-pumped operation of Nd:YLF at 1.047μm with intracavity doubling to 523.6nm, *Opt. Lett.* 11:204-206 (1986)

30. T. Baer, Large amplitude fluctuations due to longitudinal mode coupling in diode-pumped intracavity -doubled Nd:YAG lasers, *J. Opt. Soc. Amer. B* 3:1175-1180 (1986)

31. Michio Oka and Shigeo Kubota, Stable intracavity doubling of orthogonal linearly polarized modes in diode-pumped Nd:YAG lasers, *Opt. Lett.* 13:805-807 (1988)

32. A. Ashkin, G. D. Boyd, and J. M. Dziedzic, Resonant Optical Second Harmonic Generation and Mixing, *IEEE Journ. Quant. Electr.* QE-2:109-123 (1966)

33. W. J. Kozlovsky, C. D. Nabors, and Robert L. Byer, Efficient Second Harmonic Generation of a Diode-Laser-Pumped cw Nd:YAG Laser Using Monolithic $MgO:LiNbO_3$ External Resonant Cavities, *IEEE Journ. Quant. Electr.* 24: 913-919 (1988)

34. D. C. Gerstenberger, G. E. Tye, and R. W. Wallace, Efficient second-harmonic conversion of cw single-frequency Nd:YAG light by frequency locking to a monolithic ring frequency doubler, *Opt. Lett.* 16: 992-994 (1991)

35. Z. Y. Ou, S. F. Pereira, E. S. Polzik, and H. J. Kimble, 85% efficiency for cw frequency doubling from 1.08 to 0.54μm, *Opt. Lett.* 17:640-643 (1992)

36. W. J. Kozlovsky, W. Lenth, E. E. Latta, A. Moser, and G. L. Bona, Generation of 41mW of blue radiation by frequency doubling of a GaAlAs diode laser, *Appl. Phys. Lett.* 56:2291 (1990)

37. W. P. Risk and W. J. Koslovsky, Efficient generation of blue light be doubly resonant sum-frequency mixing in a monolithic KTP resonator, *Opt. Lett.* 17:707-709 (1992)

38. S. T. Yang, C. C. Pohalski, E. K. Gustafson, R. L. Byer, R. S. Feigelson, R. J. Raymakers, and R. K. Route, 6.5W, 532nm radiation by cw resonant external-cavity second-harmonic generation of an 18W Nd:YAG laser in LiB_3O_5, *Opt. Lett.* 16:1493-1495 (1991)
39. D. H. Jundt, M. M. Fejer and Robert L. Byer, Optical Properties of Lithium-Rich Lithium Niobate Fabricated by Vaport Transport Equilibrium, *IEEE Jour. Quant. Electr.* 26: 135-138 (1990)
40. D. H. Jundt, M. M. Fejer, R. L. Byer, R. G. Norwood, and P. F. Bordui, 69% Efficient continuous-wave second-harmonic generation in lithium-rich lithium niobate, *Opt. Lett.* 16: 1856-1858 (1991)
41. G. A. Magel, M. M. Fejer and R. L. Byer, Quasi-phasematched second-harmonic generation of blue light in periodically poled $LiNbO_3$, *Appl. Phys. Lett.* 56: 108-110 (1990)
42. R. L. Byer, Optical Parametric Oscillators, in *Quantum Electronics: A Treatise*, H. Rabin and C. L. Tang eds.(Academic Press, New York, 1975) Vol.1, part B, pp 587-702
43. D. C. Gerstenberger, G. E. Tye and R. W. Wallace, Optical Parametric Oscillation in MgO:$LiNbO_3$ Driven by a Diode Pumped Single Frequency Q-Switched Laser, *IEEE Photon. Technol. Lett.* 2:15-17 (1990)
44. R. C. Eckardt, C. D. Nabors, W. J. Kozlovsky and R. L. Byer, Optical Parametric oscillator frequency tuning and control, *Jour. Opt. Soc. Am. B*, 8:646-667 (1991)
45. W. J Kozlovsky, C. D. Nabors, R. C. Eckardt and R. L Byer, Monolithic MgO:$LiNbO_3$ doubly resonant optical parametric oscillator pumped by a frequency-doubled diode-laser-pumped Nd:YAG laser, *Opt. Lett.* 14:66-68 (1989)
46. C. D. Nabors, R. C. Eckardt, W. J. Kozlovsky and R. L. Byer, Efficient, single-axial-mode operation of a monolithic MgO:$LiNbO_3$ optical parametric oscillator, *Opt. Lett* 14:1134-1136 (1989)
47. C. D. Nabors, S. T. Yang, T Day, and R. L. Byer, Coherence properties of a doubly resonant monolithic optical parametric oscillator, *J. Opt. Soc. Am. B*, 7:815-820 (1990)
48. W. J. Kozlovsky, E. K. Gustafson, R. C. Eckardt, and R. L. Byer, Efficient monolithic MgO:$LiNbO_3$ singly resonant optical parametric oscillator, *Opt. Lett.* 13:1102-1104 (1988)
49. S. T. Yang, CW singly resonant KTP optical parametric Oscillator, private communication November 1, 1992
50. Robert C. Eckardt, H. Masuda, Y. X. Fan, and R. L. Byer, Absolute and Relative Nonlinear Optical Coefficients of KDP, KD*P, BaB_2O_4, $LiIO_3$, MgO:$LiNbO_3$ and KTP Measured by Phase-Matched Second-Harmonic Generation, *IEEE Journ. Quant. Electr.* 26:922-933 (1990)
51. J. A. Armstrong, N. Bloembergen, J. Ducuing, and S. Pershan, Interactions between light waves in a nonlinear dielectric, *Phys. Rev.* 127:1918-1939 (1962)
52. E. J. Lim, M. M. Fejer and R. L. Byer, Second harmonic generation of green light in periodically poled planar lithium niobate waveguide, *Electr. Lett.* 25:174-175 (1989)
53. E. J. Lim, M. M. Fejer, R. L. Byer and W. J. Kozlovsky, Blue Light generation by frequency doubling in periodically poled lithium niobate channel waveguide, *Electr. Lett.* 25: 732-732 (1989)
54. S. Matsumoto, E. J. Lim , H. M. Hertz, M. M. Fejer, Quasiphase-matched second harmonic generaton of blue light in electrically periodically poled lithium tantalate waveguides, *Electr. Lett.* 27:2040-2042 (1991)

55. M. M. Fejer, Nonlinear Frequency Conversion in periodically-poled ferroelectric waveguides, in *Guided Wave Nonlinear Optics*, pp133-145, 1992 Kluwar Academic Publishers, Netherlands
56. M. M. Fejer, G. A. Magel, Dieter H. Jundt, Robert L. Byer, Quasi-Phase-Matched second Harmonic Generation: Tuning and Tolerances, *IEEE Journ. Quant. Electr.* 28: 2631-2654 (1992)
57. M. M. Fejer, S. J. B. Yoo, R. L. Byer, Alex Harwitt, and J. S. Harris, Jr. Observation of Extremely Large Quadratics Susceptibility at 9.6 - 10.8μm in Electric-Field-Biased AlGaAs Quantum Wells, *Phys. Rev. Lett.* 62:1041-1043 (1989)
58. S. J. B. Yoo, M. M. Fejer, R. L. Byer, and J. S. Harris, Jr. Second-order susceptibility in asymmetric quantum wells and its control by proton bombardment, *Appl. Phys. Lett.* 58:1724-1726 (1991)
59. R. Lodenkamper, M. M. Fejer, J. S. Harris, Jr. Surface Emitting second harmonic generation in vertical resonator, *Electr. Lett.* 27:1882-1883 (1991)
60. R. L. Byer, Advances in Nonlinear Optical Materials and Devices, in *Nonlinear Optics, Fundamentals, Materials and Devices*, ed. by S. Miyata, pp 379-392, North Holland, 1992
61. W. S. Martin and J. P. Chernoch, "Multiple internal reflection face pumped laser," U. S. Patent 3633 126, 1972
62. J. M. Eggleston, T. J. Kane, J. Unternahrer, and R. L. Byer, Slab-geometry Nd:glass laser performance studies, *Opt. Lett.* 7:405-407 (1982)
63. J. M. Eggleston, T. J. Kane, K. Kuhn, J. Unternahrer, and R. L. Byer, The Slab Geometry Laser - Part I: Theory, *IEEE Journ. Quant. Electr.* QE-20: 289-301 (1984)
64. T. J. Kane, J. M. Eggleston, and R. L. Byer, The Slab Geometry Laser-Part II: Thermal Effects in a Finite Slab, *IEEE Journ. Quant. Electr.* QE-21: 1195-1210 (1985)
65. T. J. Kane, R. C. Eckardt, and R. L. Byer, Reduced Thermal Focusing and Birefringence in Zig-Zag Slab Geometry Crystalling Lasers, *IEEE Journ. Quant. Electr.* QE-19: 1351-1354 (1983)
66. M Reed, K. Kuhn, J. Unternahrer, and R. L. Byer, Static Gas Conduction Cooled Slab Geometry Nd:Glass Laser, *IEEE Journ. Quant. Electr.* QE-21:412-414 (1985)
67. M. K. Reed and R. L. Byer, Performance of a Conduction Cooled Nd:glass Zigzag Slab Laser, *SPIE* 1021:128-135 (1989)
68. W. F. Krupke, M. D. Shinn, J. E. Marion, J. A. Caird, and S. E. Stokowski, Spectroscopic, optical, and thermomechanical properties of neodymium-and chromium-doped gadolinium scandium gallium garnet, *J. Opt. Soc. Am. B* 3:101-113 (1986)
69. T. Y. Fan and D. F. Welch eds. special issue on Semiconductor Diode-Pumped Solid-State Lasers, *IEEE Journ. Quant. Electr.* 28: 940-1209 (1992)
70. M. Sakamoto, G. L. Harnagel, D. F. Welch, C. R. Lennon, W. Streifer, H. Kung, and D. R. Scifres, 12.5W continuous-wave monolithic laser-diode arrays, *Opt. Lett.* 13:378-380 (1988)
71. M. Sakamoto, D. F. Welch, G. L. Harnagel, W. Streifer, H. Kung, and D. R. Scifres, Ultrahigh power 38W continuous-wave monolithic laser diode arrays, *Appl. Phys. Lett.* 52:2220-2223 (1988)
72 M. Sakamoto, D. F. Welch, J. G. Endriz, D. R. Scifres, and W. Streifer, 76W cw monolithic laser diode arrays, *Appl. Phys. Lett.* 54:2299-2301 (1989)

73. M. Sakamoto, J. G. Endriz and D. R. Scifres, 120W cw output power from monolithic AlGaAs (800nm) Laser diode array mounted on Diamond heatsink, *Electr. Lett.* 28:197-198 (1992)
74. J. G. Endriz, M. Vakili, G. S. Browder, M. DeVito, J. M. Haden, G. L. Harnagel, W. E. Plano, M.Sakamoto, D. F. Welch, S. Willing, D. P. Worland, and H. C. Yao, High Power Diode Laser Arrays, *IEEE Journ. Quant. Electr.* 28: 952-965 (1992)
75. R. Beach, W. J. Benett, B. L. Freitas, D. Mundinger, B. J. Comaskey, R. W. Solarz, and M. A. Emanuel, Modular Microchannel Cooled Heatsinks for High Avaerage Power laser diode Arrays, *IEEE Journ. Quant. Electr.* 28:966-976 (1992)
76. D. Mundinger, R. Beach, W. Benett, R. Solarz, V. Sperry and D. Ciarlo, High Average power edge emitting laser diode arrays on silicon microchannel coolers, *Appl. Phys. Lett.* 57:2172-2174 (1990)
77. T. Y. Fan and R. L. Byer, Continuous wave operation of a room-temperature, diode-laser-pumped, 946nm Nd:YAG laser, *Opt. Lett.* 12:809-811 (1987)
78. T. Y. Fan and R. L. Byer, Modeling and cw operation of a quasi-three-level 946 nm Nd:YAG laser, *IEEE Jour. Quant. Electr.* 23:605-612
79. T. Y. Fan, G. Huber, R. L. Byer, and P. Mitzscherlich, Continuous-wave operation at 21.μm of a diode-laser-pumped, Tm-sensitized $Ho:Y_3Al_5O_{12}$ laser at 300K, *Opt. Lett* 12:678-680 (1987)
80. T. Kubo and T. J. Kane, Diode-Pumped Lasers and Five Eye-Safe Wavelengths, *IEEE Journ. Quant. Electr.* 28:1033-1040 (1992)
81. S. C. Tidwell, J. F. Seamans, M. S. Bowers, and A. K. Cousins, Scaling CW Diode-End-Pumped Nd:YAG Lasers to High Average Powers, *IEEE Journ. Quant. Electr.* 28:997-1009 (1992)
82. A. K. Cousins, Temperature and Thermal Stress Scaling in Finite-Length End-Pumped Laser Rods, *IEEE Journ. Quant. Electr.* 28:1057-1069 (1992)
83. W. J. Kozlovsky and W. P. Risk, Laser-Diode-Pumped 946nm Nd:YAG laser with Resonance Enhanced Pump Absorption, *IEEE Journ. Quant. Electr.* 28:1139-1141 (1992)
84. T. M. Baer, D. F. Head, P. Gooding, G. J. Kintz, and S. Hutchenson, Performance of a Diode-Laser-Pumped Nd:YAG and Nd:YLF Lasers in a Tightly Folded Resonator Configuration, *IEEE Journ. Quant. Electr.* 28:1131-1138 (1992)
85. S. Yamaguchi and H. Imai, Efficient Nd:YAG laser End-Pumped by a 1cm Aperture Laser-Diode Bar with a GRIN lens Array Coupling, *IEEE Journ. Quant. Electr.* 28:1101-1105 (1992)
86 J. Kaskinski, W. Hughes, D. DiBiase, P. Bournes, and R. Burnham, One Joule Output from a Diode-Array-Pumped Nd:YAG Laser with Side-Pumped Rod Geometry, *IEEE Journ. Quant. Electr.* 28:977-985 (1992)
87. B. J. Comaskey, R.Beach, G. Albrecht, W. J. Bennett, B. L. Freitas, C. Petty, D. VanLue, D. Mundinger, and R.W. Solarz, High Average Power Diode Pumped Slab Laser, *IEEE Journ. Quant. Electr.* 28:992-996 (1992)

GENERATION OF VISIBLE LIGHT WITH DIODE PUMPED SOLID STATE LASERS

K. - J. Boller , J. Bartschke, R. Knappe, and R. Wallenstein

Universität Kaiserslautern
Fachbereich Physik
Erwin-Schrödinger-Str.
6750 Kaiserslautern, Germany

INTRODUCTION

The rapid development of infrared diode pump lasers and new laser materials have made the diode pumped solid state lasers efficient and reliable sources of coherent light. By now, these properties have led to the commercial availability of a number of systems. Due to the restricted wavelengths of diode pump light available today with high power, all simple three or four-level diode pumped solid state lasers emit in the near IR. Unfortunately, many technical applications, as high density optical storage and precision interferometric measurements, rather require shorter wavelengths.

Currently, three principally different approaches are used to generate visible light with DPSSL's: i) There has been considerable progress in new up-conversion laser materials,[1] but still very few examples of visible emission at room temperature are known,[2] in none of these cases exclusively pumped with laser diodes. ii) In terms of wavelength flexibility, efficiency, and output power, the traditional way of second harmonic generation (SHG) in an external build-up cavity has to mentioned. An output power of 1.6Watt was reported[3] and, with a high quality crystal, the latest record in conversion efficiency was 85%.[4] The major disadvantage of this otherwise straightforeward method is the requirement for servo control of either the length of the solid state laser cavity or of the external conversion cavity. iii) Intracavity SHG takes advantage of the high power of the fundamental for efficient conversion[5] and simultaneously circumvents the servo problem. However, a non-linear crystal placed in the laser cavity causes passive losses for the fundamental by residual reflectivity, spurious absorption, and scattering. If the non-linear crystal is not of excellent optical quality the intracavity power may be severely reduced as compared to the unloaded cavity, which results in unefficient SHG. Another disadvantage should not be underestimated, that is the complexity of the two-crystal setup with its correspondingly difficult alignment.

An elegant way to avoid most of the problems listet above is to incorporate both, the properties of a laser crystal and the properties of a non-linear crystal in one crystal. A laser based on such a crystal is usually called a self frequency doubling laser.

Although the idea is more than twenty years old, only few such materials have been tested, like $Tm:LiNbO_3$, or $Nd:LiNbO_3$.[6] It turned out that strong photorefraction breaks phasematching over a substantial part of the beams cross section. An approved attempt was done with $Nd:MgO:LiNbO_3$.[7] Here, photorefraction is greatly reduced due to MgO co-doping of the host, while the lasing ion is Nd. Indeed, strong dye laser pumping led to lasing at the two Nd-fundamentals, 1085 and 1093nm, and self frequency doubling with emission at 543 and 547nm. Diode pumping, however, was not successful in the sense that emission at the fundamental was only observed at low crystal temperature while phase matching for SHG requires heating of the crystal.

To our knowledge, there is only one self frequency doubling laser material, which has defenitely overcome the first steps of fundamental research. In 1990, 10 mWatt cw emission at 531nm was reported with a NYAB laser ($Nd_x:Y_{1-x}Al_3(BO_3)_4$), pumped by a 807nm GaAlAs diode laser array.[8]

Apart from singular reports with comparable output power and varying setups,[9] no attempt was made to systematically investigate the crystal quality and the efficiency and to improve the output power. Here we would like to point out that also other even more interesting features of NYAB have not been investigated at all.

An electro-optic effect can be expected in NYAB, since this is the case for any optically non-linear crystal. The electro-optic effect could be employed for fast phase modulation, mode locking and Q-switching without additional optical intracavity components.

A Faraday effect which, in the case of Nd:YAG, led to the realization of a monolithically integrated unidirectional ring laser with unmatched passive frequency stability [10] would be of great importance also in the case of NYAB, for high frequency stability of the visible output.

The determination of the phase matching angle for sum frequency generation of tunable 2µm light to 1062nm would be an important step towards diode pumped and widely tunable 2µm radiation from an optical parametric oscillator (OPO).

Also, when high power narrowband TEM_{00} pump radiation at 807 nm is available, for example from an injection locked diode laser array, tunable blue output at 460nm would be expected. In this case, the blue radiation is the sum frequency of the tunable pump with the high power intracavity NYAB fundamental.

OVERVIEW

The above listed potential prospects of NYAB have led us to a number of investigations. First, we show how the most efficient crystals were selected out of a set of 11 crystals. We give an impression of the reproducability in NYAB manufacturing. Then, we present a systematic investigation of the power optimization of NYAB lasers pumped with multimode diode laser arrays. We describe how the emitter size of the particular pump laser array used determines the pump focusing optics and the design of the NYAB laser cavity in order to maximize the visible output power. For a comparison to array pumping, the maximum possible efficiency of a given setup was determined with a perfectly focusable pump beam from a TEM_{00} Ti:Sapphire laser at 807nm. The optimization of the diode pump optics and laser design led to an output of 130mWatt at 531nm with a diode pump power of 1.55 Watt. This is the highest power reported so far from a diode pumped NYAB laser. In this case, the efficiency approximately reaches the same value as produced by the the TEM_{00}-pump, which demonstrates the succes of the optimization. In the next step, single frequency TEM_{00} , 531nm, 535nm, and 539nm light was obtained with a micro chip laser pumped with a low power single spatial mode diode laser at 807nm. Finally, we report on measurements of new material properties of NYAB, i.e. the determination of the Verdet constant, the electro-optic coefficient, and the phase matching angle for SHG of 2.3μm to 750nm light.

CRYSTAL SELECTION

The quality of currently available NYAB varies strongly from crystal to crystal. An important step to construct an efficient NYAB laser is to select the best crystal out of a set (in our case 11). The best way, of course, is to do a test with the actual cw NYAB laser (fig.1). An advantage of this method is that also the spatial

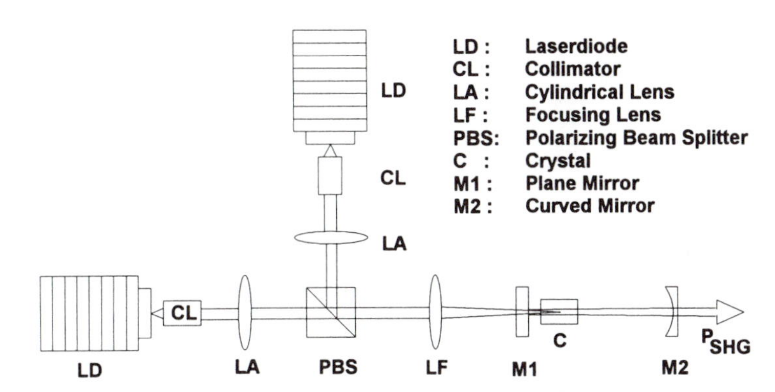

Figure 1. Schematic setup of a diode pumped NYAB laser

inhomogeneity of a crystal can be checked. Fig. 2 shows the visible and the IR output of crystal no. 11 vs. the lateral displacement of the crystal. With closer inspection of the two curves it is clear that a spatial region of efficient IR operation does not necessarily correspond to a region of high visible output. An explanation of this behaviour is that the Nd-concentration or the optical quality shows a different spatial variation than do the nonlinearity and refractive indices.

It turned out that a quick beforehand test is to measure the SHG efficiency with 1064nm pulses from a Q-switched Nd:YAG laser. The crystals (3% Nd-concentration, diameter 3mm, length 6mm, cut at an angle of 30^0 with respect to the optical axis) were mounted on a rotation stage and fine-rotated to the phase matching angle for SHG. The results are shown in fig.3.

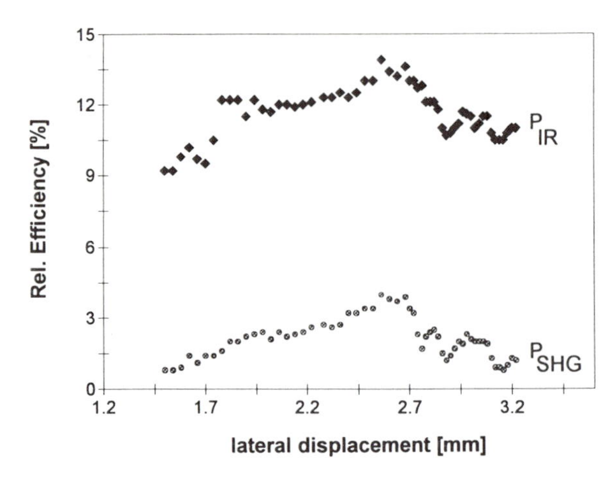

Figure 2. Relative IR output and visible output of a NYAB laser vs. the lateral displacement of the crystal

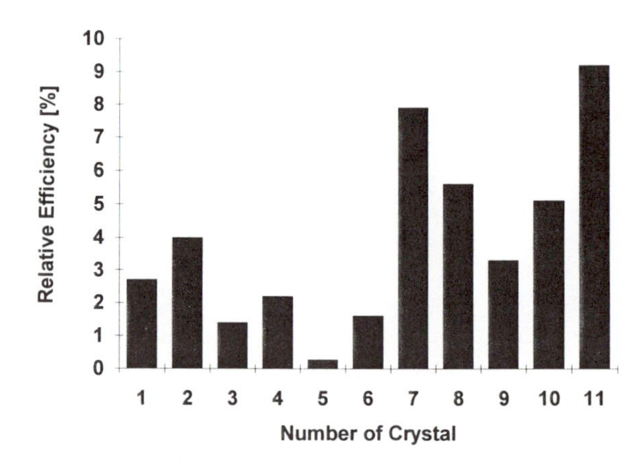

Figure 3. Relative SHG efficiency of 11 NYAB crystals tested with a Q-switched Nd:YAG-laser

The big variance of the NYAB crystal quality, on the first view, may look quite dicouraging. Nevertheless, quite a progress is made in quality and size of the crystals, compared to first reports on NYAB. The best two crystals in our set are actually of very good quality, as can be seen from the maximum output, described below. Obviously, it is the reproducability in growing the crystals which needs improvement, rather than the quality reached with selected crystals.

PUMP FOCUSING AND RESONATOR DESIGN

For maximum output power, the diode array pump optics and the resonator design of a self frequency doubling laser has to fulfill three conflicting requirements: i) The resonator beam has to posess a big cross section in order to obtain a complete overlap with the pump beam and thus a high pump efficiency. ii) The resonator beam cross section within the crystal should be small for efficient SHG efficiency, since the small-signal gain for the visible increases with the square of the fundamental intensity. iii) SHG of 531nm in NYAB can only be critically phase matched, i.e. there is a walk-off between the fundamental and the SH beam which gives rise to a minimum useful beam size.

The requirements ii) and iii) can be accounted for by the focusing theory of Boyd and Kleinman.[11] Using the NYAB refractive indices, non-linearity, [12] and the 6mm length of our crystals we calculate a beam waist of the fundamental beam of 25μm ($1/e^2$-radius of the Gaussian intensity envelope) for maximum efficient SHG.

The 25μm fundamental beam waist indeed provides the maximum SHG efficiency and the maximum visible output possible, when a TEM_{00} pump beam is used which can be perfectly mode matched to the resonator waist. A TEM_{00} beam at 807 nm from a diode laser is currently available only from single stripe low power lasers. Here, the pump radiation is typically emitted out of a 1 x 3 μm cross section. The small size of the emitting surface limits the output power of single stripe lasers to about 150mWatt. At higher values the laser is destroyed due to heating and melting of the emitting surface. This power restriction for TEM_{00} diode pump sources also limits the visible output of a NYAB laser to a few mWatt, although, concerning the SHG, an optimum output is obtained.

The situation is quite different for pumping with a diode laser array which typically consists of 10 to 40 laser stripes manufactured adjacent to each other with a distance of 10μm on the same chip. Due to this construction, an array can deliver much more output power (typically up to a few Watt cw), but the emission is spatially multimode. Perpendicular to the p-n-junction the radiation can still be matched into a TEM_{00} beam. In the other dimension, however, the emission is roughly that of an extended (i.e.spatially incoherent) light source posessing a divergence of about 8° (full beam). In this lateral direction, mode matching to a Gaussian resonator beam is not possible. Only a poor spatial overlap can be realized when an image of the emitting surface of the array (~ 1μm x 200μm) is placed around the resonator beam.

To illustrate this point more clearly let's consider the following typical setup. A short focus imaging lens (collimator with f = 6.5mm) is placed right behind the array, eventually followed by a cylindrical lens for a correction of astigmatism. The resonator beam waist is also placed as close as possible to the array optics at the location of the image of the array emission. This can be done best by using a hemispherical resonator. The smallest possible size of the array image in such an arrangement is of the order of ~200μm in the lateral direction, but behind the image the pump radiation diverges much stronger than would a TEM_{00} beam of the same beam diameter. If, on the other hand, one tries to overlap with the resonator mode over a longer distance, the diameter of the arrays image would be much bigger than the waist of a Gaussian beam, again causing a low pump effeciency.

From these considerations it is clear that finding an optimum setup has to start with the emitter size of a given diode laser array. The size of the arrays emitting surface determines the size of the arrays image at the location of the beam waist in the NYAB crystal. The corresponding fundamental beam waist of the NYAB resonator is then matched to this image.

The optimization can be restricted to the transverse dimension because, in in the other dimension, the emitter width of the diode array radiation is diffraction limited and can be tranferred into a Gaussian mode.

In the following section we give some simple geometrical considerations for maximizing the total visible output power that can be obtained with diode array pumping. Fig. 4 shows a schematic picture of the arrays lateral emitter size, the pump focusing optics, the image of the laser array in the NYAB crystal, and the beam waist of the fundamental.

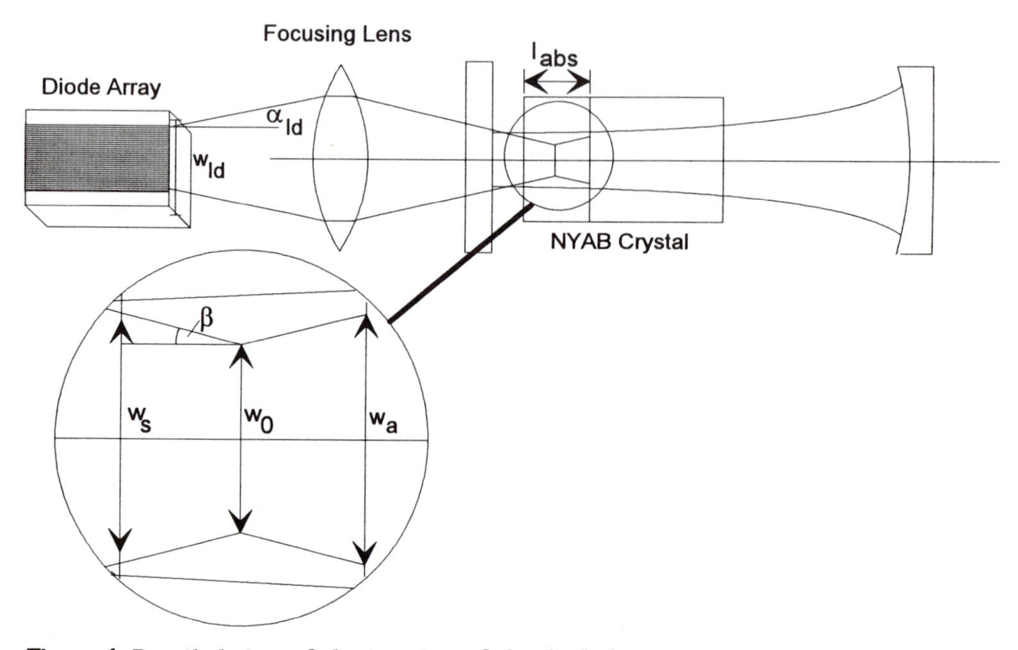

Figure 4. Detailed view of the imaging of the diode laser array into the NYAB cavity

The brightness of the array pump beam is proportional to the product of the beam divergence α_{ld} and the emitter size w_{ld}. As long as no aberrations are present, imaging does not change the brightness (see fig.4):

$$n_{air} \cdot w_{ld} \cdot \alpha_{ld} = n_{nyab} \cdot w_0 \cdot \beta \tag{1}$$

w_o is the diameter of the arrays image in the crystal, n_{air} and n_{nyab} are the refractive indices of air and NYAB ($n_{nyab} \approx 1.76$).

YAB can be doped with a very high Nd-concentration, without quenching the upper laser level lifetime and without making the crystal mechanically unstable. Therefore, in contrast to Nd:YAG, NYAB can provide very short absorption lengths. This considerably eases the pump focusing requirements. While in Nd:YAG the pump beam has to be matched over the whole length of the resonator beam, in NYAB it has only to be matched roughly over the absorption length.

Our goal is to determine β, so that the image size of the array is minimized only within the absorption length l_{abs}. Once β is determined the required focusing optics, in fig.4 just represented by a single lens, can be designed using the Gaussian lens formula and the formula for transverse magnification.

The arrays image size inside the crystal at a distance of $l_{abs}/2$ from the surface is given by eq.1:

$$w_0 = \frac{n_{air}\ w_{ld}\ \alpha_{ld}}{n_{nyab}\ \beta} \tag{2}$$

The size of the beam w_s at the crystal surface and at the absorption length w_a is:

$$w_s = w_a = w_0 + l_{abs} \cdot \tan(\beta/2), \tag{3}$$

i.e. w_s, $w_a > w_0$. The minimum value of w_s, w_a, and w_o under variation of β can be found by determining the point of zero slope of w_s with respect to β:

$$\frac{dw_s}{d\beta} = \frac{w_{ld}\ \alpha_{ld}}{n_{nyab}}\left(-\frac{1}{\beta_0^2}\right) + \frac{l_{abs}}{2\cos^2(\beta_0/2)} = 0 \text{ , i.e.} \tag{4}$$

$$\sqrt{2\ w_{ld}\ \alpha_{ld}\ /\ (\ l_{abs} \cdot n_{nyab}\ )}\ \cos(\beta_0/2) = \beta_0 . \tag{5}$$

This equation can be solved for β_0, the optimum convergence angle, for example by iteration. Eq.2 is then used to determine the corresponding minimum spotsize w_s^*. Now, a NYAB laser resonator has to be designed that posesses a beam diameter of w_s^* at the pump focus location, using conventional resonator theory . Fig. 5 shows the optimum resonator beam diameter of a NYAB laser with l_{abs} = 2mm (Nd-concentration = 3%), as a function of the pump arrays emitter size.

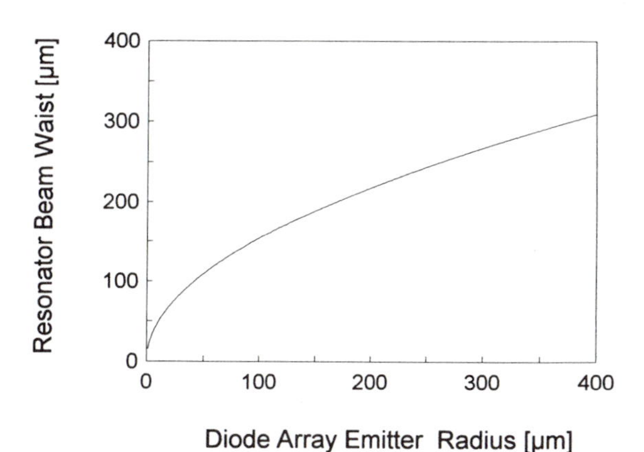

Figure 5. Optimum resonator beam waist (radius) in a NYAB laser with a 6mm crystal of 3% doping concentration (l_{abs} = 2 mm) vs. the emitter radius (=half diameter) of a diode array pump laser

Once the pump and resonator optics is selected for a specified array, the total output power at the second harmonic is maximized by that setup. This could be proofed experimentally. When the array and pump optics were replaced by a mode matched TEM_{00}-pump beam, the second harmonic output with the spatially "perfect" pump beam should not exceed the output generated with the diode array. Also, if an array with a smaller emitter (which is less powerful) is used instead, the visible output power should not depend on the type of array, but only on the actual pump power operated at.

EXPERIMENTAL SETUP

The schematic setup is given in fig.1. The beam of two diode laser arrays were collimated with short focal length lens systems (Melles Griot diode laser collimator), which are AR coated for 800nm and corrected for sherical aberration. Cylindrical lenses were used to cancel astigmatism. The two beams which had orthogonal linear polarization were combined using a polarizing beam splitter cube where the pump power of the diodes is added. A spherical lens ($f \approx$ 40mm) after the beam combiner focuses the pump light on the NYAB crystal surface.

The crystal is mounted on a stage for precise translation and rotation. Three degrees of freedom (two of transverse translation, one of rotation) are recommended for finding the spatial region in a crystal which provides the highest output power, and to rotate the crystal for phase matching of SHG. The crystal surfaces are AR coated for 1062nm and the surface facing the pump diodes is AR coated for 807nm.

We used a linear hemispherical resonator with the plane pump mirror HT coated for 807nm and HR coated for 1062nm (R~99.9%). The curved output mirror was HR coated for 1062nm and transparent for 531nm. The mirror curvature and the resonator length was varied in the range from 20 to 50mm and thereby the resonator waist in the crystal was set to the calculated optimum value.

The pump sources were diode lasers with the following emitter size and maximum power: SDL-2462 with w_{dl}= 200μm and P=1Watt, SDL-2432 with 100μm and 0.5Watt, Siemens SFH 487401 with 200μm and 0.8 Watt, and a broad area diode SFH 482201 with 60μm and 250mWatt. The pump diodes were used both singly and in combinations of two. We note that due to the non-linear gain, the output at 531nm with two diodes combined is bigger than the sum of the outputs generated with the diodes used separately.

RESULTS WITH MATCHED DESIGN

For a test of the optimization we first took three different pump diodes in single use with the emitter sizes 60μm (Siemens 250mWatt), 100μm (SDL 500mWatt), and 200μm (SDL 1Watt). Optimum pump focusing for the 60μm emitter was realized as follows: a 4.6mm collimator, a cylindrical telescope f=6.5/19mm directly following the collimator, and a spherical lens of f=40mm behind the cylindrical lens. The corresponding focusing lenses for the other diodes were 40 and 20mm for the

100μm and the 200μm emitter, respectively, while the cylindrical telescope was slightly varied to compensate for the different emission angle in the lateral direction.

The optimum resonator beam radii in the crystal were calculated to be 50μm for the 60μm emitter, 70μm for the 100μm emitter, and 120μm for the 200μm emitter. The radii were set by a cavity length of 25mm and an output mirror curvature radius of 25mm (35mm/35mm for the 100μm emitter and 50mm/50mm for the 200μm emitter).

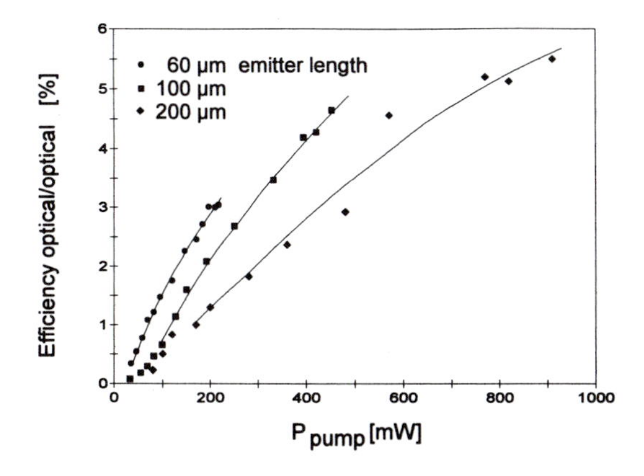

Figure 6. Efficiency of three NYAB laser setups with optimized pump and cavity design vs. the pump power. The optimization is done for three diode pump lasers with different emitter sizes

Fig.6 shows the measured optical/optical efficiency (visible output power / IR pump power emitted by the diode laser) with the three decribed setups as a function of the diode pump power. It can bee seen that the setup with the smallest emitter size has a high efficiency already at low pump power. Since in this case the pump power is limited to about 200mWatt the maximum efficiency is 3% with an output in the visible of 6mWatt. At this low pump power a setup with 100 or 200μm emitter size shows only about 2% efficiency, or 1%, respectively. In the next paragraph we show that the efficiency is reduced to the same lower values, determined by the cavity design, when a small diode is used in a setup that is optimized for a bigger diode. From fig.6 is clear that the advantage of a big diode setup lies in the range of high pump power. Here, the efficiency reaches a value of 4.5% at 400mWatt (100μm-diode) and 5.5% at 900mWatt (200μm-diode). This latter value corresponds to a 531nm output of about 50 mWatt. These measurements show that each setup was successfully optimized for maximum output power in the visible.

OPTIMIZED SETUP FOR A 200μm EMITTER

In the next step we compare the performance of a NYAB laser with single diode lasers or diode laser combinations. The resonator was optimized for a 200μm emitter size, while pumping was performed with 200μm diodes, as well as with smaller diodes. The experimental results are summarized in fig.7 .

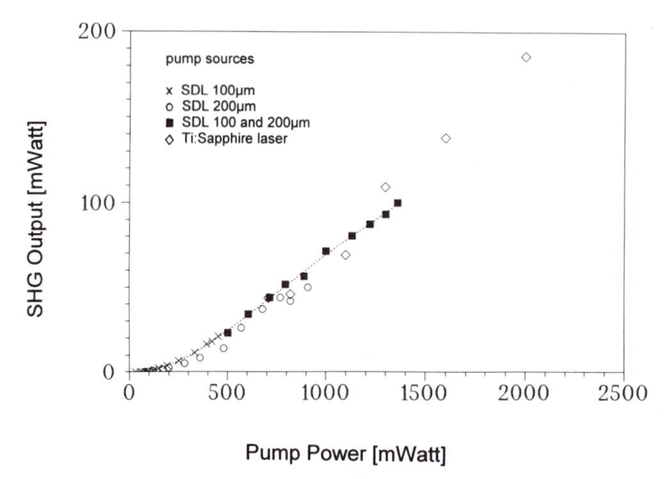

Figure 7. Visible output power from a NYAB laser, optimized for a 200μm emitter size, for a variety of diode lasers, with 200μm emitter, and smaller emitter vs. the optical pump power

All data points approximately follow the same curve. The dependence of the 531nm output from the pump power was proven to be quadratic below ~ 300mWatt and is linear above 500 mWatt. This shows that, above 500mWatt, the losses at the fundamental wave in the NYAB resonator are dominated by SHG rather than by mirror losses or passive losses. At high pump power no saturation of the output was observed. The successful optimization of pump optics and resonator for a 200μm emitter can be concluded from the data in fig.7, since also for emitters smaller than 200μm, and even for a TEM_{OO} Ti:Sapphire pump laser, the same output power is reached.

Another observation should be mentioned here. For a TEM_{OO} Ti:Sapphire pump laser the optimum beam waist is 25μm. The cavity beam waist was set to this value with a cavity length of 25 mm and an output mirror curvature of 25mm. From fig. 8 can be seen that, up to the maximum pump power from the Ti:Sapphire (≈2Watt) the 531nm output still shows a linear dependence vs. the pump power (maximum output = 350mWatt at 531nm). Obviously, at these powers, the NYAB laser does not exhibit

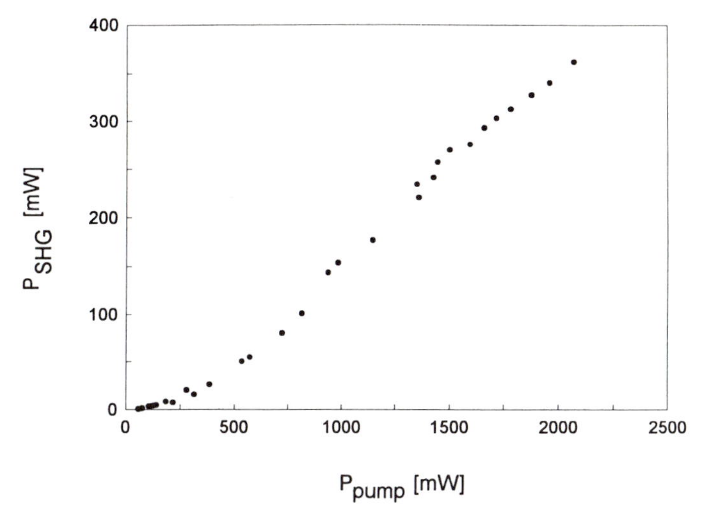

Figure 8. Visible output power from an optimized NYAB laser setup, pumped with a TEM_{OO} Ti:Saphhire laser

thermal effects, as i.e. thermally induced change of the NYAB birefringence which would disimprove phasematching, or a significant amount of thermally excited population in the lower laser state which would reduce the slope efficiency of the fundamental.

The highest visible output power from a diode pumped NYAB laser was generated with two diode arrays SFH 46758 (with integrated cylindrical lenses) with

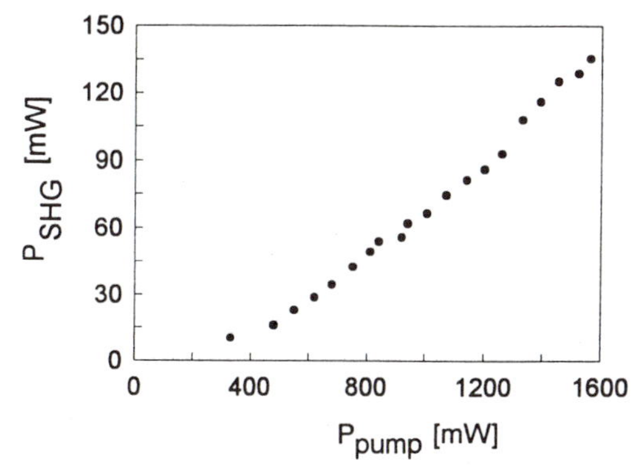

Figure 9. NYAB laser, optmized for a 200μm emitter size, pumped with the beams of two Siemens diode laser arrays having integrated cylindrical lenses, vs. the optical pump power

a total pump power of 2 x 800mWatt. The dependence of the output power from the pump power is shown in fig.9. To our knowledge, the maximum output of 130mWatt is so far the higest value from a diode pumped NYAB laser at 531nm in a unidirectional TEM_{00} beam. The optical/optical efficiency (visible power/optical pump power) reached in this case is about 8% , and the electrical/optical efficiency is about 4% (visible power/electrical power delivered to the pump diodes).

NYAB MICRO CHIP LASER

Since a NYAB laser both incorporates a Nd-based laser material and an optically non-linear material within one crystal, it is possible to reduce the dimensions of the laser cavity into the sub-milimeter range.

The above described NYAB lasers are still realized with separate components (two mirrors, one crystal) resulting in a cavity length in the order of centimeters. Such a cavity posesses a free spectral range (FSR) of a small fraction of a wavenumber. Although the 1062nm gain in NYAB is homogeneously broadened, spatial hole burning leads to oscillation of a number of longitudinal modes within the gain profile, which has a width of about 3 cm^{-1}. This is different in a short cavity laser. Here, the FSR may be on the order of the width of the gain profile so that only one mode can oscillate, and thus single frequency visible output is generated.

We realized a short cavity laser with a NYAB micro chip. In our experiment we used a platelet of 3mm diameter and 700µm thick, with plane and parallel surfaces, cut at an angle of 30° . The chip was HR coated on one side for 1062nm and partially reflective for 1062nm at the other side. The coatings were transparent for the second harmonic and the pump wavelength. A plane HR mirror for 1062nm was brought into close proximity (~10µm) with the partially transmissive facet of the chip by installing both components into one high precision mirror mount. This setup with still one separate mirror was chosen because, at the beginning of the experiment, it was not clear whether a curved mirror would be nessecary for laser oscillation.

The micro chip laser was first pumped with a diode array. Due to the comparatively low doping concentration (3% Nd) and small thickness only about 10% of the pump light is absorbed. With a pump power of 1 Watt focused to a spot size of 200µm the micro chip setup was easily aligned to laser emission and visible emission. Both the fundamental and visible beam were spatially multimode with array pumping. NYAB emission by pumping with a 1Watt from a 120µm diameter multimode glas fiber was readily obtained. In this case, the fiber end was just to bring into close contact with the front facet of the chip. This pump arrangement makes the setup rather simple since no further pump optics is required.

Depending on the tilt angle of the chips back mirror we measured three different wavelengths in the visible output. These wavelengths, 531nm, 539nm and 535nm were found to be the second harmonics and the sum frequency of the two fundamental wavelength of NYAB at 1062nm and 1078nm emitted by the crystal field split components of the laser transition $F_{3/2}$ - $I_{11/2}$. We suppose that tilting of the mirror may have resulted in an adjustable finesse of the micro chip cavity, individually for the two fundamental wavelengths.

In the next step, we tried to generate TEM_{00} single mode output. A suppression of transversal mode operation of the fundamental was realized by pumping with a single stripe diode (SDL 5422) with a maximum output power of 100mWatt. The visible emission spectrum was measured by sending the output beam through a negative short focal lens and a solid state etalon (finesse= 15, FSR= 1.6cm^{-1}). The

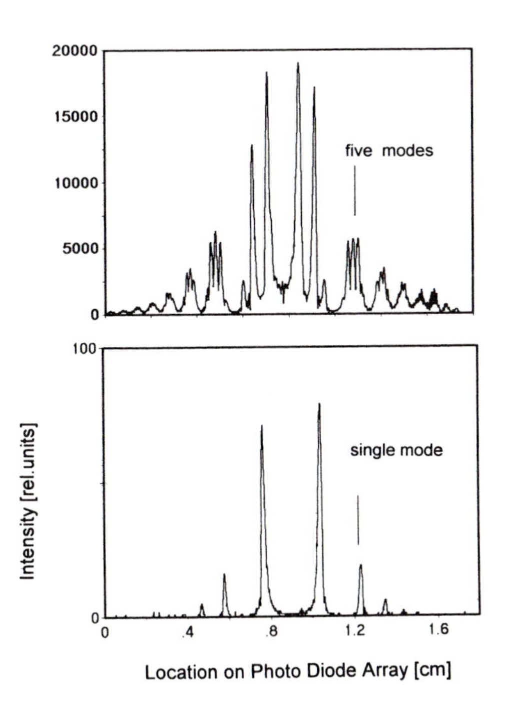

Figure 10. Emission spectra of a 700μm NYAB micro chip laser pumped with a single longitudinal mode diode laser. The spectra are recorded with a linear photo diode array of 1.8cm length in the diverging beam from the chip behind a solid state etalon. The spectrum containes a set of 5 frequencies at 100mWatt pump power (upper), and is single frequency for a pump power below 50mWatt (lower).

interference pattern behind the etalon which consists of concentric light cones was detected with a linear photo diode array and is shown in fig.10. When the maximum pump power was used the TEM_{00} beam contained 3-5 frequencies, separated by the FSR of the NYAB chip (fig. 9, upper). At a pump power below 50 mWatt the spectrum contained only one frequency (fig.9, lower), i.e. single longitudinal mode oscillation was realized.

MAGNETO-OPTICAL PROPERTIES

The Faraday effect, i.e. the rotation of light polarization caused by a magnetic field, is of big importance for the realization of compact lasers with an extremely high frequency stability. In 1985, Byer and coworkers[10] invented a monolithic Nd:YAG ring laser in which the coexistence of spatial hole burning modes is suppressed by unidirectional operation.

An intracavity optical isolator can cause unidirectional operation because it makes the cavity loss depending on the direction of propagation. An isolator usually consists of three components, that is a Faraday rotating material (in a magnetic field) placed between a conventional polarization rotator (i.e. $\lambda/2$ wave plate) and a linear polarizer. In the monolihically integrated ring laser a sufficiently strong Farady rotation was achieved by applying a magnetic field to the laser crystal. Although the Verdet constant of Nd:YAG is comparatively small ($\sim 100^{o}/T/m$) single mode operation can routinely be observed. The monolithic ring laser received big attention because of its extremely high passive frequency stability (free running bandwidth ≈ 3-10KHz), under the condition that it is pumped by a diode laser. Servo control of the cavity length has by now improved the frequency stability[13] with respect to a reference cavity up to the Townes-Schawlow quantum limit, which in this case is a bandwidth of about 50mHz. An absolute frequency stability of 150 Hz (with respect to a molecular absorption line for the second harmonic) was reported lately.[13]

These impressive data on frequency stability, obtained with a simple design, are so far unmatched by any other type of laser. The stability is the result of diode pumping, the ultra-stable monolithic setup and, finally, the sufficiently big Faraday effect of the laser crystal itself, which makes such a device possible at all.

We believe that such a monolithically integrated ring laser can be constructed with NYAB as well, provided that NYAB posesses a Verdet constant comparable to that of Nd:YAG.

Fig.11 shows the setup to measure the Verdet constant of NYAB. A HeNe laser beam is first sent through a Glan Thomson polarizer, through a NYAB crystal

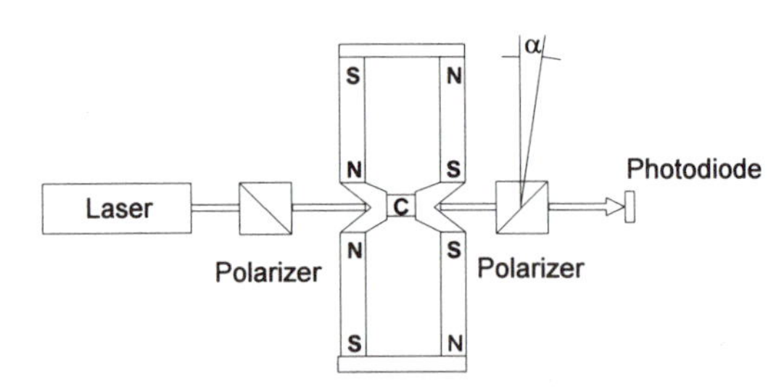

Figure 11. Experimental setup for the measurement of Faraday rotation in NYAB

(3% doping concentration, 6mm length), and then through a second crossed Glan Thomson polarizer (analyzer). The crystal is cut at 30^{0} and was rotated until the direction of the electric field was in the plane of the optical axis and the direction of propagation. In this case the laser beam is linearly polarized also behind the crystal. The analyzer was rotated until the intensity on the photo diode is minimized. When a magnetic field was applied to the crystal, the analyzer had to be rotated by a an angle of ≈ 1° in order to obtain again a minimum signal from the diode. The absolute value of the Verdet constant of NYAB was determined by comparing the rotation angle to that obtained with a Nd:YAG crystal.

We determined the Verdet constant of NYAB to be 200° +/-100°/T/m. This value is sufficiently high to realize a monolithic NYAB ring laser with visible emission.

ELECTRO-OPTICAL PROPERTIES

The electro-optical effect of NYAB is of particular interest, since it offers the possibility to construct a fast phase modulator, a loss modulator, or a Q-switch in a compact and low loss one-crystal design.

The measurement of an electro-optic coefficient was performed in the following way (see fig.12). The beam from a HeNe laser is linearly polarized and sent through

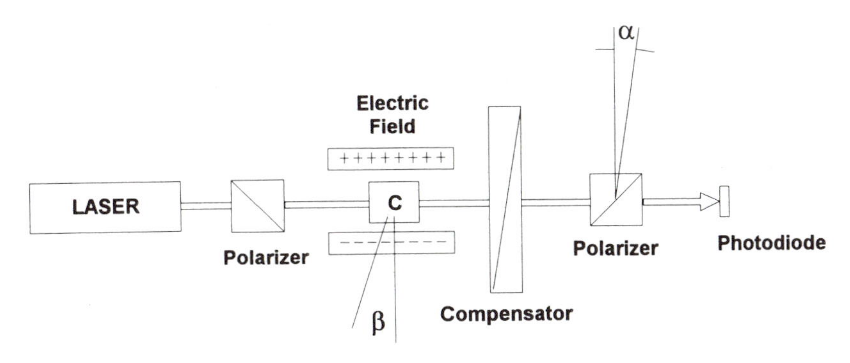

Figure 12. Setup for the measurement of the electro-optical effect of NYAB

the NYAB crystal. The crystal was rotated around the propagation axis to split the beam into an ordinary and extra-ordinary beam by approximately equal amounts (electric field vector rotated to 45° with respect to the plane of the propagation direction and the optical axis). The resulting elliptical polarization was transformed back into linear polarization with a Soleil Babinet compensator and then sent through a Glan Thomson analyzer. In contrast to the experiment for the Faraday effect, now a stripe shaped interference pattern was visible behind the analyzer when it was rotated to minimum transmitted intensity. The center minimum of the pattern was sent onto a photo diode. In this configuration a static electric field of several hundred V/mm was applied vertical to the direction of propagation and at 45° with respect to the direction of the linear polarization of the probe laser.

We observed a shift of the intensity pattern perpendicular to the stripes, with the shift beeing proportional to the applied voltage. The electric field necessary to shift the pattern by one half of a spatial period (phase shift of $\lambda/2$) was measured to be about 500 Volt/mm with a crystal of 6mm length. This value is of the same order as that of KDP. Thus we expect that NYAB can be succesfully used as an electro-optic crystal, simultaneously to its use as a laser crystal.

SECOND HARMONIC GENERATION AT 2μm

So far, the optical non-linearity of NYAB has only been discussed to generate shorter wavelengths from the fundamental. Nevertheless, NYAB might also be used to generate tunable IR light around 2μm, operating as an optical parametric oscillator (OPO).

In such a device, the intense intracavity fundmental at 1064nm would serve as the OPOs pump field of frequency ω_p. Depending on the rotation angle of the crystal, a signal at ω_s and idler field at ω_i could be emitted with $\omega_p = \omega_s + \omega_i$. Although we are presently far from realizing such a device, a first step of investigation towards such a diode pumped OPO was done.

The realization of an OPO from NYAB would require the knowledge of the phase matching angle at which a certain pair of IR frequencies is emitted. Instead of directly observing OPO emission we demonstrated the inverse process to OPO, that is sum frequency generation. Currently, our experiments are limited to the degenerate case, i.e. SHG using wavelengths in the range from 2.3 to 0.75μm.

The third harmonic of a Q-switched Nd:YAG laser was used to pump an OPO from ß-Barium borate[14] (BBO) and to generate nanosecond light pulses (pulse energy about 10mJ), continously tunable from 2.3μm to 0.75μm. The pulses were focused through a NYAB crystal mounted on a rotation stage. SHG was easily observed when the crystal was rotated to the correct phase matching angle. The rotation angle was measured with respect to vertical incidence of the pump light to the crystal.

Fig.13 shows the measured phase matching angle for SHG in NYAB vs. the pump wavelength, varied in the range of interest for NYAB OPO operation. The Sellmeier equations which are the basis for the proper construction of a NYAB-OPO were approximately derived from fig.13. A comparison to refractive index data that were

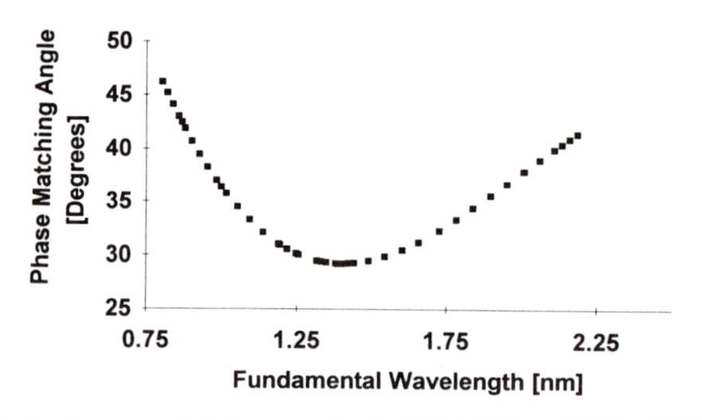

Figure 13. Measured phase matching angle in NYAB for SHG of 2.3 to 0.75μm light from a pulsed OPO

previously derived by other groups[12] by extrapolation from measurements in the visible and at 1064nm show considerable discrepancy. Especially, the minimum at 1.4μm in fig. 13 is not contained in previous data.

Although the measurements presented are quite preliminary, we are very much encouraged to proceed in the realization of a NYAB-OPO.

SUMMARY

First, we gave a short review on the various very succesfull attempts to generate visible light with diode pumped solid state lasers. Then we presented new measurements with self frequency doubling NYAB lasers which are promising candidates to efficiently generate visible light from high power diode lasers. We demonstrated the optimization of the pump optics and the design of the laser cavity for maximum visible output power. We report the so far highest 531nm output power of 130 mWatt generated with 1.55 Watt of diode pumping. A NYAB micro chip laser was constructed and successfully operated with diode array pumping, pumping through a multimode glass fiber, and with a low power single stripe diode laser. In the latter case, single longitudinal mode operation was achieved. Three different wavelengths in the visible were generated, 531nm, 535nm, and 539 nm. The magneto-optical and electro-optical properties of NYAB were investigated for the first time. The Verdet constant was measured to be about 100°/T/m and the electro-optical cooefficient was about 500V/mm^2/(λ/2). Finally, Sellmeier coefficients for the refractive indices of NYAB were determined in the new and important wavelength range from 2.3 to 0.75μm, where a diode pumped optical parametric oscillator based on NYAB seems feasable.

REFERENCES

[1] B. P. Scott, F. Zhao, R. S. F. Chang, and N. Djeu, Opt. Lett.**18**, 113 (1993)

[2.] W. Lenth and R. G. McFarlane, Opt. Phot. News, March 1992, p8

[3] D. H. Jundt, M. M. Fejer, R. G. Norwood, P. F. Bordui, and R. L. Byer, Opt. Lett.**16**, 1856 (1991)

[4] Z. Y. Ou, S. F. Pereire, E. S. Polzik, and H. J. Kimble, Opt. Lett.17, 640 (1992)

[5] D. W. Anthon, D. L. Sipes, T. J. Pier, and M. R. Ressl, IEEE J. Quantum Electron. **28**, 1148 (1991); L. R. Marshall, A. D. Hays, A. Kaz, and R. L. Burnham, IEEE J. Quantum Electron. **28**, 1158 (1991)

[6] L. F. Johnson and A. A. Ballman, J. Appl. Phys. **40**, 297 (1969); V. G. Dimitrev, E. V. Raevskii, N. M. Rubina, L. N. Rashkoich, O. O. Silichev, and A. A. Fomichev, Sov. Tech. Phys. Let. **5**, 590 (1979)

[7] T. Y. Fan, A. Cordova-Plaza, M. J. F. Digonnet, R. L. Byer, and H. J. Shaw, J. Opt. Soc. Am. B **3**, 140 (1986)

[8] I. Schütz, I. Freitag, and R. Wallenstein, Opt. Commun. **77**, 221 (1990)

[9] S. G. Grubb, R. S. Cannon, and G. J. Dixon, Conference on Lasers and Electro-Optics CLEO 1991, paper CFJ6, p 106

[10] T. J. Kane and R. L. Byer, Opt. Lett. **10**, 65 (1985); T. J. Kane, A. C. Nilsson, and R. L. Byer, Opt. Lett. **12**, 175 (1987)

[11] G. D. Boyd and D. A. Kleinman, J. Appl. Phys. **39**, 3597 (1968)

[12] B. S. Lu, J. Wang, H. F. Pan, and M. H. Jiang, J. Appl. Phys. **66**, 6052 (1989)

[13] see the contribution of R. L. Byer, **ibid.**

[14] A. Fix, T. Schröder, and R. Wallenstein, Laser and Optoelektronik **3/91**, 106 (1991)

HIGH POWER DIODE LASERS

Christian Hanke

Siemens AG
Corporate Research and Development
Munich
FRG

INTRODUCTION

The unique features of semiconductor laser diodes open a wide field of applications. The high power capability of these devices in combination with high efficiency, small volume, a wide range of available wavelengths and the cost reduction by mass production are the main factors for the increasing use of laser diodes. The purpose of this article is to give an overview of the recent developments in the field of high power laser diodes. The scope is from single stripe lasers with diffraction limited beams to stacked laser arrays with kW-power capability.

There are many excellent textbooks[1,2,3] dealing with all fundamental aspects of semiconductor lasers, so there is no special chapter on the basics. The outline of this article, which mostly deals with GaAs-based laser diodes, is as follows. In the first chapter the material systems and the range of possible wavelengths are discussed. The next chapter describes the fundamentals limiting the output power of semiconductor lasers and the means to increase it. Laser diodes with diffraction limited beam quality for applications like optical recording or frequency doubling are discussed in the following chapter. The laser diodes suitable for the wide field of pumping solid state laser are descibed. The issue of reliability is discussed and in the last chapter concluding remarks and an outlook into the future will be given.

Solid State Lasers: New Developments and Applications
Edited by M. Inguscio and R. Wallenstein, Plenum Press, New York, 1993

MATERIALS

The light emitting process in laser diodes is due the radiative recombination of excess carriers in a direct bandgap semiconductor material. Holes and electrons are injected into the so-called active region through a pn-junction. Therefore the bandgap energy of this semiconductor material essentially determines the wavelength of a laser. There are a lot of direct semiconductors, so a variety of laser wavelengths ranging from approximately 400 nm up to 30 μm can be achieved[4] in principle. Because today the research and development in the field of laser diodes is mostly initiated by the applications, two classes of semiconductor materials are of main interest; on one hand the GaAs- and on the other the InP-based systems. Fig. 1 shows the bandgap energy versus the lattice constant for semiconductor alloys with elements of group III and V of the periodic system. Lasers based on InP-substrates use lattice matched layer structures of InGaAsP or InGaAlAs and can in principle cover the range from 1100 nm up to 1650 nm. In this band there are the two dominant wavelengths (1300 and 1550 nm) for optical fibre communication systems. The lasers usually operate in the low power regime of several milliwatts. Because there are only a few applications of high power lasers with an outputpower of 100 mW and more, e.g. for pumping erbium-doped fibre amplifiers at 1480 nm, this category of lasers is not described in detail in this article.

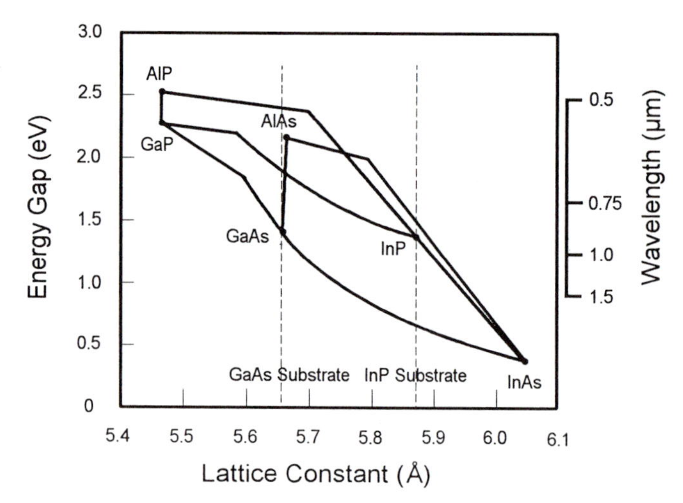

Figure 1. Bandgap energy versus lattice constant for the most important semiconductor alloys used for laser diodes

Most high power laser diodes are based on GaAs-substrates. With lattice matched AlGaAs, InGaAlP or InGaAsP layers the wavelength range from 630 nm up to 880 nm can be covered. Most of the developments were done with the AlGaAs/GaAs-system, where the first semiconductor laser was demonstrated. Lasers with wavelength from 780 nm to 880 nm are commercially available. The most prominent application of high power laser diodes is pumping

Nd-YAG solid state lasers. In this case an emission wavelength of 807 nm corresponding to the absorption peak is required. In double heterostructure lasers this wavelength can be achieved either by an active layer of $Al_{0.1}Ga_{0.9}As$ or by the use of so called quantum well active layers of GaAs. Extremely thin active layers in the range of a few nanometers shift the energy levels to 1.58 eV due to the quantum confinement of the carriers. Sophisticated epitaxial growth techniques like MOCVD (Metal Organic Vapour Phase Deposition) or MBE (Molecular Beam Epitaxy) are required, to fabricate these layer structures with high uniformity and reproducibilty.

Since some years there is considerable interest in so called strained layers, which are not lattice matched. Under the condition that the layers are below a critical thickness[5] the mismatch is elastically accommodated and no dislocations deteriorating the laser performance are generated. By adding a certain amount of In to GaAs the wavelength can be shifted from 880 nm up to 1100 nm. This opens a new field of applications like pumping Er-doped fibre amplifiers and lasers at 980 nm. There are additional features using InGaAs-quantum wells instead of GaAs. The biaxial strain changes the band structure[6] and in consequence lower threshold current densities can be achieved. On the other hand the strain reduces the growth of so called dark line defects[7], which are a main source for degradation and sudden failure in conventional AlGaAs-lasers. This improvement is described further in the chapter about laser reliability. It is interesting to note, that above 900 nm the GaAs-substrate is transparent. Therefore new laser structures emitting perpendicular to the substrate can be realised in this wavelength range.

The optical storage technique with its demand for shorter wavelengths to increase the recording density of optical discs has initiated the development of visible lasers[8] with an emission wavelength from 630 to 700 nm, using (Al)InGaP-layers lattice matched to GaAs-substrates. The introduction of quantum wells and strain has greatly improved the performance of these lasers[9]. The same application also leads the research to even shorter wavelengths in the green or blue regime. With laser structures of group II and VI like ZnSe or ZnS room temperature operation of diode lasers with an emission wavelength of 480 nm has been demonstrated just recently[10]. With the occurrence of high power lasers in this wavelength range new diode pumped solid state lasers can be envisioned.

The properties of semiconductor lasers depend on the design of the transversal and lateral structur, on the quality of the epitaxial layers and the fabrication process. The main guidelines for the design of high power laser diodes are the maximum achievable power and the optimization of the overall efficiency η, which is described by formula 1.

$$\eta = \frac{P_{opt}}{P_{el}} = \frac{\eta_d (I - I_{th})}{I\frac{h\upsilon}{e} + I^2 R_s} \qquad (1)$$

I is the operating current for obtaining the output power P_{opt}, P_{el} is the total electrical power and R_s is the series resistance of the device. The following formulae describe the threshold current I_{th} and the differential efficiency η_d.

$$\mathrm{I_{th}} = \frac{eV}{\tau}\left\{n_{tr} + \frac{1}{\Gamma A}\left(\alpha + \frac{1}{2L}\ln\left(\frac{1}{R_f R_b}\right)\right)\right\} \tag{2}$$

$$\eta_d = \eta_i \frac{h\nu}{e} \frac{\frac{1}{2L}\ln\left(\frac{1}{R_f R_b}\right)}{\frac{1}{2L}\ln\left(\frac{1}{R_f R_b}\right) + \alpha} \tag{3}$$

The parameters are the transparency carrier density n_{tr}, the internal efficiency η_i, the confinement factor Γ, the gain coefficient A, the total internal losses α, the front and back mirror reflectivities R_f, R_b, the cavity length L, the photon energy $h\nu$, the electron charge e and the volume of the active layer V. For an optimized overall efficiency a low threshold current, low internal losses and a low series resistance are essential.

Parameters affecting the threshold current and the efficiency had been collected from literature for the main semiconductor laser materials and are summarized in table 1.

Table 1. Characteristic parameters for various semiconductor laser material systems

Material/Substrate	AlGaAs/GaAs	InGaAlAs/GaAs	InGaAlP/GaAs	InGaAsP/InP
Wavelength λ / µm	0.72 - 0.88	0.8 - 1.1	0.63 - 0.7	1.2 -1.65
Internal efficiency η_i	$\leq$ 1.0	$\leq$ 1.0	$\leq$ 0.95	$\leq$ 0.80
Internal loss α_i / cm^{-1}	4 -15	2 - 10	~ 10	5 - 10
Threshold current density i_{th} /Acm^{-2}	80 - 700	50 - 400	200 - 3000	200 - 1500
External efficiency η_d / W/A	$\leq$ 1	$\leq$ 1.1	$\leq$ 1	$\leq$ 0.5
Characteristic Temperature T_o / K	120 -200	100 - 200	60 - 100	50 - 70

The internal efficiency and the internal losses are mostly related to the quality of the epitaxial layers whereas the external efficiency, the threshold current density and the series resistance are more influenced by the laser design and -technology and by the mirror coatings (R_f, R_b).

POWER LIMITATIONS IN SEMICONDUCTOR LASER DIODES

There are two fundamental limits for the output power of semiconductor laser diodes as indicated in the P/I-characteristic in fig. 2.

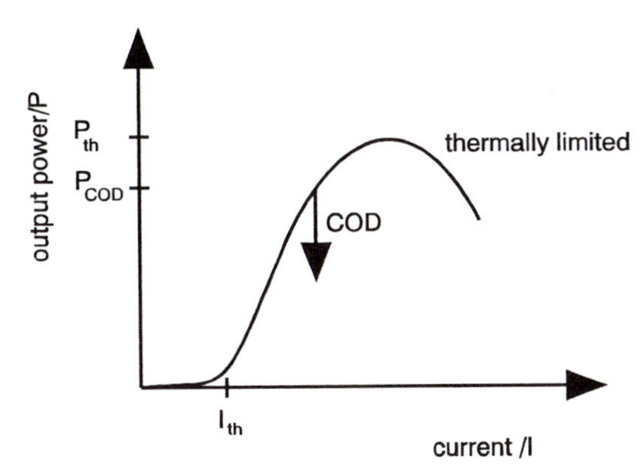

Figure 2. Schematic representation of the power limitations in the P/I-characteristic of a semiconductor laser diode

The thermal limit occurs because the laser characteristics are temperature sensitive. The temperature dependence of the threshold current can be approximated by equation (4).

$$I_s(T) = I_s(T_r)\exp\left(\frac{T - T_r}{T_o}\right) \qquad (4)$$

The so-called characteristic temperature T_O depends on the material system for the laser and on the design of the laser. Typical values are listed in table 1.

The thermal limitation or roll over which is an reversible phenomena is due to the heating of the semiconductor chip by the supplied electrical power. This leads to an increase of the threshold current with increasing bias current. The difference between injection current and threshold current decreases and therefore above a certain output power a rollover occurs. The maximum output power depends on the quality of heat sinking, the heat sink temperature, the total electrical power, the duty cycle and the characteristic temperature of the laser. The thermal rollovericreases with the characteristic temperature.

Especially in GaAs-based lasers with active layers of InGaAlAs or InGaAlP there is an second important power limitation the so called **c**atastrophic **o**ptical **d**amage (COD). In this case the output power breaks down abruptly above a certain level and the laser is irreversibly damaged. This phenomenon is related to an heating effect in a small volume at the laser mirror. Due to the surface recombination velocity, which is higher in the systems mentioned above compared to InGaAsP, the injected excess carriers in the active layer recombine nonradiatively delivering their energy via phonons to the lattice. The temperature rise lowers the bandgap locally and leads to an increased absorption of the laser radiation at the mirror. The higher carrier density in combination with a diffusion of the carriers to the semiconductor surface leads to an effective power transfer into a small volume. Above a critical value there is a positive feedback and a thermal runaway occurs with a temperature rise up to the melting point of the material. The semiconductor crystal is then permanently damaged. Due to the very small volume involved and the effective absorption the COD can occur on a very short time scale in the range of a few nanoseconds. Therefore these lasers are prone to uncontrolled surge currents. The typical power density for COD depends on the laser structure, the material in the active layer and on the operation conditions like pulswidth or cw(continuous wave)-operation. Typical values are in the range from 1 to 10 MW/cm^2. There are several means to lower the mirror temperature. Mirror coatings with e.g. Al_2O_3[11] or special chemical treatments[12] lower the recombination velocity and thus increase the COD level. The most effective way is to lower the excess carrier density at the mirror either by special contact configurations[13] or by the introduction of a so-called nonabsorbing mirror[14] which can be realised by a mirror region of a material with a bandgap energy higher than the lasing photon energy. By this method the COD-level can be increased two- to fourfold[14]. But the manufacturing of lasers with nonabsorbing mirrors is technologically challenging. Due to the low surface recombination velocity at the mirror InGaAsP-lasers for 1.3 μm and 1.55 μm do not suffer from COD.

The COD is an effect of the optical power density impinging on the mirror in conjunction with the excess carriers. The easiest way to increase the maximum output power is to increase the emitting area. In the transversal region the mode size depends on the layer structure which can be designed for a wide width like in LOC-(large optical cavity) or in SCH(separate confinement heterostructure)-designs. On the other hand the low confinement factor in high power designs leads to higher threshold currents for this type of laser. There are many ways to increase the lateral mode size. These laser structures like broad area lasers and laser arrays are described later.

For lasers with diffraction limited beam quality only one lateral and transversal mode may oscillate which in turn limits the lateral nearfield size in the range of 4 to 6 μm. Therefore these lasers are limited to powers up to a few hundred milliwatts. This will be discussed in the following chapter.

LASER DIODES WITH HIGH BEAM QUALITY

In fiber telecommunication systems laser diodes usually operate in the low power regime (e. g. 2-5 mW). But there are a lot of other applications requiring high power in the range of 100 mW and more in a diffraction limited beam. Free space communication between satellites, high speed optical recording, pumping Er-doped fibre amplifiers and optical frequency doubling for the generation of coherent blue light are the most prominent examples for the application of these specially designed laser diodes. Often these lasers must also emit in a single logitudinal mode. The direct current modulation of laser diodes is an additional advantage.

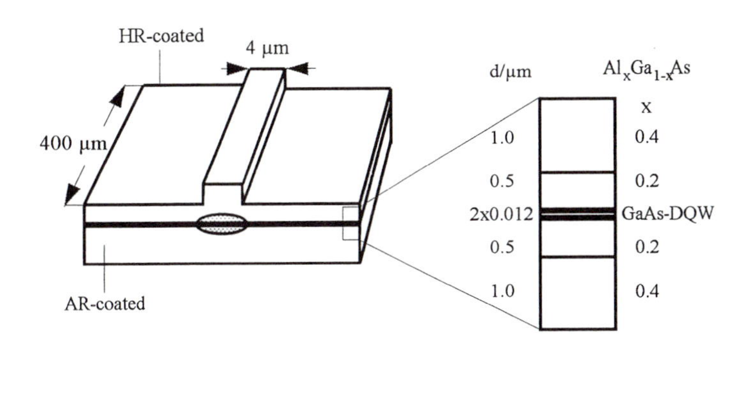

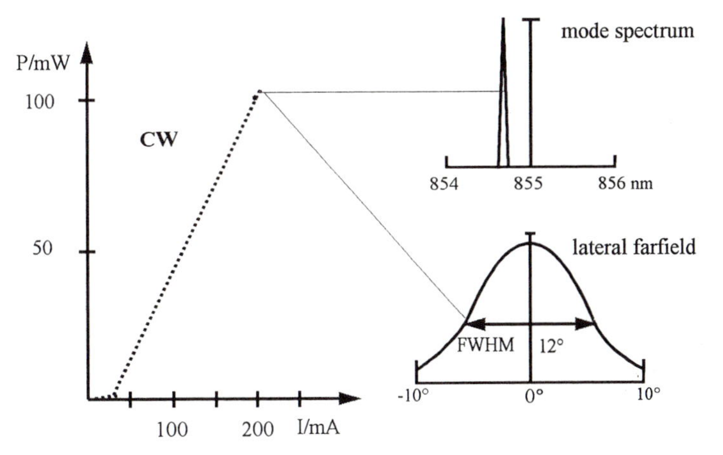

Figure 3. Structure and characteristics of an AlGaAs/GaAs index guided quantum well laser. (For explanation see text.)

High power lasers with broad stripes or with array configurations support a large number of lateral modes due to the big dimension of the lateral cavity compared to the small wavelength in the high index semiconductor material. For high beam quality an index guided structure is needed, providing fundamental mode operation both in the transversal (perpendicular to the epitaxial layers) and in the lateral (parallel) direction.

An example of an AlGaAs/GaAs quantum well high power laser[15] is given in fig. 3. The active layer consists of an double quantum well (L_z=12 nm) embedded in a waveguiding layer ($Al_{0.2}Ga_{0.8}As$) with a total thickness of 1 μm. This design leads to a rather low confinement factor Γ of the guided wave of 1.5% in one quantum well. This increases on one hand the threshold current of the laser but on the other hand the power density for COD is essentially increased, because it is determined by the overlap of the optical wave with the carriers in the active layer at the mirror plane. In the design of high power lasers there is always a tradeoff between a high COD-level and a low threshold current. The transversal optical waveguide can support more than one mode, but only the fundamental mode will oscillate because the higher modes don't reach threshold due to their low confinement. In the lateral direction the optical wave is guided by a ridge type waveguide. The distance between the active layer and the surface outside the ridge determines the effective lateral index step due to the different boundary conditions for the optical wave under or outside the ridge. A careful design of both the ridge width and the effective lateral index step ensures the fundamental lateral mode operation up to high power levels. The laser chip is asymmetrically coated with two λ/4-pairs of alumina and silicon on the back mirror (R=90%) and a λ/4-layer of aluminia on the front mirror (R=5%) to enhance the light emission from one side. The P/I-characteristic shows a threshold current of 35 mA and a differential efficiency up to 0.9 W/A. The characteristic is kinkfree up to 100 mW in cw-operation. The lateral farfield is single lobed and stable up to 100 mW indicating the fundamental mode operation over the whole range. In addition the spectrum shows a single mode even at the highest output power, making this laser an excellent candidate for frequency doubling.

Laser arrays usually operate in a multitude of lateral modes. In the past there have been many efforts to force an array structure to emit only one lateral mode to get a diffraction limited high power beam. Many of the approaches failed due to the complex interaction of temperature gradients and inhomogeneous carrier density on the waveguiding properties. Only recently a structure with specially designed antiguide waveguides has been reported[16]. It has very high intermodal selectivity yielding a diffraction limited output of 1W in pulsed and 0.5 W in cw operation.

LASER DIODES FOR PUMPING SOLID-STATE LASERS

The field of application for high-power diode lasers covers areas where the thermal effect of the radiation is used like in contactless soldering systems or in ophtalmology. But the most rapidly growing field of application is that of pumping solid state lasers, where the small bandwidth of the radiation is dominant. There were early attempts[17] to provide the pump power by light emitting diodes or by lasers diodes but only with the availability of reliable high power diodes there was a renaissance in that field. The following table compares the main characteristics of flash lamps and laser diodes as pump sources.

The most attractive features of laser diodes are their high efficiency reducing the cooling requirements. The narrow spectral width fitting to the small absorption bands of the solid state laser material enables an efficient power transfer and reduces the thermal loading of the crystal. The small volume enables compact light weight laser systems, which are even suitable for space-borne applications. The low driving voltage is an additional advantage. The high reliability reduces the maintainance costs of the laser. Depending on the operation conditions of the solid state laser (high/low power, pulsed, cw) a wide variety of laser diodes in various pump module configurations are needed, whose characteristics must match the requirements of the whole system.

Table 2. Comparison between flash lamps and laser diodes for pumping solid state lasers

	Flash lamp	Laser diodes
Lifetime	500-1000 h	10 000- 50 000 h
Supply voltage	2 - 5 kV	2 - 4 V
Power supply	big	compact
Modulation and control	difficult	easy
Spatial emission	nearly isotropic	focusable
Spectral emission	broad	narrow, tunable
Efficiency P_{light} / P_{el}	60 - 80 %	30 - 50 %
Transfer efficiency	4 - 8 %	~ 100%
Laser efficiency P_L / P_{el}	0.5 - 2 %	5 - 15 %
Cooling	easy	sometimes complex

Diode pumped solid state laser systems

There is a wide variety of solid state laser materials suitable for diode pumping. The most important ones are listed in the table 3 in addition with the emission wavelengths, the pump bands and the suitable semiconductor lasers.

The Nd-Yag laser is the most advanced diode pumped system for which (In)AlGaAs/GaAs-lasers with a wavelength of 807 nm perfectly match the absorption band. Because of their significance the development of these pump lasers is described in more detail in the following chapters. A few remarks on the other diode lasers will be given.

Judging the maximum reported output power, one has to distinguish between the record high values, either limited thermally or by COD, and the specified maximum power for commercially available devices, which is fairly lower and guarantees a resonable lifetime.

Broad-area lasers and laser arrays

Due to the COD-power limit the most effective way to increase the total output power is to increase the lateral emission area at the mirror. Broad area lasers have typical widths from 50 to 400 μm which are uniformly pumped. The homogeneity in the epitaxial layers grown by MOCVD or MBE is so high that the lasing modes are evenly distributed over the whole area without filamentation. The other solution is the lateral integration of many (up to 40) small stripe lasers with a typical width of 3-5 μm on one chip. In these laser arrays the optical fields

Table 3. Diode laser pumped solid state lasers

Solid state material	Emission wavelength/μm	Pump band / μm	Semiconductor laser
Nd-YAG	1.06 ; 1.3	0.807	(In)GaAlAs/GaAs
Ho-YAG	2.1	0.785	AlGaAs/GaAs
Tm-YAG	2.02	0.78	AlGaAs/GaAs
Er-YAG	1.65	0.98	InGaAs/GaAs
Yb-YAG	1.03	0.93 -0.98	InGaAs/GaAs
LiCaF	0.72 - 0.84	0.63 -0.69	InAlAsP/GaAs
LiSaF	0.75 - 0.98	0.63 -0.69	InAlAsP/GaAs

of the individual emitters can either be coupled or uncoupled. For the same active lateral width there is no big difference in the lasing characteristics between broad area lasers and laser arrays. Despite the fact that cw-power levels of more than 4 W can be obtained from devices with a lateral active width of 100 μm, the typical recommended output power levels in cw-operation are in the range between .7 to 1.5 W. The total conversion efficiency lies between 30 and 45%. The lateral farfield patterns are double lobed for laser arrays and more or less single lobed (but not diffraction limited) for broad area devices. Due to the lateral multimode behaviour the typical FWHM-values are between 10° and 30°. The transversal farfield pattern is diffraction limited and the FWHM depends on the design of the transversal waveguide yielding values between 20° and 40°.

For an 100 μm aperture InGaAlP-laser with an emission wavelength of 685 nm a maximum cw-power of 1 W has been reported[18].The highest reported cw-value for the strained InGaAs-system (λ=1.06 μm) for the same aperture width is 5.25W[19].

Laser diode bars

A further increase in output power is possible by the integration of groups of broad area lasers or laser arrays on long bars. These devices with a usual bar length of 1 cm are suitable for side pumping schemes. A bar like this represents a line source of pump radiation. The number of individual lasers on a bar and the filling factor dependend on the operation condition. In cw-operation a great steady-state heat flow must be removed from the bar; therefore the packing density is low and depends on the quality of the heatsink. Especially for pumping pulsed Nd-YAG lasers the so called quasi-cw-mode has been established, where 200 μs long pulses corresponding to the lifetime of the upper state are applied at low repetition rate. In this case the heat can be stored at an intermediate level and removed in the pulse pause.

For a 1 cm long bar the typical specified cw-output power is 10 W to 20 W depending on the filling factor. The overall efficiency approaches 30 %. By improving the thermal resistance using a diamond heatsink a record cw-value of 120 W is reported[20] for an 1 cm long bar with an effective aperture width of 0.72 cm operated at a temperature of 2 °C. In the quasi-cw mode the higher packing density enables specified powers up to 60 W, whereas record values go up to 300 W[21] before the COD occurs. The highest reported cw-power[18] for an 8 mm long InGaAlP-laser bar emitting at 690 nm is 8.5 W, the quasi cw-value approaches 60 W.

Two-dimensional laser stacks

For efficient side pumping configurations especially in high power slab designs, two-dimensional pumps sources are needed, which can deliver high optical power over a wide emitting area. This can be achieved by stacking many pretested bars mounted on subheatsinks in the vertical direction. The typical distance between the bars is 0.2 to 0.5 mm. The number of laserbars ranges from 3 to 42 in one unit[22]. Due to the very high power density in the stack configuration these two-dimensional arrays usually have a forced cooling system either by air- or liquidflow. The topic of integrated microcoolers is beyond the scope of this review, but there are several articles[22] to which the reader can refer. Usually the laser diode stacks are operated in the quasi-cw-mode at a low duty cycle of typically 0.2 to 8%. Commercially available stacks with an emitting area of 1 cm^2 deliver a specified power from 1.5 up to 2.4 kW. A laserstack[23] with an improved heatsink has a power capability of 1.5 kW at a very high duty cycle of 25%. The highest reported power[24] from an assembly of 900 bars aligned in three rows in an area of 4.5x9 cm^2 is 45 kW. Due to the modular assembly this system can easily be expanded up to 18000 bars so that a projected power of 700 kW is expected.

Surface emitting lasers

The rack and stack technology has the advantage, that the individual laser bars can be tested, on the other hand this technique requires many production steps, especially the cleaving into individual bars, their coating and the mounting on sub heatsinks, which are very cost sensitive. A fabrication process of the lasers on a wafer scale promises a more cost effective technology. Monolithic two-dimensional laser arrays require the implementation of surface emitting lasers, which emit perpendicular to the substrate surface. There are several approaches like vertical cavity lasers[25], lasers with horizontal cavities in which the light is deflected either by second order gratings[26], by internal[27] or external[28] 45°-mirrors. Vertical cavity lasers are restricted to low power applications due to the high threshold current density, the high series resistance and the high thermal resistance. Because lasers with grating couplers, external deflectors or with integral turning up mirrors cannot be mounted junction down, the maximum output power will be limited thermally and a high output power can only be generated in short puls operation.

We favour a laser structure with a folded cavity, where a 45° turning mirror deflects the light through the substrate. The basic structure with the advantage of junction down mounting is shown in fig 4.

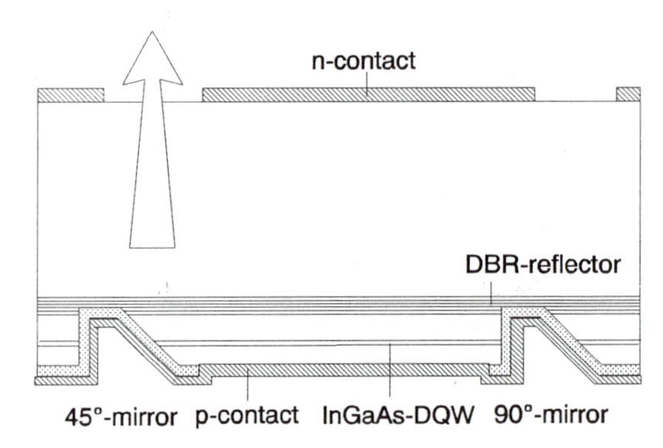

Figure 4: Structure of a surface emitting laser with an integral 45°-mirror and a DBR-reflector

Six pairs of epitaxially grown $GaAs/Al_{0.6}Ga_{0.4}As$-$\lambda/4$-layers form a DBR-reflector designed for a reflectivity of 30% at a wavelength of 960 nm. The transversal laser structure consists of an $In_{0.2}Ga_{0.8}As$-double quantum well (DQW) active layer with a nominal thickness of 8 nm embedded in $Al_{0.2}Ga_{0.8}As$-$Al_{0.4}Ga_{0.6}As$ confinement and waveguiding layers. The 45°- and the 90°-mirrors are etched by argon ion beam milling in a single step. The cavity length is 450 μm. After the etching process all mirrors are in situ coated with Al_2O_3. Broad area contacts with a width of 50 μm are opened in the oxide and the p-metallization is applied which also covers the vertical mirror. After thinning the wafer the n-contact with windows for the laser emission is evaporated and tempered. After dicing individual laser chips are mounted junction-down on copper heatsinks.

The room temperature cw-threshold current is 110 mA and a characteristic temperature T_0 of 200 K is measured. The slope efficiency is 0.42 W/A for the device with an anti reflective coated GaAs-substrate. The absorption loss in the substrate is estimated to be 10 %. A thermally limited cw-output power of 0.85 W is obtained. This corresponds to a cw-line density of 124 W/cm, which is to the best of our knowledge the highest value obtained for an InGaAs surface emitting laser. This high value reflects the good quality of both etched mirrors. The lasers are located on 500 μm centres so that a power-density of 250 W/cm^2 can be estimated for an 2D-array. Due to the fresnel reflection at the substrate surface the farfield patterns show interference fringes. The longitudinal mode spectrum is centred at 952 nm and shows unusual large mode spacing due to the back reflections from the surface. A very first attempt for an two-dimensional array (1x1 mm^2) with 2x2 broad-area lasers showed a threshold current of 450 mA and a maximum output power of 1.6 W. By improving the thermal resistance and by increasing the packing density a higher power can be realised. A maximum cw-power of 50 W for an 1cm^2-array with a similar structure is reported[29].

This laser concept with an integral 45°-mirror is not restricted to wavelengths where the substrate is transparent. It is also possible to remove the substrate at the emission area[30], so that for instance the pump wavelength for Nd-YAG-lasers can also be realised with this laser structure.

LASER RELIABILITY

The lifetime of laser diodes has continuously increased from the very first devices, which often degraded during the first measurements. On the other hand a gain-guided GaAs-double heterostructure laser operates for more than ten years in our laboratories[31] at an output power of 10 mW at an elevated temperature of 80 °C without degradation. This result shows that there is in principle no inherent degradation process associated with the lasing operation. Nevertheless many lasers show degradation. Therefore the investigation on the failure modes and on the physics of degradation is still under progress. With new material systems, epitaxies like MOCVD or MBE and new technologies the question of reliability has to be answered again, because the requirements for the lifetime of laser diodes are challenging especially in space-borne applications. In this review not all aspects of laser reliability can be stressed, but there are several comprehensive review articles[32,33].

There at least three failure modes for semiconductor lasers:

a) Rapid degradation and infant failures

b) Gradual degradation

c) Sudden failure

The rapid degradation often occurs in the early stage of operation and is often related to process induced damage of the crystal. The defect generation and motion leads to the failure of the device. This kind of damaged lasers can easily selected by an burn-in test over a few hundred hours.

The gradual degradation is characterised by a decrease of the optical output power at a rate of -0.1 to -10 %/khr, which is usually combined with an increase of the threshold current and a decrease of the slope efficiency. Point defect generation in the crystal and/or changes at the interfaces between the mirror and the coating are believed to be responsible for this degradation mode.

The sudden failure is characterised by an abrupt change in output power down to zero in a very short period of time. This failure mode, mostly observed in AlGaAs/GaAs-based lasers, can occur even after thousands of hours of operation. This fact excludes a selection by burn-in methods. The examination of failed devices often reveals so called dark line defects (DLD), which cause a local detoriation of the crystal leading to areas of enhanced nonradiative recombination. These DLDs can be initiated by epitaxial crystal defects or by process induced damage. Under operation they can grow even in areas where no current is injected, because photons generated by spontaneous emission are distributed over the whole chip. Due to the interaction of photons and defects the DLD-formation can start from defects everywhere in the chip[7]. The DLDs predominantly grow in the ⟨100⟩-crystal direction with a velocity from 0.1 to 10 μm/h depending on the current- and photon density[34]. In contrast to AlGaAs-based lasers which are very sensitive to DLDs, in strained layer InGaAlAs-lasers and in InGaAsP/InP-lasers for the optical communication this type of failure is nearly completely suppressed. The ⟨100⟩-growth velocity of DLD is dramatically reduced[7,33,34] and a substantial increase in reliability can be expected from strained InGaAlAs-lasers. The inhibited growth of DLDs in InGaAlAs-layers is possibly due to an dislocation pinning mechanism. Similar to alloy-hardening the incorporation of the relatively large In-atoms leads to an effective suppression of glide or climb motions of defects[35].

In the literature there are many reports on lifetime and degradation of high-power lasers. Because there is a wide variety as well of laser structures (arrays, broad-area-lasers, bars, stacks) as of operating conditions (cw, quasi-cw, heatsink temperature), it is impossible to compare all these results. Only a few outstanding results will be presented here. A constant cw-power lifetest with 20 W for a 1cm long bar over 2000 h is reported[36] with a projected lifetime of up to 15000 h. Operation in the quasi-cw-mode (100 μs/50% duty cycle) exceeding $3 \cdot 10^9$ pulses with little degradation has been performed[37] for a linear bar. To the best of my knowledge no lifetime data on two-dimensional arrays either stacks or monolithic versions have been published yet.

CONCLUDING REMARKS

The status of AlGaAs-lasers has reached a level of maturity, where no quantum leap in efficiency and output power density can be expected. By applying the rack and stack technology powerlevels of several hundred kilowatts can be delivered from large area modules, which seemed to be unattainable years ago. Also the reliability of high power laser diodes has reached a high level. Nevertheless the introduction of strain by implementing Indium in the active layer of the lasers will improve the resistance again sudden failures. This is especially important for densely packed devices like two-dimensional arrays where a large area of semiconductor is consumed. New material systems (InGaAlP/GaAs,InGaAlAs/GaAs) have appeared, offering an enlarged choice of wavelengths from 650 nm up to 1100 nm, which render new diode pumped solid state lasers possible. With the advent of the roomtemperature operation of II-VI semiconductor laser diodes, also the blue-green spectral range can be assessed. The issue of high power operation and reliability of these lasers has to be solved in the future. Laser diodes in the near infrared regime with diffraction limited beam quality have surpassed the 100 mW-barrier and special designed arrays touched the 1 W-level. The question of even higher power-levels is still open.

For a real breakthrough in mass applications, especially for pumping purposes, the crucial question is that of the costs of laser diode modules. There must be a real mass production to reduce not only the costs for the semiconductor laser itself but for the whole module including testing. The monolithic two-dimensional laser array is a promising approach to reach that goal, because it enables waferscale testing and mounting. On the other hand in the field of packaging and thermal management there is space enough for improvements by new engineering solutions.

Acknowledgement: The contributions by J. Luft, Siemens Components Division and J. Jansen, Heimann Optoelectronics are gratefully acknowledged. I thank L. Korte and B. Stegmüller for carefully reading the manuscript.

REFERENCES

1. G. H. B. Thompson, Physics of Semiconductor Laser Devices, John Wiley & Sons, 1980
2. G. P. Agrawal, N. K. Dutta, Long-wavelength lasers, Van Nostrand Reinhold Company, New York 1986
3. P. K. Cheo (editor), Handbook of Solid-State Lasers, Marcel Dekker Inc. 1989
4. H. C. Casey Jr., M. B. Panish, Heterostructure Lasers, Academic, New York 1978
5. J. W. Mattews, A. E. Blakeslee, Defects in Epitaxial Multilayers; Journal of Crystal Growth, 27 (1974), pp. 118-125
6. S. W. Corzine, R. H. Yan, L. A. Coldren, Theoretical gain in strained InGaAs/AlGaAs quantum wells including valence band mixing effects; Appl. Phys. Lett. 57 (26), Dec 1990, pp. 2835-2837
7. R. G. Waters, D. P. Bour, S. L. Yellen, N. F. Ruggieri, Inhibited Dark-Line Defect Formation in Strained InGaAs/AlGaAs Quantum Well Lasers; IEEE Photonics Technology Letters; Vol. 2, No. 8, Aug. 1990, pp. 531-5332
8. G. Hatakoshi, K: Itaya, M. Ishikwa, M. Okajima, Y. Uematsu, Short-Wavelength InGaAlP Visible Laser Diodes, IEEE Journal of Quantum Electronics, Vol. 27, No. 6, June 1991, pp. 1476-1482

9. T. Katsuyama, I. Yoshida, J. Shinkai, J. Hashimoto, H. Hayashi, High temperature (>150 °C) and low threshold current operation of AlGaInP/Ga_xIn_{1-x} strained multiple quantum well visible laser diodes, Appl. Phys. Lett. 59 (26), Dec. 1991, pp. 3351-3353

10. H. Cheng, J. M. Depuydt, ZnSe-based Blue/Green Light Emitting Diodes, paper TuA.2 LEOS 1991 Summer Topical Meeting on "Epitaxial Materials and In-Situ Processing for.Optoelectronic Devices", July 1991, Newport Beach, USA

11. D.J. Webb, H.-P. Dietrich, F. Gfeller, A. Moser, P. Vettiger, Passivation of GaAs Laser Mirrors By Ion-Beam Deposited Al_2O_3 , Mat. Res. Soc. Symp. Proc. Fall Meeting 1988 Boston; Vol 128, pp. 507-512

12. J. S. Yoo, H. H. Lee, P. Zory, Enhancement of Output Intensity Limit of Semiconductor Lasers by Chemical Passivation of Mirror facets, IEEE Photonics Technology Letters; Vol 3 , No. 3, March 1991, pp. 202-203

13. F. Herrmann, S. Beeck, G. Abstreiter, C. Hanke, C. Hoyler, L. Korte, Reduction of mirror temperature in GaAs/AlGaAs quantum well laser diodes with segmented contacts; Appl. Phys. Lett. 58 (10); March 1991, pp. 1007-1009

14. R. L. Thornton, D. F. Welch, R. D. Burnham, T. T. Paoli, P. S., Cross, High Power (2.1W) 10-stripe AlGaAs laser array with Si disordered facet windows, Appl. Phys. Lett. 49 (23); Dec. 1986, pp. 1572-1574

15. C. Hanke, unpublished results

16. D. Botez, M. Jansen, L. Mawst, G. Peterson, T. J. Roth, Watt-range, coherent, uniphase powers from phase-locked arrays of antiguided diode lasers, Appl. Phys. Lett. 58 (19); May 1991, pp. 2070-2072

17. M. Ross, YAG laser operation by semiconductor laser pumping, Proc. IEEE,56, p. 196

18. D. F. Welch, D. R. Scifres, High power, 8.5 W, Visible Laser Diodes, Electronics Letters, Oct. 91, vol. 27, No. 21 pp. 1915-1916

19. R. F. Murison, N. Holehouse, S. R. Lee, A. H. Moore, Strained Layer Quantum Well Lasers at 1.06 µm with up to 5.25 W from single 100 µm Stripe, Paper SDL 1.7, LEOS '91, Annual Meeting 1991, San Jose, CA USA

20. M. Sakamoto, J. G. Endriz, D. R. Scifres, 120 W cw Output Power From Monolithic AlGaAs (800 nm) Laser Diode Array Mounted On Diamond Heatsink, Electronics Letters, Jan. 92, vol. 28, No. 2 pp. 197-198

21. J. G. Endriz , M. Vakili, G. S. Browder, M. A. DeVito, J. M. Haden, G. L. Harnagel, W., E. Plano, M. Sakamoto, D. F. Welch, S. Willing, D. P. Worland, H. C. Yao: High Power Laser Arrays, IEEE Journal of Quantum Electronics, Vol. 28, No. 4, April 1992, pp. 951-965

22. R. Beach, W: J: Bennet, B. L. Freitas, D., Mundinger, B. J. Comaskey, R. Solarz, M. A. Emanuel: Modular Microchannel Cooled Heatsinks for High Average power Laser Arrays; IEEE Journal of Quantum Electronics, Vol. 28, No. 4, April 1992, pp. 966-976

23. R. Solarz, High Power laser arrays, Paper SDL 1.1 , LEOS '91, Annual Meeting 1991, San Jose, CA USA

24. Photonics Spectra, Vol. 25, Issue 10, October 1991, p. 50

25. J. L. Jewell, A. Scherer, S. L. McCall, Y. H. Lee, S. Walker, J. P. Harbison, L. T. Florez: Low-Threshold Electrically Pumped Vertical-Cavity Surface-Emitting Microlasers, Electronics Letters, Aug. 89, vol. 25, No. 25, pp. 1123-1124

26. J. M. Hammer, N. W. Carlson, G. A. Evans, M. Lurie, S. L. Palfrey, C. J. Kaiser, M. G. Harvey, E. A. James, J. B. Kirk, F. R. Elia: Phase-locked operation of coupled pairs of grating-surface-emitting diode lasers, Appl. Phys. Lett. 50 (1); March 1987, pp. 659-661

27. W. D. Goodhue, J. P. Donnelly, C. A. Wang, G. A. Lincoln, K, Rauschenbach, R. J. Bailey, G. D. Johnson: Monolithic two-dimensional surface-emitting strained-layer InGaAs/AlGaAs and AlInGaAs/AlGaAs diode laser arrays with over 50% differential efficiency, Appl. Phys. Lett. 59 (6); Aug. 1991, pp.632-634

28. S. S. Ou, M. Jansen, J. J. Yang, M. Sergant: High power cw operation of GaAs/GaAlAs surface-emitting lasers mounted in the junction-up configuration, Appl. Phys. Lett. 59 (9); Aug. 1991, pp. 1037-1039

29. D. W. Nam, R. G. Waarts, D. F. Welch, D. R. Scifres: High-Power cw Monolithic 2-D arrays of surface-emitting lasers, paper CWN8, CLEO 92, Anaheim CA USA

30. M. Jansen, J. J. Yang, S. S. Ou, M. Sergant, L. Mawst, J. Rozenbergs, J. Wilcox, D. Botez: Monolithic two-dimensional surface-emitting diode laser arrays mounted in the junction down configuration, Appl. Phys. Lett. 59 (21); Nov. 1991, pp. 2663-2665

31. H.-D. Wolf, private communication

32. O. Ueda, J. Electrochem. Soc. 135, 11C, 1988

33. R. G. Waters, Diode Laser Degradation Mechanisms: A Review, Prog. Quant. Electr. 1991, Vol. 15, pp. 153-174

34. K. Fukukai, S. Ishikawa, K. Endo, T. Yusua, Current density dependence for dark-line defect growth velocity in strained InGaAs/AlGaAs quantum well laser diodes, Japanese Journal of Applied Physics, Vol. 30, No. 3A, March 1991, pp. L 371-L 373

35. P. A. Kirkby, IEEE Journal of Quantum Electronics QE-11, 572 (1975)

36. M. Sakamoto, J. G. Endriz, D. R. Scifres: 20 W CW Monolithic AlGaAs (810 nm) Laser Diode Arrays, Electronics Letters, Vol. 28, No. 2, pp.178-179 (1992)

37. R. Beach, D. Mundinger, W. Benett, V. Sperry, B. Comaskey, R. Solarz: High-reliability silicon microchannel submount for high average power laser diode arrays, Appl. Phys. Lett. 56 (21), May 1990, pp. 2065-2067

EFFICIENT NARROWBAND OPTICAL PARAMETRIC OSCILLATORS OF BETA–BARIUM–BORATE (BBO) AND LITHIUM–TRIBORATE (LBO)

A. Fix, R. Feldbausch, T. Schröder, R. Urschel,
R. Wallenstein

Fakultät für Physik
Universität Kaiserslautern
6750 Kaiserslautern, Germany

INTRODUCTION

Since the first demonstration of an optical parametric oscillator (OPO) by Giordmaine and Miller[1] in 1965 the OPO has been subject to detailed theoretical and experimental investigations[2,3]. The OPO is considered as a source of powerful, broadly tunable coherent radiation. The development and the scientific application has been hampered, however, by the scarcity of nonlinear optical materials with suitable optical and mechanical properties.

The present renaissance in the research and development of optical parametric oscillators is based on new nonlinear materials and on advanced powerful pulsed solid state lasers. Crystals like beta–Bariumborate (BBO) and Lithiumborate (LBO) provide phase–matched OPO operation in a wide spectral range from the near ultraviolet to the near infrared [4,5]. Because of the high damage threshold and the high effective nonlinearity[6,7] these crystals are appropriate materials for the development of highly efficient reliably operating OPO devices. Advanced Q–switched single–mode solid state lasers (like Nd:YAG or Nd:YLF) generate high peak power light pulses which are well–suited for pumping efficiently narrowband ns–OPO systems.

Solid State Lasers: New Developments and Applications
Edited by M. Inguscio and R. Wallenstein, Plenum Press, New York, 1993

Aspects which are of special interest are design criteria, tuning range, efficiency, output power, pulse shape and the spectral properties of the generated coherent light.

BASIC PROPERTIES OF THE BBO–OPO

In principle the BBO–OPOs are of simple design. In typical systems, 12 mm, 15 mm or 18 mm long BBO crystals were placed in a 20 – 30 mm long resonator consisting of two flat mirrors. The signal wave reflectivity of the input mirror was 96 – 99 %, the one of the output coupler in the range of 50 – 75 %. Both mirrors were transparent for pump and idler wave to ensure singly resonant operation.

Pumped by the second, third, fourth or fifth harmonic of the 1.064 μm radiation of an injection–seeded Nd:YAG laser the BBO–OPO is widely tunable from the ultraviolet (230 nm) to the near infrared (3.036 μm). In Table 1 values of the signal (λ_s) – and idler (λ_i)–wave (calculated from the Sellmeier equations reported by Kato[8]) are compared with experimental results[4]. The width of the total tuning range is limited by the increasing absorption of the idler radiation at 2.2 – 3.0 μm. For 355 nm pump light the type–I phase matching angle ϑ changes from 23° to 33°. Operation in the whole tuning range (400 – 3000 nm) thus requires only one crystal cut at 30° rotated by about 10°.

Table 1

	Theory			Experiment	
λ_p [nm]	ϑ [deg]	λ_s [nm]	λ_i [nm]	λ_s [nm]	λ_i [nm]
532	20.7 - 22.8	647 - 1064	1064 - 3000	667 - 1064	1064 - 2628
355	23.1 - 33.1	403 - 710	710 - 3000	402 - 710	710 - 3036
266	26.7 - 47.6	292 - 532	532 - 3000	302 - 532	532 - 2248
213	31.3 - 72.6	229 - 426	426 - 3000	—	—

The threshold of the 355 nm–pumped BBO–OPO (crystal length ℓ = 12mm, reflectivity of the output coupler R = 70 %) is in the range of 20 – 40 MW cm^{-2} and thus well below the BBO damage threshold of several GWcm^{-2} [4,5]. The power densities at threshold are easily achieved, for example, in laser beams with 2.5 mm diameter and pulse energies of 5 – 7 mJ[4].

Because of these low threshold energies efficiencies of 30 – 40 % are obtained with pump energies of less than 60 mJ in 7 ns long pulses. With higher pulse energies the OPO output easily exceeds 100 mJ (see Fig. 1) with total efficiencies of more than 50 % which corresponds to crystal internal efficiencies exceeding 70 %.

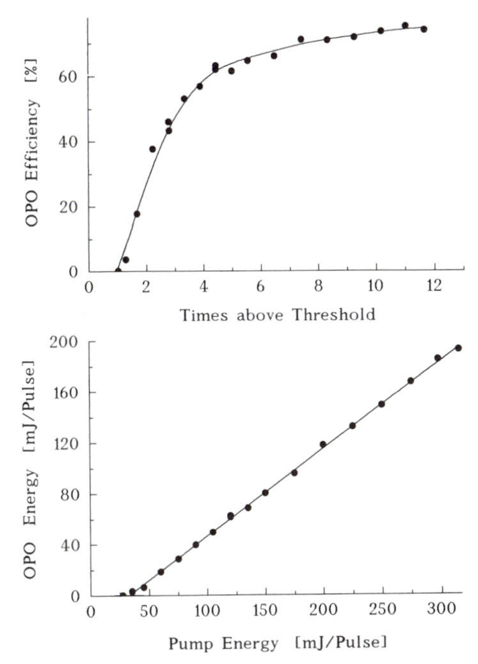

Fig. 1. BBO–OPO pulse energy and crystal internal conversion efficiency as function of the pump pulse energy (length of the BBO crystal: 15 mm; wavelength of the signal wave: $\lambda_s = 484$ nm)

The bandwidth of the BBO–OPO output depends on the wavelength, the length of the crystal, the properties of the resonator, the linewidth of the pump radiation and on the pump power. If the OPO is pumped three times above threshold the measured linewidth increased with wavelength from about 0.2 nm at 410 nm to almost 4 nm at 650 nm with further increase at wavelengths close to degeneracy (at 710 nm). This broadband radiation is of advantage for multiplex spectroscopy with optical array detection of the dispersed radiation (as demonstrated by multiplex CARS spectroscopy[9,10] and intracavity absorption spectroscopy[11]).

Many applications require, however, narrowband operation. The traditional approach to develop a pulsed OPO with narrow bandwidth is to incorporate wavelength–selective elements such as gratings and etalons in the OPO cavity[12]. The losses introduced into the cavity usually cause additional cavity losses which increase the threshold and reduce the efficiency.

An alternative straightforward technique for obtaining single–mode operation is injection seeding. This method – first demonstrated by Bjorkholm and Danielmeyer[13] – has meanwhile been used successfully in the development of single–mode OPOs. BBO–OPOs have been seeded with the radiation of a single– mode pulsed Nd:YAG laser[14], a cw single–mode dye laser[4] or a pulsed narrowband dye laser[15]. Using narrowband radiation from a low intensity tunable pulsed dye laser for injection seeding the OPO provided narrowband (0.1 cm^{-1}) signal and idler outputs with optical frequencies continuously tunable over a wide tuning range (> 100 cm^{-1}). As demonstrated in early experiments[14] the injection seeding not only provides single–mode operation but also reduces the pump power at threshold and the oscillator build–up time.

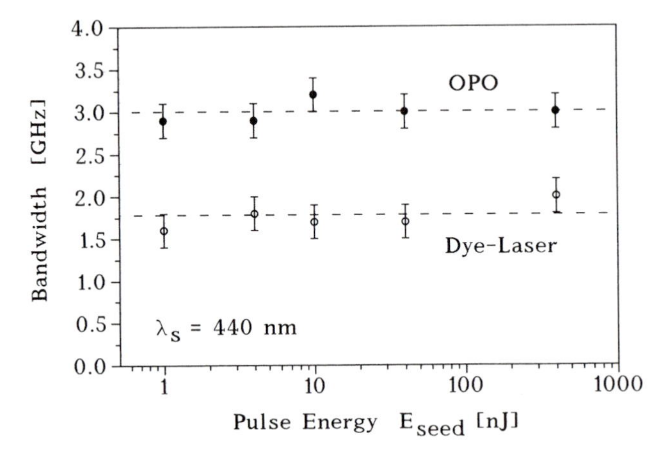

Fig. 2. Bandwidth of the BBO–OPO (seeded by a pulsed dye laser) as function of the pulse energy of the seed–radiation (wavelength of the signal wave: $\lambda_s = 440$ nm)

Detailed measurements[16] demonstrated that the OPO–bandwidth ($\approx$ 0.1cm^{-1}) is broader than the dye laser linewidth (0.06 cm^{-1}). Powers of the seed laser as low as 1nJ were sufficient for a precise spectral control (see Fig. 2). The bandwidth of the seeded OPO was independent of the pump power.

For most precise frequency control tunable cw seed sources are of particular advantage. In this respect compact single mode diode lasers are of special interest. In our investigations[18] a 355 nm–pumped BBO–OPO (with a 30 mm long linear or a

3–mirror–ring cavity) was injection seeded at the wavelength of the nonresonant idler wave with the radiation of single–stripe GaAlAs diode lasers with an external grating as reflective wavelength selector for feedback and tuning.

At room temperature the used diode lasers generated light at 810 nm or 830 nm with a spectral bandwidth of less than 1 MHz and an output power of about 20 mW. The OPO resonators were resonant for visible signal radiation in the range of 600 – 700 nm. As demonstrated in the experiment idler seed powers below 100 μW are sufficient for stable single–mode OPO operation. The spectral bandwidth of the signal wave is less than 600 MHz (see Fig. 3). In principle diode lasers are available at wavelengths of 665 nm to 1.6 μm. If a 355 nm–pumped OPO is seeded at the idler wavelength (0.71 – 1.6 μm) the signal wave provides narrowband radiation in the visible at 456 – 710 nm. Seeding a 532 nm–pumped OPO with diode laser light at 665 – 1064 nm will produce narrowband infrared idler radiation at 1.064 – 2.66 μm.

These data clearly demonstrate that single–mode BBO–OPO systems seeded by a tunable diode laser will combine the advantageous properties of wide operating range, compact solid–state design, reliability and high efficiency.

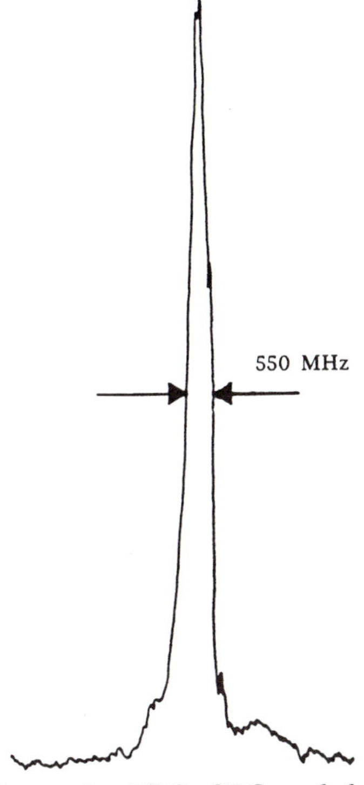

Fig. 3. Linewidth of a ring cavity BBO–OPO seeded at the idler wavelength (λ_i = 810 nm) by a single–mode diode laser

For further investigations of the spectral properties and quantum fluctuations of seeded and unseeded OPOs the pulse–to–pulse variations of the intensity of the individual OPO modes has been measured[19]. For this investigation we used an OPO with sufficiently large mode–spacing. The BBO–OPO consisted of a 3.6 mm long flat–flat mirror resonator and a 2.5 mm thick BBO crystal. The cavity is resonant for wavelengths at 490 – 600 nm. Despite the short crystal this OPO provides a conversion efficiency of about 2.5 % at a pump energy of 60 mJ (1.5 times above threshold).

The mode spectra are recorded with a 1 m spectrometer and a 1728–element photodiode–array as detector. The resolution of this device is 0.25 cm^{-1} which exceeds about four times the OPO modespacing of 0.96 cm^{-1}. From pulse to pulse the intensity of the OPO modes varies considerably. This is shown in Fig. 4A where 10 arbitrarily chosen (successively detected) pulse spectra are displayed. The average of 1000 single pulse spectra is displayed in Fig. 4B. The mode intensities are modulated due to an etalon effect caused by the uncoated BBO crystal. Fig. 5A shows the intensity distribution for modes with highest mean intensity. This distribution is close to an exponential function. An analysis of the mode–intensity cross correlation coefficient for different modes reveals that mode coupling is negligible. An interesting question is of course the statistical behaviour of the mode intensities when the OPO is exposed to seed radiation. Fig. 5B shows, as an example, an averaged spectrum of the BBO–OPO output seeded at 532 nm by the second harmonic of a seeded Nd:YAG laser. As expected only one mode is present in the output spectrum.

The detection of the pulse intensity in the individual modes provides a very sensitive means for a detailed study of the seeding of the OPO. The dependence of the intensity in the seeded and unseeded modes as function of the pump and seed power will provide reliable information on the energy in the seeded and unseeded part of the output. Besides the seeding by narrowband laser radiation the possible influence of intense incoherent light or of spectrally broad coherent (fs) light pulses on the OPO output spectrum are of special interest. At present these investigations as well as a detailed theoretical analysis of the experimental results are in progress.

THE LBO–OPO

In addition to BBO, crystals of LBO are available in sufficiently large size and with high optical quality. Compared to BBO, LBO offers the advantages of a small

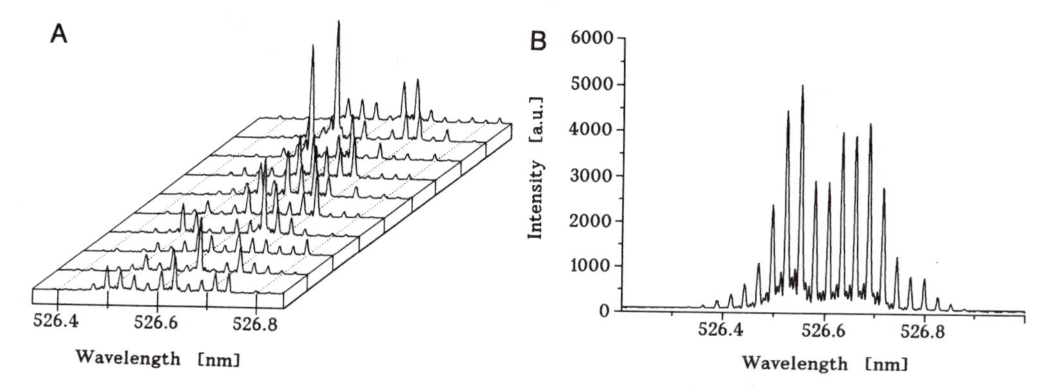

Fig. 4. A: Successively recorded single pulse spectra of the BBO–OPO.
B: Averaged spectrum of 1000 successive pulses of the BBO–OPO at a signal wavelength of 526 nm. The mode spacings are 0.96 cm^{-1}. The resolution of the spectrograph is 0.25 cm^{-1}.

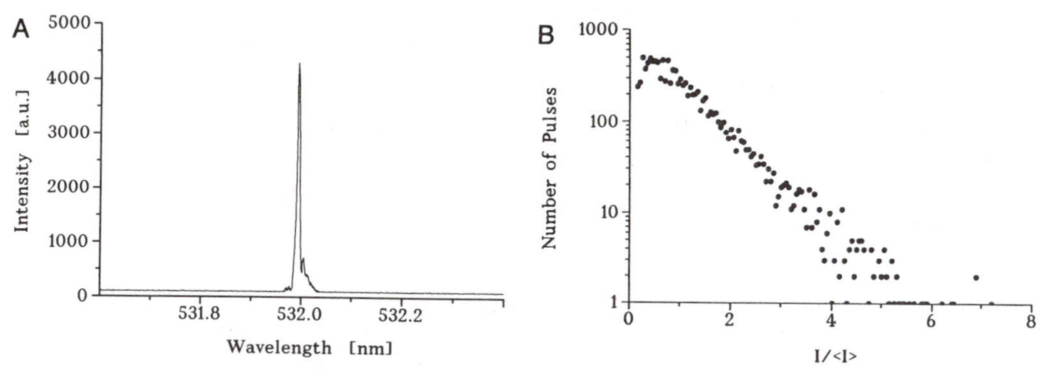

Fig. 5. A: OPO mode intensity distribution of 1000 pulses at a signal wavelength of 526 nm.
B: Average spectrum of 100 pulses of the BBO–OPO seeded at 532 nm. The seed energy is 40 μJ/pulse.

walk–off (ρ[LBO] ≈ 0.125 ρ[BBO] and a large acceptance angle ($\Delta\vartheta$[LBO] ≈ 10 $\Delta\vartheta$ [BBO])[6,7]. The LBO material should thus provide longer effective crystal lengths and output beams of very good spatial quality.

The LBO–OPO is tunable in about the same wavelength range as the BBO–OPO. This was confirmed experimentally for pumping with 266 nm, 355 nm and 532 nm radiation. Also, the linewidth of the unseeded LBO–OPO is comparable to that of the BBO–OPO.

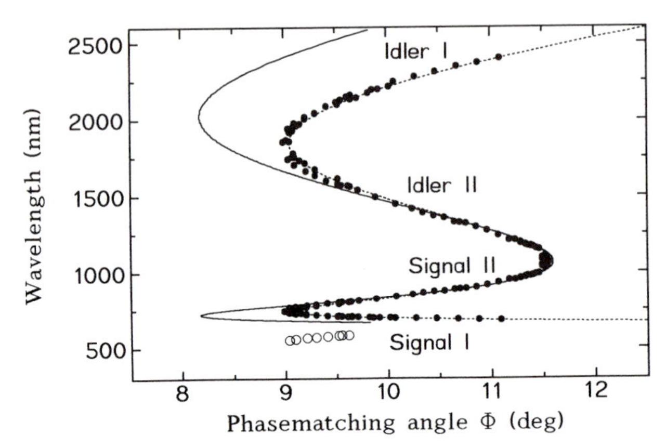

Fig. 6. Multiple frequency operation of the 532 nm pumped LBO–OPO. Besides the observed two pairs of signal– and idler–waves the sum frequency ($\omega = \omega_s^{II} + \omega_i^{I}$) generates visible light at 556–586 nm. The solid curve represents the tuning calculated from the Sellmeier equations reported by Kato[8].

The LBO has, however, the disadvantage of a larger thermal birefringence (η[LBO] ≈ 8η[BBO]) and a smaller effective nonlinearity (d_{eff}[LBO] ≈ 0.66 d_{eff}[BBO]). Because of the smaller nonlinearity the power density J_0 at threshold is higher for the LBO–OPO than for the BBO–OPO. A comparison of the values of J_0 measured – for identical experimental conditions – in the wavelength range of 500–600 nm with an 11 mm long LBO crystal and a 12 mm long BBO crystal indicated that the threshold of the LBO–OPO is higher by a factor of about 2.5.

At a pulse energy of J=170 mJ (J/J_0=5) the conversion efficiency for a 15 mm long crystal exceeded 45%. The energy density of threshold was about 0.2 J cm^{-2}.

One of the most interesting features of the LBO–OPO is the multiple colour operation if pumped by 532 nm radiation[20]. As seen from Fig. 6 the wavelength tuning shows three turning points. For a phase matching angle θ in the range of 9° – 11.6° the LBO–OPO generates simultaneously two pairs of signal– and idler waves. At $\theta = 9.1° - 9.6°$ the OPO–crystal generates in addition visible light at 556 – 586 nm, by type–I sum frequency mixing of the signal wave II and idler wave I. The results of a detailed investigation of the properties of the LBO–OPO[21] will be published elsewhere[21].

SUMMARY

In summary OPOs of BBO and LBO are highly efficient devices for the generation of narrowband ns–light pulses. Of particular advantage are the large tuning range, the high efficiency, and the ease of operation. With diode–laser–pumped solid state lasers as pump source these OPOs will be the base for powerful broadly tunable, all–solid–state coherent light sources.

REFERENCES

1. J.A. Giordmaine and R.C. Miller, Phys. Lett. **14**, 973 (1965)
2. S.E. Harris, Proc. IEEE **57**, 2096 (1069)
3. R.L. Byer, in Treatise in Quantum Electronics, edited by H. Rabin and C.L. Tang (Academic, New York, 1973) pp. 587–702, see also: Y.X. Fa and R.L. Byer in SPIE Proceedings Vol. **461**, 27 (1984)
4. A. Fix, T. Schröder and R. Wallenstein, Laser und Optoelektron. **23**, 106 (1991), and references therein.
5. C.L. Tang, W.R. Bosenberg, T. Ukachi, R.J. Lane, and L.K. Cheng, Proc. IEEE **80**, 365 (1992) and references therein
6. A. Borsutzky, R. Brünger, Ch. Huang, and R. Wallenstein, Appl. Phys. **B 52**, 55 (1991) and references therein
7. A. Borsutzky, R. Brünger, and R. Wallenstein, Appl. Phys. **B 52**, 380 (1991)
8. K. Kato, IEEE J. Quant. Electron. **QE–22**, 1013 (1986)
9. M.J. Johnson, J.G. Haub, H.–D. Barth, and B.J. Orr, Opt. Lett. (in press)
10. D. Brüggemann, J. Hertzberg, B. Wies, Y. Waschke, R. Noll, K.F. Knoche, and G. Herziger, Appl. Phys. B (in press)
11. K.J. Boller and T. Schröder, International Conference on Quantum Electronics, OSA Technical Digest Series **9**, 562 (1992)

12. See for example: S.J. Brosnan and R.L. Byer IEEE J. Quantum Electron. **QE–15**, 415 (1979); T.K. Minton, S.A. Reid, H.L. Kim and J.D. McDonald, Opt. Commun. **69**, 289 (1989)
13. J.E. Bjorkholm and H.G. Danielmeyer, Appl. Phys. Lett. **15**, 171 (1969)
14. Y.Y. Fan, R.C. Eckardt, R.L. Byer, J. Nolting, and R. Wallenstein, Appl. Phys. Lett. **53**, 2014 (1988)
15. J.G. Haub, M.J. Johnson, B.J. Orr, and R. Wallenstein, Appl. Phys. Lett. **58**, 1718 (1991)
16. R. Urschel, Diplom–thesis 1992, University of Kaiserslautern (to be published)
17. C.E. Hamilton and W.R. Bosenberg, Conference on Lasers and Electrooptics, OSA Technical Digest Series, **12**, 370 (1992)
18. A. Fix, R. Feldbausch, M. Inguscio, G.M. Tino, and R. Wallenstein International Conference on Quantum Electronics, OSA Technical Digest Series 9, 528 (1992)
19. A. Fix, PhD–thesis 1993, University of Kaiserslautern (to be published)
20. T. Schröder, A. Fix, and R. Wallenstein, Conference on Lasers and Electrooptics, OSA Technical Digest Series, **12** (1992)
21. T. Schröder, PhD–thesis, University of Kaiserslautern 1993 (to be published)

TRAVELLING WAVE PARAMETRIC GENERATION OF HIGHLY COHERENT FEMTOSECOND PULSES TUNABLE FROM 750 TO 3000 nm

G.P. Banfi[(1)], R. Danielius[(2)], P. Di Trapani[(1)], P. Foggi[(3)],
A. Piskarskas[(2)], R. Righini[(3)] and I. Sa'nta[(3)]

[(1)] Dipartimento di Elettronica, Universita' di Pavia, 27100 Pavia, Italy
[(2)] Laser Research Center, Vilnius University, Vilnius 232054, Lithuania
[(3)] European Laboratory for Nonlinear Spectroscopy (LENS), Firenze, Italy

We report here some preliminary results of an extensive investigation devoted to the generation of tunable pulses through parametric conversion of a 200 fs, 100 μJ pulse at 600 nm delivered by an amplified dye laser. Different crystals and combinations, all capable of achieving the same pulse duration of the pump, have been tested in the travelling wave configuration adopted. We achieved for the first time in this regime pulses with a time-bandwidth product as small as 0.7, the limit being dictated by the pump itself. To this end, confirming our expectations, BBO in type II phase matching definitely outperforms type I, still allowing the same 800-3000 nm tuning range and comparable conversion efficiencies. When bandwidth is not a request, LBO appears the best candidate for the minimal energy requirements.

1- Introduction

Powerful femtosecond pulses, tunable in a wide range are strongly required for many application, in particular for time resolved and nonlinear spectroscopy. A great improvement has been achieved recently in the direct generation of tunable pulses with the Ti-saffire lasers: beside very short pulses, they offer the practical advantages of solid state, possibility of large energy through amplification, and a much broader tunability when compared to the few tens of nanometers of a dye laser. But even with this source, the tunability is limited to a fraction of the central wavelength and the infrared region of the spectrum behind 1 μm is totally uncovered.

The demands of a broad tunability within a single device and of IR pulses can both be satisfied through parametric conversion from a source at fixed wavelength. Parametric

Solid State Lasers: New Developments and Applications
Edited by M. Inguscio and R. Wallenstein, Plenum Press, New York, 1993

generation finds favourable condition with short pulses since the high intensities one can handle allow to achieve more easily sufficient gains. We notice that new materials, such as BBO and LBO, with high second order susceptibility and high damage threshold, have also contributed to the recent revival in the all the field of parametric generation. A high nonlinearity is desired also with ultrashort pulses, since one can better face the problems of pulse broadening, by employing short crystals, and the effects of group velocity dispersion (GVD).

With short pulses parametric conversion can be obtained either by a synchronously pumped optical parametric oscillator (OPO), eventually followed by an optical parametric amplifier (OPA) to boost single pulse energy, or by a travelling-wave optical parametric converter (TOPC). This last one (Fig.1) is essentially an OPA driven at very high gains, so that it can amplify in just two or three stages (each involving a pass in the crystal) the initial parametric fluorescence to intensities comparable to that of the pump beam. For most applications the practical advantages of TOPC are undoubtable. Easiness of adjustment and tuning, reliability, and the fact that a single energetic tunable pulse is immediately available once a fixed frequency pump pulse is provided, make it the ideal candidate in many complex spectroscopic arrangements. What in general TOPC doesn't provide is the intrinsic beam conditioning (bandwidth, spatial profile, and more subtle effects leading to self-compression) offered by the many passes in an OPO configuration. However, we anticipate that the spectral and spatial quality of the beam can be controlled in the design of the set-up and by a proper choice of the nonlinear crystals.

During the years, as soon as sources of shorter pump pulses became available, parametric conversion has been successfully operated in the new regime. With femtoseconds, which we here consider, significant results have been achieved with synchronously pumped OPO's: with the second harmonic of a ML (mode-locked) train generated by a ML Nd:glass laser, widely tunable fs pulses, each about 100 nJ in energy and shorter than those of the pump itself, have been obtained;[1] while OPO have been recently reported to operate also in the continuous fs regime , employing either a ML dye laser [2] or a ML Ti-saffire laser as a pump source.[3]

The same trend toward shorter pulses has been pursued with TOPC. The high gain required in this arrangement benefits even more substantially than OPO's from the higher intensities that usually come along with short pump pulses, and successful operation becomes easier going from the ns to the ps regime. However, moving down to the fs regime, while the possibility to obtain transform limited pulses increases, new difficulties , mainly connected with the GVD, are encountered. Beside an eventual broadening in time, the separation of pump, signal and idler pulses might cause the parametric gain to decrease at the desired wavelength. This fact, often complicated by competing processes, can deny the achieve the conversion or the tunability in all the potential range of wavelengths.

The investigation on which we here report was aimed to find the criteria to be followed in designing an TPOC operating in the 100 fs regime, and to have at least a qualitative understanding of the role of the different mechanism which lead to the aforementioned difficulties. Keeping in mind the final use of the device, the points addressed have been tunability range, pump energy required, conversion efficiency, time duration of the tunable converted pulses and especially their time-bandwidth product. As pump beam we employed a 200 fs, 0.1 mJ pulse at 0.6 μm. With a similar pump, a 15 % conversion to 830 nm pulses has been reported with a BBO in type I p-m (hereafter denoted BBO I).[4] Always with a BBO I, but with the higher energy of 600 μJ, pulses of 150 fs were obtained at 830 nm and a potential tunability between 775 and 3000 nm has been indicated.[5] In both cases the output pulses were far from being transform-limited. Subpicosecond transform-limited pulses have been obtained by some of present authors pumping TOPC with the second harmonic of a 1.5 ps pulse generated by a Nd-glass laser. To this end different crystals and cuts have been investigated, the best choice turning out in that context a KDP II in the first seeding stage and BBO in the second one.[6] In aiming to transform-limited pulses, these previous findings in the 1 ps regime best served as guide-lines for present experiment.

2-The experimental set-up

The pump source is based on a CW-ML dye laser (Spectra Physics model 3695) synchronously pumped by 4 ps pulses in the green. These last ones are obtained from a frequency doubled CW-ML Nd:YAG laser (Spectra Physics model 3800), followed by a fiber/grating compressor. The pulses from the dye laser are also compressed and finally amplified, at 10 Hz repetition rate, in a three stage dye amplifier (Quanta Ray PDA1). More details on the pump source are given in ref 7. After a spatial filter, the pulse at 605 nm had, in the best conditions, an energy of 130 μj. Pulse duration, from correlation trace, was 200 fs and its FWHM spectral-width about 130 cm^{-1}, which means a time-bandwidth product $\Delta\nu\Delta\tau$ almost twice the transform-limited value for a gaussian pulse.

An unfolded representation of the OPO set-up is depicted in Fig.1. The system is composed by two stages, the "seed generator" and the " amplifier", pumped respectively by p1 and p2, the two beams into which the laser pump has been splitted. Their energy ratio E_{p1}/E_{p2} was 1: 3.5, and taking into account the losses of a few uncoated optics, we had 20 μj and 70 μj for the first and the second stage, respectively (in the real lay-out the seed was generated via a double pass in the same crystal, with a broadband dielectric mirror serving as total reflector between the the two passes) . The synchronisms between the two pulses was controlled by a proper delay line. The spot sizes were estimated by transmission through a set of pin-holes of different diameter. After collimation, d_2, the diameter (FWHM of the intensity) of p2, was set to 1.25 mm from which we estimate a peak intensity in space and time of 20 GW/cm^2 in front of the amplifier. Focusing with an 1/2 meter lens, a rough "power in the bucket" method suggested the pump beam divergence to be sligthly larger than 2 times the diffraction limit. Different focusing conditions have been tried out for d_1; most data here reported have been taken collimating p1 to $d_1 \approx 0.35 - 0.4$ mm on the first crystal, with a maximum peak pump intensity that could reach 100 GW/cm^2. After some propagation the generated seed is collimated to a large spot before being injected together with p2 in the amplifier. The system has always been operated imposing collinear phase-matching.

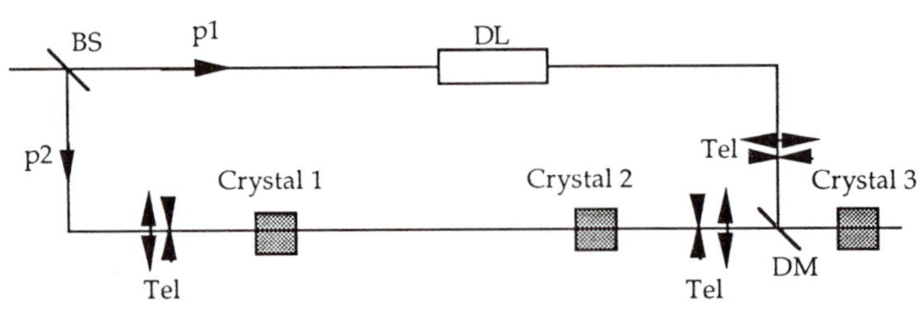

Fig.1

Unfolded lay-out of the TOPC. Tel: telescope; BS; beam splitter; DM: dichroic mirror; pl and p2: pump beams; DL: delay line.

The following crystals have been used: BBO I, BBO II, and LBO, with this last one cut for 90-degree non-critical p-m. All BBO's were 8 mm long, while LBO 's length was 15 mm. A few measurements have also been performed with a 4 mm long BBO II in the seeder.

Beside a sensitive bolometer and photodiodes, the diagnostics included a spectrometer which allowed to record the spectral distribution of the signal pulses up to $\lambda_s \approx 1$ µm (in the following we call arbitrarily signal and idler the modes generated at shorter and longer wavelength respectively, and λ_s and λ_i will denote the signal and idler central wavelengths). For time measurements a single shot autocorrelator developed in Vilnius University proved most useful during the adjustment procedure. Final data have been taken through a conventional scanning autocorrelator equipped with a thin KDP crystal.

Tab. I

Summary of the results obtained in different configurations

Generator & Amplifier	**BBO II BBO II**	**BBO I BBO I**	**LBO LBO**	**LBO BBO II**
Tuning range	750-3000 nm	750-3000 nm	840-2200 nm	840-2200 nm
Conversion Efficiency (last stage)	30%-15% 70mj pump	33%-18% 70mj pump	15 % 25mj pump	30 % 30mj pump
Pulse Duration (fs)	240 @0.8mm 195 @ 1mm		150-200	
Spectral Bandwith (cm-1)	100-200 (200 0.8mm)	200-over 1000	200-over 1000	100
DtDn	0.7 -1.3 (1.3 @800 nm)	>>1	>>1	

3-Results

a) BBO crystals

One of the main results of the present work is the demonstration of the complete tunability obtainable with the BBO OPO, also in the case of a relatively weak pump pulse. It should be noticed that the totally available pump energy is in our case five time lower than in ref.5, and even lower figures do apply when we consider the energy fraction devoted to the seeder. The best results in terms of minimum energy required, and wide tunability in all the 0.75-3 μm range have been obtained with a small spot and high intensities in the first stage, this being true both for BBO in type I and type II p-m. Some data which summarize the performance obtained are given in table I, while in Fig.2 we have reported for BBO II the conversion efficiency versus wavelength. In the following η will denote the conversion conversion efficiency to signal+idler in the amplifier, and , being the seed input energy negligible, $\eta = (E_{sign} + E_{idl})/E_{p2}$.

For BBO II, the easiest condition of operation and maximum conversions have been observed for λ_s around 800 nm, in a region 100 nm wide close to the edge of the tuning curve. In this region, conversion η larger than 20% are easily achieved with much lower pump energy, and the results do not depend crucially on the focussing choices and on the position of crystal 2. The conversion efficiency decreases moving away from this 800 nm region, but can be maintained over 15 % in all the tuning range, degeneracy included, with proper choice of d1 and of the distance among crystals 1-2. We notice that the conditions for conversions become more crucial moving away from the more favourable wavelengths around 800 nm, and indeed with large d1 the tunability ceases. Also weakening the gain in crystal 2 by translating it causes a drop of η. At the same time we note that the parametric emission after the first crystal in the seeder occurs in a wide cone with brigth ring some degrees off-axis. As it will be discussed later, we attribute all these findings to the group velocity dispersion, which changes significantly in the tuning range. In the 800 nm region,

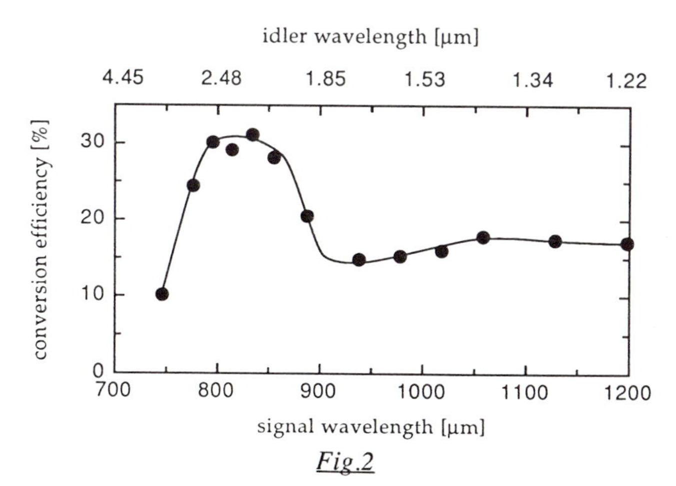

Fig.2

Energy conversion efficiency in the final amplifier stage. Pump energy: E=70mj.

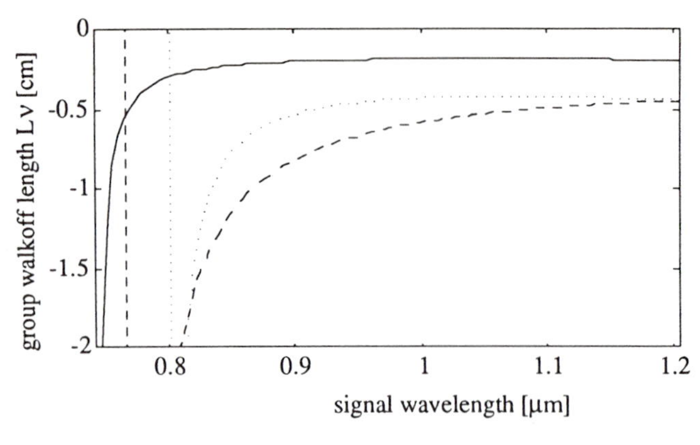

Fig.3

Group walkoff length $L_{n\ s/i} = t\ (1/u_{s/i} - 1/u_p)$, where t is the pulse duration and $u_{s/i/p}$ are the signal / idler / pump group velocities. Full line: L_{ni} for the II p-m type BBO; dashed and dotted lines: L_{ns} and L_{ni} for the I p-m type BBO. Pulse duration: t=200 fs. (L_{ns} for the II p-m type BBO is much larger than 1 cm in the whole tuning range).

the GVD (of the idler for BBO II, of both signal and idler for BBO I) for collinear p-m is small so that the three pulses - pump, signal and idler- stay together and interact for maximum parametric yield almost throughout all the crystal length. But tuning for collinear phase-matching to shorter signal wavelengths, the GVD become larger (Fig.3), the three pulses separate and the gain at the set λ_s decreases. At the same time, however, parametric action can find more favourable condition in a non collinear p-m at different λ's, for which the GDV are smaller. This explains the cone and side-emission at undesired λ's in the first crystal. Attempts to compensate the reduced gain at the set collinear λ_s by increasing the pump intensity do help as far as the competing side emission doesn't became so strong to cause pump depletion. It is our opinion that some off-axis emission is practically unavoidable; all one can do is to contain it, and waist as small as possible pump energy in it. In our arrangement we leave the job to select and amplify the collinear tunable λ_s to crystal 2. This requires a small gain in this stage, care must only be taken to ensure the proper distance among the two crystals in the seeder in order to select the collinear λ_s.

A trend similar to that of BBO II was observed for the conversion efficiency of BBO I, installed both in the seeder and amplifier. Also magnitude of η were comparable in the two cases. It appears that BBO I achieves stable operating conditions at a slightly lower pump energy , a fact we attributed to the higher gain associated with a longer group walkoff length of the idler (Fig.3).

We tried also BBO I as a further amplifier after the BBO II configuration. This increased somewhat the output energy out of the 800 nm region, making conversion more smooth in the tuning range. The practical advantages of this further pass in a crystal are still to be assessed.

In BBO crystals the tuning range is limited by IR absorption : for BBO II this absorption hindered the parametric process for $\lambda_i > 3200$ nm. As expected from the higher gains/unit length one imposes in the fs regime respect to longer pulses, one can push a little more the tunability into the range of IR absorption. In the transparency range the conversion efficiency to the idler mode can be simply calculated from the total one and from the ratio of photon energies. In presence of idler absorption, which becomes significant for $\lambda_i > 2.5$ μm, directly measurement with bolometer, after filtering out from the output the signal beam, have been preferred. We had a 9 % conversion to the pulse at $\lambda_i \approx 2$ μm, while a still detectable signal of 0.64 μJ, is still available at $\lambda_i \approx 3$ μm, for a corresponding 0.9% conversion rate to this wavelength. Being at this tuning $\eta \approx 12\%$ and the photon energy ratio ≈ 5, we estimated the pump energy fraction going to the IR pulse to be 2.4 % . Then, at 3 μm, about 2/3 of the generated idler pulse is lost due to the IR absorption.

In order to maximize the conversion efficiency, the synchronism of p2 must be slightly adjusted during the tuning. We just mention here that such adjustment is not delicate at all. From the plot of the conversion rate versus p2 delay at fixed λ's, we saw that a precision of 50 μm in the delay line was more than sufficient. This plot exhibits the curious feature of being asymmetric (a feature already noted by the authors of ref 5), a behaviour that we attribute to the fact that the amplifier works either seeded by the signal or by the idler pulse. The asymmetry disappears replacing BBO II with BBO I in the amplifier, since in this case the idler beam generated in the BBO II seeder has the wrong polarization to be amplified.

The fluctuations of the energy of the output pulse are essentially determined by the seed brightness and by the operating conditions of the amplifier. As always in TOPC the amplifier must be driven to the saturation regime in order to dump the unavoidable fluctuation of the seeding pulse. We remind that the seeder works by amplifying, practically in the "small signal " regime, the initial parametric fluorescence, and then carries all the fluctuations due to the quantum noise of the initial process. When the amplifier has a gain of 10^4 -10^5, driving it to saturation requires a hundred of nJ energy in the seed. For $\eta >$ 15-20 % we measured shot-to-shot energy fluctuation around 15%, which must be compared with 10 % variations of the pump pulse. Taking into account the necessity of having a decent energy stability, we would fix, conservatively, to 100 μJ the minimum pump energy required for properly operating the BBO II converter in all the tuning range, while somewhat lower figure could be worked out with BBO I.

One of the main results of present investigation concerns the spectral quality of the generated pulses. Some spectral distribution can be seen in Fig.4. As we expected, BBO II showed a much smaller bandwidth than BBO I, and actually this is the reason of our efforts with this crystal cut. For BBO II, $\Delta\nu_s$ is about 110-140 cm^{-1} from 1 μm to 0.8 μm, increasing to 200 cm^{-1} just at the very edge of the tuning curve at 750 nm (by $\Delta\nu$ we denote the FWHM of the spectral distribution expressed in wavenumbers). Direct bandwidth measurements are limited to $\lambda_s < 1$ μm, due to the detector sensitivity, but by frequency doubling the signal pulse we could anyway estimate that the bandwidth stays below 150 cm^{-1} up to degeneracy (λ_s =1.2 μm).

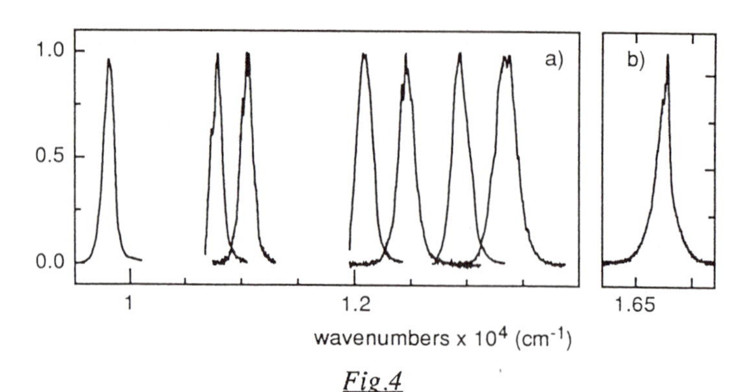

Fig.4

a) Spectra of the II p-m type BBO for signal wavelengths from 1020 nm (left) to 750nm (rigth) ; b) the pump spectrum.

We notice that the spectral width is comparable to that of the pump, and this let us conjecture that smaller $\Delta\nu$ can be obtained with a narrower bandwidth pump, yielding fully transform-limited 200 fs pulses . Actually the conjecture is supported by the simple calculations one can do with longer pulses, which suggest for a narrow spectrum pump an intrinsic $\Delta\nu_s \approx 50$ cm^{-1} for BBO II. The bandwidth is much larger for BBO I. This is obvious on approaching degeneracy, since the first order term in $(\Delta\nu)^{-1}$ vanishes in a type I p-m, and $\Delta\nu_s \approx 1000$ cm^{-1} have been estimated from the angle one could rotate a crystal generating the SH of the output. But also at 900 nm, the $\Delta\nu_s = 350$ cm^{-1} we measured is still 3 times larger than for BBO II. As shown also by simple estimations, the group velocities of signal and idler in the two different cuts are such to make type II slightly harder to be pumped, but more selective in wavelength.

All the effort in narrowing the bandwidth is aimed to transform-limited pulse, and it would be of no advantage if pulse duration were to become longer in so doing. Autocorrelation traces of the signal pulses have been taken at two wavelengths, $\lambda_s = 1$ µm and at 0.8 µm. For BBO II, the FWHM of the autocorrelation traces were all between 260 and 340 fs, with lower values occurring at 1 µm, while traces of the pump taken in the same experimental run were ≈ 280 fs wide. These results show that parametric conversion without lengthening of the pulses can be accomplished with BBO II in the seeder (with the advantage of a narrow spectral width) and with the same crystal also in the amplifier. There is then no need to replace BBO II with BBO I in the amplifier as we were prepared to, and actually such replacement brought no significant changes in present 200 fs regime.

The time-bandwidth product decreases to 1.3 at $\lambda_s = 0.8$ µm . In view of the regular behaviour exhibited by τ_p and $\Delta\nu_s$, and by the fair agreement of $\Delta\nu_s$ with the predicted trend, we feel confident to state that the best figure showed at $\lambda_s = 1$ µm should hold in the whole tuning for λ_s ranging from 850 nm up to degeneracy, even if no direct measurements of bandwith could be taken in this region.

The signal pulse after the amplifier has a spot size comparable to the pump, while its divergence is about 1.3 x larger (measurements at 820 nm). Taking also into account the differences in wavelength, the spatial quality of the pump appears to be not worsened in the conversion. With no focusing a 10% conversion of the signal to SH was obtained in a BBO I doubler, while by focussing with a 200 mm lens, continuum was produced in a 3 mm thick plate of fused silica (estimated intensity 200 GW/cm^2). We notice that the spatial beam characteristics are essentially defined during the second pass in the seeder.

b) LBO crystal

LBO is a biaxial crystal, with magnitude of the second-order susceptibility comparable to that of BBO. The samples we tested were cut for non-critical 90 degree p-m, and they were quite long (15 mm), a circumstance this last one which might at first side induced scepticism on their possibility to be used in conjunction with short pulses. Tuning is in this case accomplished by changing the temperature T of the crystal. With a 600 nm pump, temperatures lay in the comfortable range 30-90 C°. Temperature here reported must be regarded only as indicative; the simple arrangement we used was only aimed to guarantee repeatibility rather than giving the exact T values. Degeneracy is at $T \approx 30$ C°, while at $T \approx 90$ C°, λ_s and λ_i are 840 and 2220 nm, respectively.

We could tune the crystal in the range 840-2200 nm. The main advantage of LBO stems from the much lower energy needed by the seeder, and in the smaller overall energy requirements for a stable operation. Some figures are reported in Table 1. Actually we estimated that the seeder could operate with only a few μJ and that a TOPC with an overall demand of 20 μJ pump would be feasible. Beside the reduced tuning range, the draw-back for this LBO configuration is of course the bandwidth, which is huge close to degeneracy. On the contrary, we found pulse duration to be be rather short, with τ_p as low as 150 fs when the average conversion was kept to 5%.

4 -Discussion

We shall now briefly discuss some criteria which led us to the design of the arrangement adopted, and which appears to be confirmed by the results obtained. Two main problems are to obtain spectra narrow enough for a transform limited pulse, and a tunable seed in spite of the competing non-collinear generation, as it occurs with BBO. Quantitative prediction would require a sophisticated numerical modelling, but a qualitative understanding on these questions can come from more handwaving arguments.

Let's first consider the bandwidth. Indications on how a given crystal will perform can be had from the simple expressions of the parametric gain in presence of a narrow pump, according to which the gain-bandwidth $\Delta\nu$ is given by :

$\Delta\nu = \frac{4\sqrt{\ln 2}}{2\pi c} \frac{\sqrt{\Gamma/L}}{|\nu_{si}|}$ (for $\nu_{si} \neq 0$), or by $\Delta\nu = \frac{4}{2\pi c} \left[\frac{\ln 2\, \Gamma/L}{(g_s+g_i)^2}\right]^{1/4}$ when $\nu_{si} \approx 0$, such as occurs at degeneracy. Here $\nu_{si} = 1/u_s - 1/u_i$, u_s and u_i are the group velocities of signal and idler, respectively; while $g = \partial^2 k/\partial\omega^2$. We notice that, up to a certain extend, it is easier to achieve transform-limited pulses as τ_p decreases. To this end consider for example, a transform limited 30 ps pump pulse in the visible generating a signal wave at 1μm: whichever the crystals employed, the above expressions show is impossible to obtain a transform limited output since the wavelength selectivity imposed by the p-m condition during parametric amplification is not sufficient for the $\Delta\nu \approx 1\ cm^{-1}$ required for the purpose. The request relaxes to $20\ cm^{-1}$ when $\tau_p \approx$ 1ps, and indeed we succeeded in obtaining transform limited pulses in these regime.[6] However, in the first stage we had to employ a long KDP crystal, type II p-m, in order to have a narrow bandwidth seed, while amplification - and eventual time shortening - of this seed was provided by the wide-bandwidth second stage employing BBO I. In the first stage the advantages of the higher nonlinearity and the broader tunability of BBO had to be given up due to its insufficient spectral selectivity. As expected, the situation improved by employing in the seeder BBO in type II pm.[8] This last experiment showed that the spectral width was still larger than needed for a transform limited 1 ps pulse, but it clearly indicated that it would be perfect for the 200 fs regime.

The other relevant question in TOPC is to obtain a proper tunable seed, possibly with the minimum cost in pump energy. We note that this last request becomes especially important for applications which do not require much energy in the tunable pulse, say 1 μJ, and for which one would then desire to scale down the pump energy eventually in favour of an higher repetition rate. And while scaling appears feasible in the amplifier, this is not the case with the seeder , where gains of $\approx 10^{15}$ must be provided in any case.We believe that a small spot in the first crystal is the best choice to achieve full tunability and to cut the energy demand. The minimum spot size compatible with the crystal walk-off angle has been suggested for d_1 by previous 1 ps experiments in order to maximize the ratio between the spatial brightness of the seed and the pump energy lost in its generation.[8] Even somewhat smaller spots can be adopted in present fs regime, and to this end one must consider the gains at the set λ_s and at the competing non-collinear generation. We shall denote them G_c and G_{n-c}, respectively. In presence of GVD, one defines the pulse splitting length $l_{g,m}$ among fields g, m -any two of the three involved in parametric interaction- as $l_{g,m} = \tau_p/\nu_{g,m}$. Two things then might hamper exponential gain on the whole crystal length L : longitudinal detachment of pulses due to the small l's, and transverse detachment due to walk-off and/or non-collinear generation. Both act as damping mechanism on the 3-wave interaction, but while the first one is fixed for a given crystal and cut, the second one can be very sensitive on beam size. As it can be seen in Fig.3 , for BBO I the l's are $\approx$ 4 mm, and, with d_1 =300 μm, GVD remains the main damping factor when one consider that the walk-off angle is $\approx$ 3 degrees. This is even more true for BBO II, in view of its smaller $l_{pump\text{-}idler}$ and walk-off angles. On the contrary, the disturbing non-collinear processes, which gain in view of the favourable long l's they encounter in generating in off-axis directions, find in a reduced spot-size the only damping mechanism. As we already mentioned, non-collinear generation is a serious trouble, especially when it comes to deplete the pump. The accurate description of the features of this generation with BBO I given in ref.4 let us infer that the problem was significant also in that case, and actually no indication of direct tunability out of the 830 nm region were reported. As already reported, enlarging d1, while keeping a constant energy for p1, we failed to achieve tunability, while side emission was clearly present.

We notice that chosing a small d1 gave also the side effect of reaching almost far field condition for both pump and seed within a modest propagation distance, an fact that definitely advantages the spatial quality of the beam.

5 -Conclusion

We have shown that travelling wave parametric conversion can be operated in the 100 fs regime, preserving - at least for signal wave we directly tested- not only the pulse duration, but also the spatial and the spectral quality of the pump. All these desirable features can be maintained together with full tunability, which, with BBO II, ranged from 750 nm to degeneracy (at 1200 nm) for the signal wave, while the accompanying idler pulse extended to 3 μm. The demand on pump energy are relatively modest, 100 μJ being sufficient for a stable output and an overall conversion efficiency (including the seeder) of 10-20 % according to the chosen wavelengths, and much lower energy is needed if one relaxes the request of an almost transform-limited signal pulse.

Present results have been obtained with a 0.6 μm pump, but the criteria we have here outlined can as well be applied with other pump sources, in particular with the promising Ti-saffire laser.

References

[1] R.Laenen, H.Graener, and A.Lauberau, Optics Letters 15, 971 (1990)

[2] G.Mak, Q.Fu, and H.M.Van Driel, Applied Phys. Lett., in press

[3] W.S. Pelouch, P.E.Powers, and C.L.Tang, Int. Quant. Electr. Conf., Vienna, June 1992

[4] W. Joosen, H.J.Bakker, l.D.Noordam, H.G.Muller and H.B.van Linden, JOSA B, 8 (1991) 2087

[5] W. Joosen, P.Agostini, G.Petite, J.P.Chambaret, and A. Antonetti, Opt.Lett. 17, 133 (1992)

[6] R. Danielius, A. Piskarskas, D. Podenas, P.Di Trapani, A. Varanavicius, and G.P.Banfi, Opt. Comm. 87, 23 (92)

[7] P.Foggi, R.Righini, R.Torre, L.Angeloni, and S.Califano, J. Chem. Phys. 96, 110 (1992)

[8] R. Danielius et al., Internal Report, Vilnius University 1992

FREQUENCY CONVERSION OF PS-PULSES: TUNABLE PULSES FROM THE UV TO THE NEAR INFRARED

R. Beigang and A. Nebel

Fachbereich Physik
Universität Kaiserslautern, FRG

INTRODUCTION

Titanium doped sapphire (Ti:Al_2O_3) lasers have turned out to be efficient, reliable tunable laser systems in the visible and near infrared spectral region[1,2]. Due to their large gain bandwidth they can produce transform limited mode locked pulses with sub-picosecond pulselength and considerable peak power[3-7]. In conjunction with efficient nonlinear crystals they are well suited for frequency conversion processes which should lead to tunable sub-picosecond pulses in wavelength ranges so far not easily accessible with lasers. The large tuning range of the Ti:Al_2O_3 laser makes such a system in particular very attractive for the generation of tunable sub-picosecond pulses in the **blue, uv and even the vuv** spectral region.

Resonant frequency doubling of cw active and passive mode-locked Ti:Al_2O_3 laser radiation in an external cavity using $LiJO_3$ crystals was reported recently [8,9]. The maximum usable average output power was 125 mW at 381 nm with a repetition rate of 240 MHz and a pulse length of 6 - 15 ps. The use of an external enhancement cavity requires an active length stabilization scheme as the free spectral range of the cavity has to be matched exactly to the repetition rate of the laser. Here we report on efficient external frequency conversion, including second, third and fourth harmonic generation, without an enhancement cavity in three different nonlinear materials. The properties and behaviour of the various crystals are investigated in detail in order to find the optimum configuration for the different nonlinear processes.

Optical parametric oscillators (OPO), on the other hand, are very attractive light sources for the generation of tunable ultrashort light pulses in the **near infrared** spectral region. In particular, cw mode locked systems synchronously pumped by a mode locked pump source can produce tunable picosecond and subpicosecond pulses with high repetition rate. The high repetition rate is of particular importance for high signal-to-noise-ratio time-resolved spectroscopy. The first synchronously pumped OPO was a double resonant $BaNaNbO_5$ OPO around 1 µm pumped by a frequency doubled mode locked Nd:YAG laser. It was characterized by a very low threshold, pulse length around 35 ps, and average powers of 56 mW at a repetition rate of 140 MHz [10]. Owing to the resonance at signal and idler wave simultaneously the emission spectrum consisted of several clusters leading to highly unstable output pulses. To avoid these instabilities singly resonant operation is preferable where only the idler or signal wave is made reso-

Solid State Lasers: New Developments and Applications
Edited by M. Inguscio and R. Wallenstein, Plenum Press, New York, 1993

nant inside the cavity. These systems show a dramatically improved stability compared to double resonant OPOs. However, the threshold will increase by at least one order of magnitude.

The advent of powerful mode locked pump sources like dye lasers, Nd:YAG, Nd:YLF and, in particular, Ti:Sapphire lasers allows for the operation of singly resonant OPOs despite the fact of the high pump power at threshold. The combination of synchronous pumping at a high repetition rate and a singly resonant OPO resonator leads to stable, widely tunable and low-noise ps- and fs-systems for a wide variety of applications.

Femtosecond pulses from a cw mode locked OPO were first obtained by Edelstein et al. in an intracavity pumped $KTiPO_4$ (KTP)-OPO [11, 12, 13]. Inside a colliding pulse mode locked dye laser 105 fs pulses tunable from 0.755 μm to 1.04 μm and 1.5 μm to 3.2 μm with average powers up to 3 mW were achieved.

Recently Mak et al. [14] first synchronously pumped an external cavity OPO with a cw mode locked dye laser at a repetition rate of 76 MHz. With a pump pulse length of 105 fs tunable 220 fs pulses were produced in a wavelength range from 1.20 μm to 1.34 μm by rotating the nonlinear crystal. Maximum average output powers reached 30 mW for the signal wave. Using the same critical phase matching arrangement with a Ti:Sapphire pump laser they could increase the average output power up to 185 mW for 149 fs pulses [15]. With external pulse compression the pulse length was reduced to 62 fs at an average power of 175 mW.

Using a diode laser pumped mode locked and frequency doubled Nd:YLF laser as the pump source an all solid state picosecond OPO was demonstrated in KTP [16, 17]. The OPO was tunable around the degeneracy wavelength from 1002 nm to 1096nm.

Here the demonstration of a ps-OPO with a KTP nonlinear crystal synchronously pumped by a cw mode locked Ti:Sapphire laser will be reported. In contrast to all other cw mode locked OPOs noncritical phase matching was used giving rise to a high conversion efficiency. Tuning was accomplished by tuning the wave length of the pump laser without changing the critical OPO resonator configuration. In this way large tuning ranges can easily be obtained without realigning the OPO resonator.

We will first report results from experiments dealing with the generation of tunable blue and uv ps-pulses and then discuss the performance of noncritically phasematched optical parametric oscillators for the generation of tunable ps-pulses in the near infrared spectral region. As we will show the whole wavelenght range from 190 nm down to 4 μm can be covered with tunable ps-pulses by nonlinear processes in crystals with a high efficiency starting from a mode-locked Ti:sapphire laser.

CRYSTAL CONSIDERATIONS

For efficient frequency conversion the choice of the nonlinear material is of particular importance. Because the transparency range of most nonlinear materials is limited at shorter wavelength only a few crystals are suited for harmonic generation in the UV and VUV. In addition to a high nonlinear coefficient, a relativly large acceptance bandwidth is required in order to sustain the short pulselength in the conversion process. For higher order processes like frequency tripling it is advantageous to have a good beam quality after the first step in order to obtain an optimum overlap and focus in the subsequent steps.

The most efficient nonlinear crystals which provide phase-matched second harmonic generation of $Ti:Al_2O_3$ laser radiation are $LiJO_3$, β-Barium-borat (BBO) and Lithium-triborat (LBO)[18, 19]. $LiJO_3$ has the highest nonlinear coefficient but its transparency range extends down to 300 nm, so that it can be used only for frequency doubling. Due to the large birefringence Dn and the large walk off angle a a poor beam quality is to be expected which is a disadvantage for subsequent mixing processes. The high nonlinear coefficient d_{eff}, on the other hand, should guarantee a high conversion efficiency.

le, however, the nonlinear coefficient is smaller by almost a factor of 2. In LBO the nonlinear coefficient is reduced again by a factor of 1.5. The walk-off, however, is smaller by a factor of 4. The acceptance bandwidth is given by $\Delta\nu \approx 1/\Delta n\, l$, so that for a 10 mm long crystal for all three materials a sufficient acceptance bandwidth is guaranteed for transform limited pulses with pulse lengths of appproximately $\Delta t \approx 1.5$ ps.

SECOND HARMONIC GENERATION

The $Ti{:}Al_2O_3$ laser used in all experiments reported here is a commercial laser system (Spectra Physics, "Tsunami") with a maximum average power (repetition rate 82 MHz) of 1.75 W at 790 nm, a tuning range from 720 nm to 850 nm and a pulse length between1.35 ps and 1.5 ps. Assuming a $sech^2$ pulse shape the time-bandwidth product of $\Delta\nu\,\Delta t \approx 0.32$ indicates transform limited pulses. For the second harmonic generation the fundamental radiation of the $Ti{:}Al_2O_3$ laser is focussed into the nonlinear crystal and recollimated with an r = 150 mm mirror. The second harmonic radiation is separated from the fundamental with a quartz prism and monitored with a power meter. The focal length of the focussing lens was varied to adjust the confocal parameter of the focussed beam to the crystal length and acceptance angle.
For a 5x5x10 mm^3 long 43° cut $LiJO_3$ crystal and a focal length of 80 mm a maximum average output power at the second harmonic of $P_{SH}(LiJO_3) = 700$ mW was measured at the power meter. Accounting for the Fresnel losses at the uncoated crystal surface a conversion efficiency of nearly 50 % was obtained. In the case of BBO a 5x5x8 mm^3 long 30° cut crystal and a focal length of 60 mm produced an output power of $P_{SH}(BBO) = 450$ mW which corresponds to a conversion efficiency of 27 %. With the same focal length and a 6x6x8 mm^3 long $\Phi=32^o$ ($\Theta=90^o$)cut LBO a maximum output power of $P_{SH}(LBO) = 350$ mW was generated. In all cases the average fundamental power was $P_o = 1.75$ W at 790 nm. The complete tuning curves for all three crystals are shown in Fig. 1. The pulse length of the second harmonic pulses was determined with an autocorrelation measurement of the the second harmonic. A typical autocorrelation trace is shown in Fig. 2. Deconvolution of the shape and assuming again a $sech^2$ pulse shape results in a pulse length of 1.1 ps which corresponds to a reduction by a factor of 1.4 which is slightly smaller than the theoretically expected reduction of 1.55 for a $sech^2$ pulse. Together

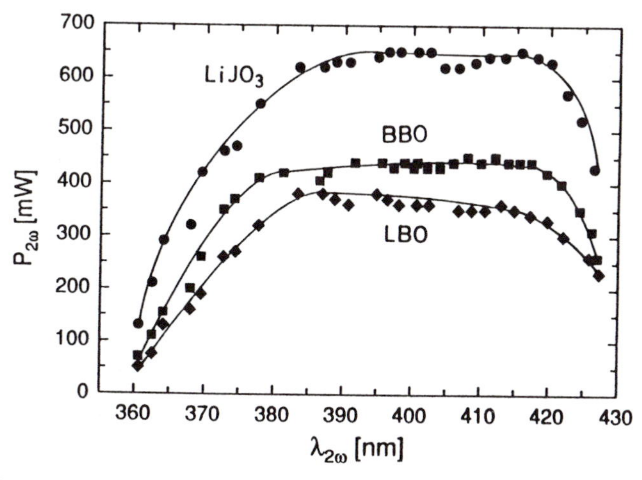

Figure 1. Second harmonic tuning curves for $LiJO_3$, LBO and BBO crystals

with the measured spectrum a time-bandwidth product of $\Delta\nu\,\Delta t = 0.33$ was obtained. The pulse width was also determined with a crosscorrelation measurement between the fundamental and the second harmonic giving the same results. As expected from the crystal data the beam quality of the second harmonic radiation generated with LBO was far superiour compared to BBO and $LiJO_3$.

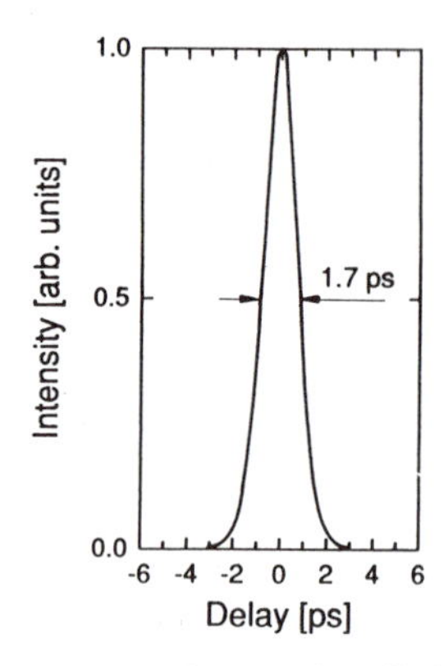

Figure 2. Autocorrelation trace of the second harmonic rafiation at a wavelength of 400 nm.

THIRD HARMONIC GENERATION

For the sum frequency generation of the fundamental and second harmonic radiation the difference in group velocity (group velocity mismatch GVM) between the fundamental and second harmonic has to be taken into account. After the doubling crystal the pulses at the second harmonic are delayed with respect to the pulses at the fundamental. Results of calculations of the group velocity dispersion for the crystals used in these experiments show that after the doubling process there is only a small temporal overlap between fundamental and second harmonic pulses for a pulse lenght of 1.5 ps.

The experimental set up for frequency tripling is shown in Fig.3. For all tripling experiments the second harmonic was generated using LBO in order to obtain a good beam profil. After separation of the fundamental and second harmonic radiation with a dichroic mirror the time delay is compensated in a variable delay line for the fundamental radiation. The beams are combined before the mixing crystal with a second dichroic mirror, focussed into the crystal with a f= 50 mm mirror (spot size approximately 15 µm) and recollimated with an uncoated f = 60 mm quartz lens. A quartz prism is again used for separation of fundamental, second and third harmonic. LBO allows for phase matched frequency tripling down to 785 nm . With a $\Phi=70^o$ ($\Theta=90^o$) cut, 7 mm long LBO a maximum average power of 35 mW at 275 nm was obtained. Fig. 4 displays the tuning curve extending from 266 nm to 283 nm. The short wavelength cut off is caused by the decreasing nonlinear coefficient d_{eff} when reaching the phase match angel of $\Phi = 90^o$. Dividing the third harmonic power $P_{3\omega}$ by $P_{2\omega}P_{\omega}$ gives the dependence of d_{eff} on the wavelength. The measured dependence is in good agreement with theoretical calculations of d_{eff}.

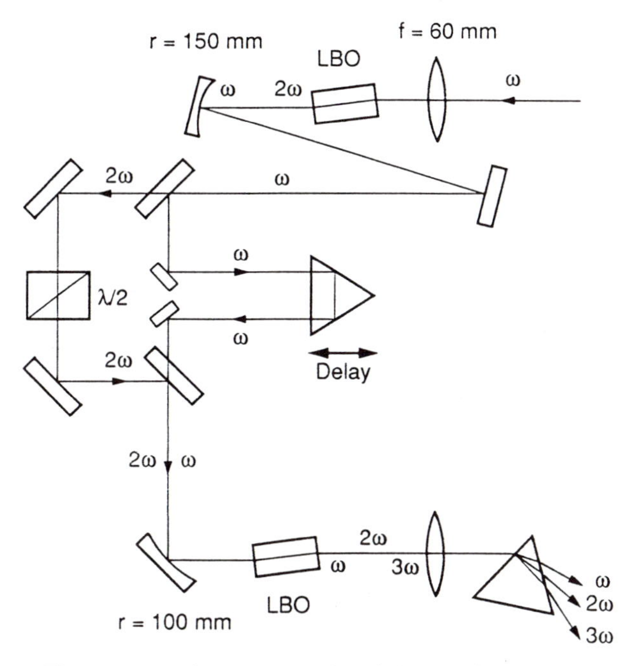

Figure 3. Experimental set-up for third harmonic generation.

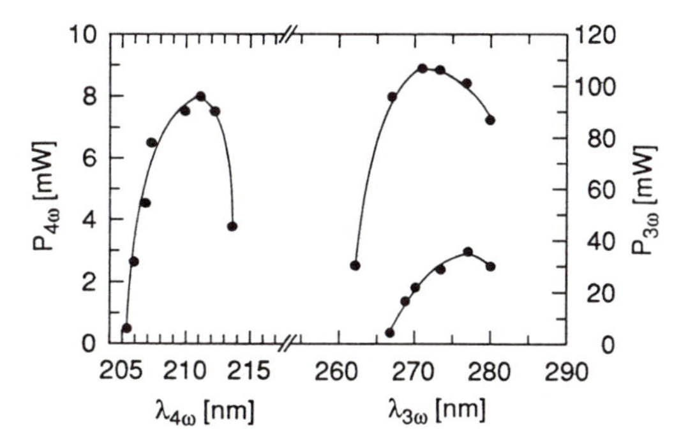

Figure 4. Third and fourth harmonic tuning curves for LBO and BBO crystals

Due to the higher nonlinear coefficient of BBO a much higher uv power was obtained with a 50^ocut 12 mm long BBO crystal. The maximum average power was 120 mW at 272 nm. The tuning range is also shown in Fig. 4. The short wavelength cut off at 262 nm was caused by the limited reflectivity range of the optics used in the optical delay line (see Fig. 3). The higher walk off angle compared to LBO resulted in an elliptical beam profil of the uv radiation.

FOURTH HARMONIC GENERATION

The shortest wavelength which can be generated by frequency tripling of the $Ti:Al_2O_3$ radiation is 240 nm. Shorter wavelenghts are generated by the fourth harmonic using a 75° cut 8 mm long BBO crystal. BBO allows for phase matched second harmonic generation down to 410 nm, so that uv radiation from 205 nm to 213 nm can be obtained with the mirror set used in the $Ti:Al_2O_3$ laser. Utilizing the whole possible tuning range uv radiation up to approximately 260 nm can be generated.

In our experimental set up the second harmonic radiation produced either with a LBO, BBO or $LiJO_3$ crystal is focussed directly into the BBO to a spot size of about 13 μm with a f = 50 mm dichroic mirror and recollimated with a f = 60 mm uncoated quartz lens. The fourth harmonic is separated from the second harmonic and the fundamental with a quartz prism. A maximum average power of 10 mW was generated with LBO as doubling crystal. Although the average po wer at the second harmonic was higher for BBO and $LiJO_3$, the power at the fourth harmonic did not reach the same level compared to LBO due to the poor beam profile. A typical tuning curve is shown in Fig. 4. Assuming the same reduction factor for the pulse length as in the case of the second harmic generation a pulse length at the fourth harmonic well below one picosecond can be expected.

In order to generate tunable pulses below 205 nm sum frequency mixing between the fundamental ω and the third harmonic 3ω can be applied in BBO. This is described in more detail in reference 22.

OPTICAL PARAMETRIC OSCILLATOR

A schematic diagram of the OPO resonator is shown in Fig. 5. The KTP crystal (3x3x6 mm^3) is placed at the waist of a 4 mirror ring cavity. The curvature of the two spherical mirrors was varied from 75 mm to 150 mm. The output coupler transmission was changed from 0.5% to 15%. The pump beam was focussed collinearly through one of the spherical mirrors with lenses of different focal length (75 mm < f < 200mm). Optimum performance of our system was obtained with spherical mirrors of 100 mm radius, a focal length of 100 mm for the pump beam focus and a transmission of 7.5% of the output coupler. The cavity length was matched to the length of the pump laser as required for synchronous pumping. The geometrical length of the pump pulses is in the order of 300 μm. As there will be no parametric generation if the round trip time of the OPO pulses does not coincide with the repetition rate of the pump pulses the length has to be matched with an acccuracy of at least the pump pulse separation divided by the maximum length mismatch. With a repetition rate of 82 MHz and a maximum mismatch of 14 μm an accuracy of 2.6 10^5 is required to see parametric oscillation. Stable operation with constant pulse length and small peak-to-peak amplitude fluctuations (< 10%) was obtained within a resonator length intervall of approximately 5 μm.

This behavior is in contrast to synchronously pumped lasers. Owing to the lifetime in the upper laser level the gain medium provides gain also for a finite cavity mismatch between pump and laser pulses resulting in a distorted pulse shape. In the case of an OPO there will be no oscillation outside a small length intervall given, in principle, by the pump pulse length. For precision adjustment of the cavity length the plane high reflector was mounted on an piezoelectric transdu-

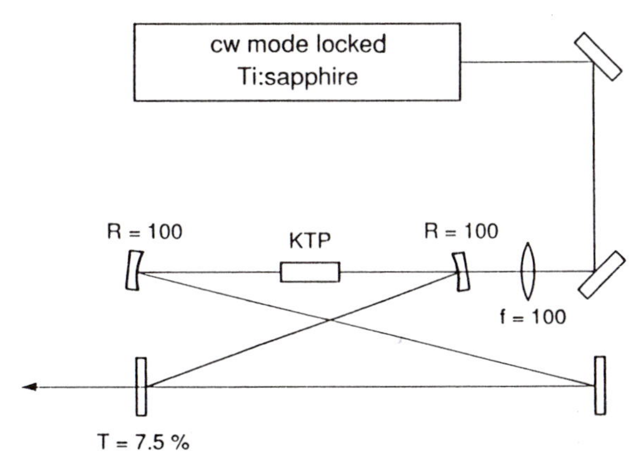

Figure 5. Resonator set-up for the optical pareametric oscillator.

cer and a precision translation stage. Even with such stringent requirements for the accuracy of the resonator length no active stabilisation of the OPO resonator length was necessary to guarantee stable OPO operation.

The KTP crystal was cut for type II phase matching in the xy-plane (e -- e + o). In this plane and for the pump wave length used noncritical phase matching becomes possible ($\phi = 0^o$, $\theta = 90^o$) with a maximum figure of merit compared to all other phase matching conditions. There is no walk off between the three interacting beams inside the crystal, namely, pump, signal and idler wave. Therefore the maximum length of the crystal is mainly limited by the difference in group velocity (GVM) of the three wavelengths. Calculations of the difference in group velocity show that crystal lengths larger than 5 mm are not reasonable for ps pulses as there will be no temporal overlap of all three interacting pulses inside the crystal after several round trips in the resonator. The crystal was antireflection coated on both sides for a center wavelength of 1150 nm. The residual losses from the crystal surfaces were less than approximately 0.5% for both sides. The extraordinary (e) polarized signal wave was resonated.

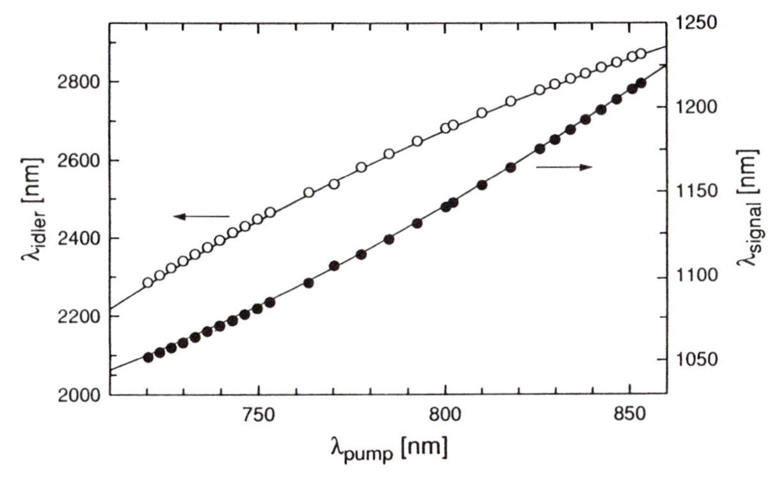

Figure 6. Tuning curve of the OPO as a function of pump wavelength.

The tuning curve of the OPO as a function of pump wavelength is shown in Fig. 6. With the Ti:Sapphire laser used in these experiments a tuning range from 1.052 μm to 1.214 μm and from 2.286 μm to 2.871 μm was achieved for the signal and idler wave, respectively, limited only by the mirror set of the Ti:Sapphire pump laser. The length of the OPO resonator was optimized at the center of the tuning curve and had to be readjusted slightly only at the ends of the tuning range caused by the change of the index of refraction. The experimental values (circles in Fig. 6) show excellent agreement with the calculated tuning curve using a Sellmeier equation and constants as given by Bierlein et al. [20]. Utilizing the whole tuning range of a Ti:Sapphire laser from 700 nm to 1100nm the OPO can, in principle, be tuned from 1.03 μm to 1.6 4 μm for the signal wave and from 2.16 μm to 3.34 μm for the idler wave.

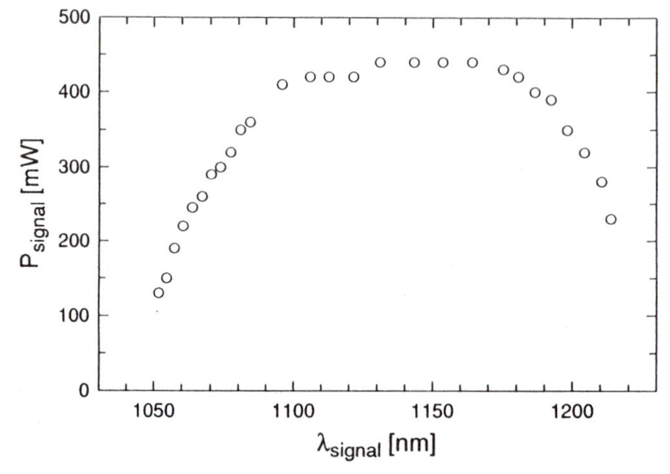

Figure 7. Signal wave tuning curve

A typical tuning curve of the KTP OPO for the signal wave as a function of signal wave length is shown in Fig. 7. Average output powers of up to 440 mW were achieved over a wide tuning range. The maximum average output power at 1140 nm was 510 mW, corresponding to a peak power of 5.2 kW or 6.2 nJ per pulse. The tuning curve of the OPO follows the power dependence of the pump laser. The power efficiency as defined by $\eta = (P_S + P_i)/P_P$ reached 38% around 1.1 μm with the optimized output coupler of 7.5 %. Pump power at threshold was as low as 390 mW resulting in a slope efficiency of 44%. The threshold pump intensity reached 178 MW/cm^2. The power dependence was linear and no saturation effects were observed.

The pulse width was determined with an intensity autocorrelation measurement. Assuming a sech^2 pulse shape the pulse length was 1.15 ps at the center of the tuning range. The spectral bandwidth of 303 GHz as measured with a monochromator resulted in a time-bandwidth product $\Delta\nu\ \Delta t = 0.35$ clearly indicating transform limited pulses. An interferometric autocorrelation measurement supported this assumption. At the very wings of the tuning curve the time-bandwidth product was increased to about 0.53. The pulses were very stable with peak-to-peak amplitude fluctuations in the order of 10% without an active length stabilization of the OPO resonator.

It should be pointed out that the OPO can also be used to generate tunable pulses in the visible spectral range by frequency doubling in $LiJO_3$ e. g. Preliminary experiments indicate that average output powers up to 50 mW can easily be obtained in the wavelength range from 525 nm to 605 nm. Thus the OPO greatly enhances the wavelength range where tunable ps pulses can be generated. Together with second, third and fourth harmonic generation of mode locked

Ti:Sapphire radiation in LBO and BBO crystals [21, 22] almost the complete wavelength range from 190 nm down to 4.5 µm can be covered with tunable ultrashort pulses starting from a Ti:Sapphire master oscillator.

SUMMARY

In summary we have demonstrated efficient external frequency conversion of cw mode locked Ti:Al_2O_3 laser radiation using LBO, BBO and $LiJO_3$ crystals. The generation of the second, third and fourth harmonic radiation leads to tunable picosecond pulses in the uv and blue spectral range. Utilizing the whole tuning range of Ti:Al_2O_3 lasers the spectral region from 205 nm to 525 nm can be covered. Maximum average output powers at 410nm, 272 nm and 210 nm of 700 mW, 120 mW and 10 mW have been obtained, respectivly. Auto- and cross-correlation measurements clearly demonstrated transform limited pulses with a pulse lengths around one picosecond for the second harmonic and sub-picoseconds for the third and fourth harmonic.

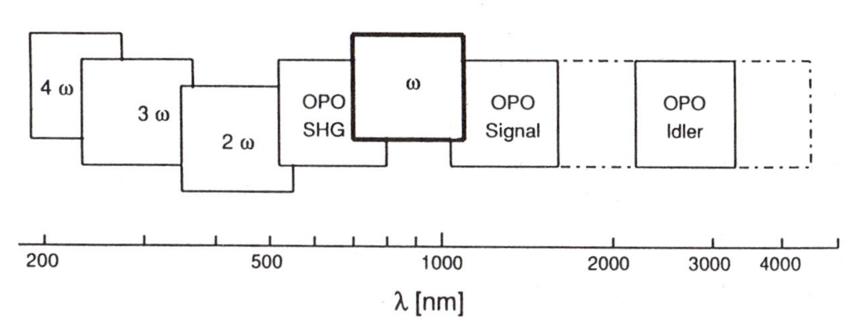

Figure 8. Total tuning range of ps-pulses generated from a cw mode-locked Ti:sapphire laser by frequency conversion in nonlinear crystals and optical parametric oscillation.

Efficient operation of a noncritically phase matched cw mode locked OPO for the generation of tunable ps and fs pulses in the near infrared spectral region has been demonstrated in KTP. The noncritical type II phase matching arrangement caused the highest figure of merit for this crystal, no walk off between the three interacting beams and, as a consequence, a high power conversion efficiency of upt o 38%. The OPO was pumped with a cw mode locked Ti:Sapphire laser (repetition rate 82 MHz). Fourier limited tunable 1.2 ps pulses with average powers of up to 510 mW for the signal wave have been obtained. Tuning was easily accomplished by tuning the pump laser without changing the OPO resonator alignment. The observed tuning range from 1.052 µm to 1.214 µm and from 2.286 µm to 2.871 µm was limited by the tuning range of the pump laser and can, in principle, be extended from 1.03 µm to 1.64 µm and from 2.16 µm to 3.34 µm for signal and idler wave, respectively.

ACKNOWLEDGEMENTS

This work was supported by the Deutsche Forschungsgemeinschaft, DFG. Helpful discussions with R. Wallenstein and his continuous support are thankfully acknowledged. The Ti:Al_2O_3 laser used in these experiments was a loan from the Spectra Physics GmbH, Germany.

REFERENCES

1. P. F. Moulten, J. Opt. Soc. Am. **B3**, 125 (1986)
2. P. Albers, E. Stark, and G. Huber, J. Opt. Soc. Am. **B3**, 134 (1986)
3. P. M. W. French, J. A. R. Williams, and J. R. Taylor, Opt. Lett. **14**, 686 (1989)
4. N. Sarukura, Y. Ishida, H. Nakano, Y. Yamamoto, Appl. Phys. Lett. **56**, 814 (1990)
5. P. M. W. French, S. M. J. Kelly, and J. R. Taylor, Opt. Lett. **15**, 378 (1990)
6. J. Goodberlet, J. Jacobsen, J. G. Fujimoto, P. A. Schulz, and T. Y. Fan, Opt. Lett. **15**, 504 (1990)
7. D. E. Spence, P. N. Kean, and W. Sibbett, Opt. Lett. **16**, 42 (1991)
8. P. F. Curley and A. I. Ferguson, Opt. Lett. **16**, 321 (1991)
9. P. F. Curley and A. I. Ferguson, Opt. Commun. **80**, 365 (1991)
10. A. Piskarskas, V. Smil´gyavichyus, and R. Umbrasas, Sov. J. Quantum Electron. **18**, 155 (1988)
11 D. C. Edelstein, E. S. Wachmann, and C. L. Tang, Appl. Phys. Lett. **54**, 1728 (1988)
12. E. S. Wachmann, D. C. Edelstein, and C. L. Tang, J. Appl. Phys. **70**, 1893 (1991)
13. E. S. Wachmann, D. C. Edelstein, and C. L. Tang, Opt. Lett. **15**, 136 (1990)
14. Q. Fu, G. Mak, and H. M. van Driel, Opt. Lett. **17**, 1006 (1992)
15. G. Mak, Q. Fu, and H. M. van Driel, Appl. Phys. Lett. **60**, 542 (1992)
16. M. Ebrahimzadeh, G. P. A. Malcolm, and A. I. Ferguson, Opt. Lett. **17**, 183 (1992)
17. M. J. McCarthy and D. C. Hanna, Opt. Lett. **17**, 402 (1992)
18. C. Chen, Y. Wu, A. Jiang, B. Wu, G. You, R. Li, S. Lin, J. Opt. Soc. Am. **B6**, 616 (1989)
19. A. Borsutzki, R. Brünger, Ch. Huang, and R. Wallenstein, Appl. Phys. **B 52**, 55 (1991)
20. J. D. Bierlein and H. Vanherzeele, JOSA **B6**, 622 (1988)
21. A. Nebel and R. Beigang, Opt. Lett. **16**, 1729 (1991)
22. A. Nebel and R. Beigang, Opt. Commun. **94**, 369 (1992)

QUASI-THREE-LEVEL LASERS

T. Y. Fan

Lincoln Laboratory, Massachusetts Institute of Technology
Lexington, Massachusetts 02173
U. S. A.

INTRODUCTION

One of the key developments in the renaissance in solid-state laser technology has been the rapid improvement of diode laser pump sources.[1-3] Some of the advantages of diode-pumping relative to lamp-pumping of solid-state lasers include higher overall efficiency, reduced thermal loading of the gain medium, higher reliability, and reduced size. In addition, there are other important differences between these two types of pump sources. For example, diode lasers have higher spectral and spatial brightness, in other words, the diode laser output is narrowband spectrally, and it is directional. The consequence is that much higher volumetric pumping density can be achieved using diode-laser pumps even with low power single-stripe diode lasers relative to high-power lamps. The high pump densities has led to the demonstration of good laser performance at room temperature of several transitions that performed only poorly or not at all at room temperature under lamp pumping. These transitions in rare-earth ions include the $^4F_{3/2}$ - $^4I_{9/2}$ near 0.94 μm in Nd^{3+},[4-6] the $^5I_7$ - $^5I_8$ near 2.1 μm in Ho^{3+},[7-10] the $^3F_4$ -$^3H_6$ near 2.0 μm in Tm^{3+},[11-13] the $^4I_{13/2}$ - $^4I_{15/2}$ near 1.5 μm in Er^{3+},[13,14] and the $^2F_{5/2}$ - $^2F_{7/2}$ near 1.0 μm in Yb^{3+} (refs. 15, 16) as shown in Fig. 1. The common element in these laser transitions is that the lower laser levels are in the ground-state multiplet which means that the lower laser levels are only a few hundred cm^{-1} above the ground state. Thus the lower levels have significant population in thermal equilibrium at room temperature since kT is 207 cm^{-1} at 300 K. This is in contrast to the common four-level Nd^{3+} transition near 1.06 μm which has a lower level about 10kT above the ground-state at 300 K thus can be considered unpopulated in thermal equilibruim, or ruby in which the lower laser level is the ground-state. With lamp-pumping, these lasers with lower levels in the ground-state manifold were typically operated at cryogenic temperatures to reduce the lower-level population in thermal equilibruim; at sufficiently low temperature these lasers become four-level lasers and efficient, low threshold performance was obtained. The

Solid State Lasers: New Developments and Applications
Edited by M. Inguscio and R. Wallenstein, Plenum Press, New York, 1993

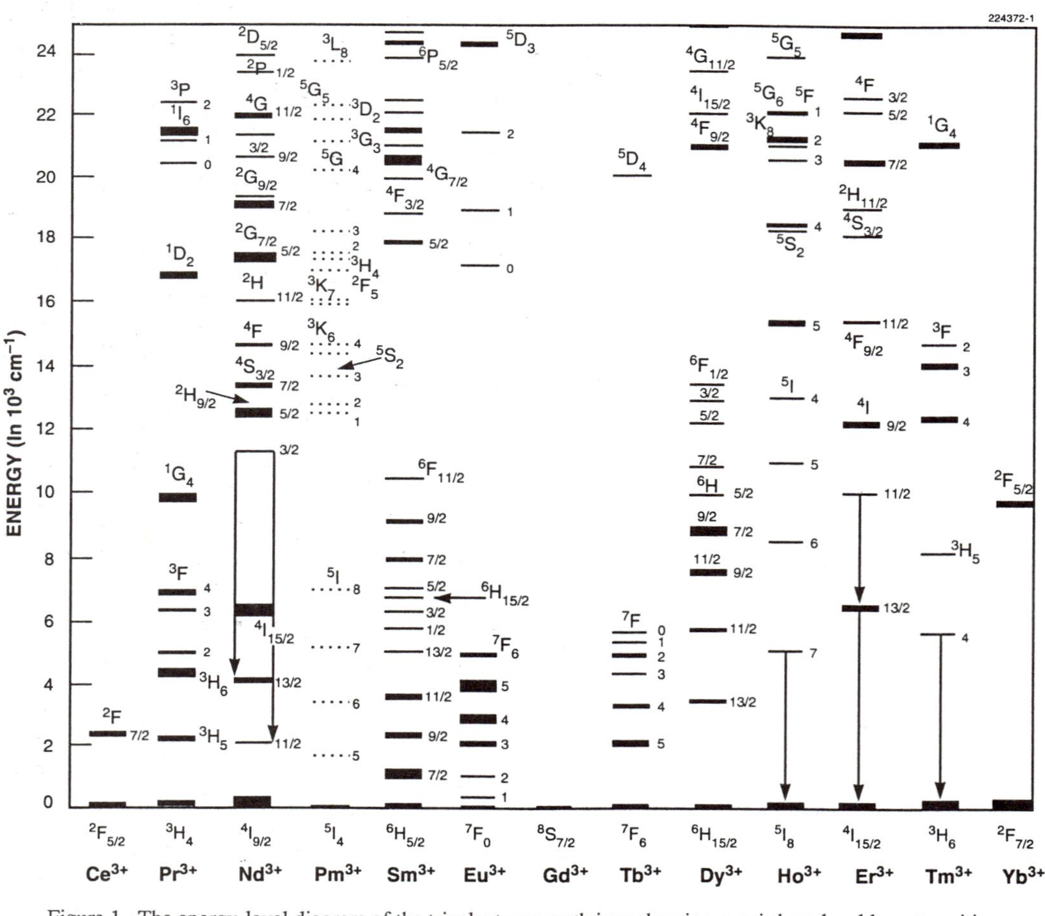

Figure 1. The energy-level diagram of the trivalent rare earth ions showing quasi-three level laser transitions. From ref. 1.

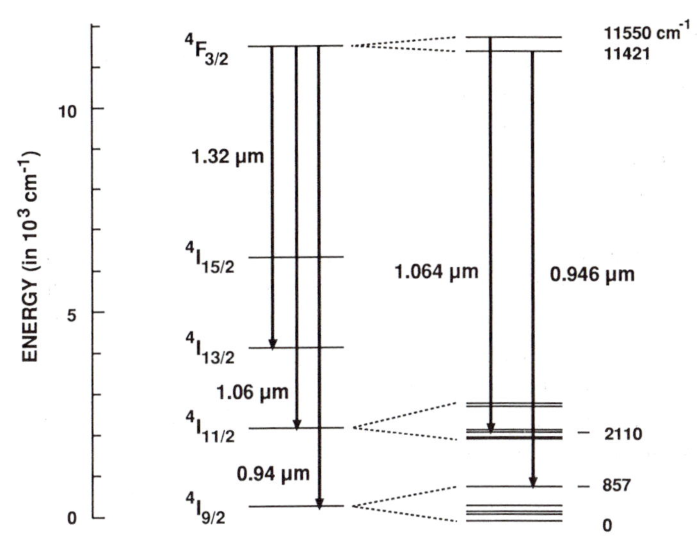

Figure 2. Energy-level diagram for Nd:YAG showing the manifold positions and the crystal-field splittings.

typical idealized three- and four-level laser models are adequate, for the most part, in describing the behavior of 1.06-µm Nd^{3+} and ruby lasers; however, they are inadequate for describing the performance of the other laser transitions mentioned above which are called quasi-three-level lasers.

Here, we show how quasi-three-level lasers are modeled and show that they can be efficient. First, the energy level structure and relaxation rates of rare-earth ions will be introduced so that the assumptions used in quasi-three-level laser models can be understood. This will be followed by models of three-level, four-level, and quasi-three-level lasers. Finally, an example of a quasi-three-level laser, Yb:YAG, will be given.

ENERGY LEVELS OF RE^{3+} IONS AND RELAXATION RATES

Energy Level Diagrams

It is useful to understand the energy-level diagrams and relaxation mechanisms in rare-earth ions since this is the most common type of laser center in solid-state lasers. Rare-earth atoms have outer electron shells with a configuration of $4f^N6s^2$. When rare-earths are doped into solids, the two 6s electrons and one of the 4f electrons go into bonding. Figure 1 shows the energy-level diagram of 4f shell of the trivalent rare earth ions.[17] The positions of the energy levels are determined by three factors: the Coulomb interaction between 4f electrons, spin-orbit coupling, and the crystal field of the host material. The Coulomb interaction splits the 4f configuration into terms that are separated typically by 10,000 cm^{-1} while the spin-orbit interaction splits the terms into manifolds that are separated by typically 3,000 cm^{-1}. The manifold positions in Figure 1 are determined by the Coulomb interaction and spin-orbit coupling; the crystal-field interaction is not included. The interaction with the crystal field is the weakest effect and only splits the energy levels within a manifold by typically 200 cm^{-1}. The small splitting is a consequence of the fact that the 4f electrons are well shielded from external interactions because their radial extent is small compared with the other electron shells.[18] One implication of the weak crystal field is that the positions of the manifolds only vary slightly as rare-earth ions are doped from host to host since neither the Coulomb interaction or spin-orbit coupling are changed by the crystal field. The other implication is that the 4f energy levels can be treated as if the rare-earth ions are isolated impurities; even at relatively high concentrations of rare earth ions, 1 x 10^{22} cm^{-3}, the 4f energy levels do not form a band structure since effectively, the 4f-shell electrons do not interact with the nearest neighbor rare-earth ion. This means that the spectra of doped rare-earth ions appear more like line spectra than band spectra. As an example, Figure 2 shows the energy-level diagram of Nd:YAG. The ^4I term is split from the ^4F term by the Coulomb interaction by ~10,000 cm^{-1}. The ^4I term is split into four manifolds by spin-orbit coupling, and the manifolds are separated by ~2,000 cm^{-1}. Finally, the crystal field splits a manifold into levels separated by ~100-200 cm^{-1}. The energy-level diagram is more complicated than that assumed for the ideal three or four-level laser model, and it would be undesirable to need to have separate rate equations to describe the populations of each of these many levels. However, there are simplifying assumptions that can be made by comparing typical radiative, nonradiative, and stimulated emission rates.

Radiative and Nonradiative Relaxation Rates

Now, let us address the lifetimes of these states. The oscillator strengths of transitions between these 4f energy levels is relatively low, typically 10^{-4} - 10^{-6}; highly allowed electric-dipole transitions have oscillator strengths near 1. The low oscillator strength in intra-4f-shell transitions of rare-earth ions is a consequence of the electric-dipole selection rules. Energy levels within an electron shell have the same parity, thus intrashell electric-dipole transitions are not allowed; an electric dipole transition is not allowed if the initial and final states have the same parity, although magnetic-dipole transitions are allowed, and these have oscillator strengths ~10^{-7}. However, if the rare-earth ion is doped into a site that is not centrosymmetric then parity is broken and electric-dipole transitions become allowed. But as already noted, the effect of the crystal field is small thus the electric-dipole transition is only weakly allowed. The radiative lifetime of the 4f manifolds is typically on the order of 50 μs - 5 ms while a highly allowed transition would have a lifetime ~10 ns.

The nonradiative relaxation rate from a manifold for an isolated rare-earth ion in a host is given by the energy-gap law. This law states that

$$\frac{1}{\tau_{\mathrm{NR}}} = B\exp(-\alpha\Delta E) \tag{1}$$

where τ_{NR} is the nonradiative lifetime, B and α are constants that are material dependent and ΔE is the energy between the manifold of interest and the next lower lying manifold. In YAG B is approximately 1 x 10^8 s^{-1} and α is about 3.1 x 10^{-3} cm (ref. 19) thus the nonradiative lifetime for the $^4F_{3/2}$ manifold in Nd:YAG is on the order of 5 ms based on an energy gap of 5,000 cm^{-1} to the next lower lying manifold $^4I_{15/2}$. This compares with the radiative lifetime of the $^4F_{3/2}$ manifold of 240 μs. The variation of the nonradiative relaxation rate with host material is mostly dependent on the effective phonon energy of the host lattice; for host lattices with high energy phonons the nonradiative relaxation rate for a given energy gap is higher than for a host with low energy phonons. The effective phonon energy is approximately equal to the highest phonon energy in the lattice. In YAG this effective phonon energy is ~700 cm^{-1}. As a general rule, oxide host materials tend to have greater nonradiative relaxation rates than fluorides.

The energy-gap law is not valid for all values of ΔE. For ΔE less than twice the effective phonon energy, the energy-gap law breaks down, and the nonradiative lifetime increases faster with decreasing ΔE than the energy gap law predicts.[20] At the current time, there is no equivalent simple law to the energy gap law for small ΔE; however it has been estimated that the relaxation times for energy level spacings within a manifold are in the picosecond range.[21] This relaxation rate is much faster than any other rates that are typical in solid-state lasers. This rate is many orders of magnitude greater than the radiative relaxation rate and a few orders of magnitude greater than typical intermanifold nonradiative relaxation rates. It is also much greater than typical stimulated emission rates in Q-switched and cw solid-state lasers. The great difference between the intramanifold relaxation rate and other important rates leads to the quasi-thermal equilibrium approximation. This approximation states that the relative populations of crystal field splittings within a manifold can be treated as if the levels are in thermal equilibrium with each other at all times.

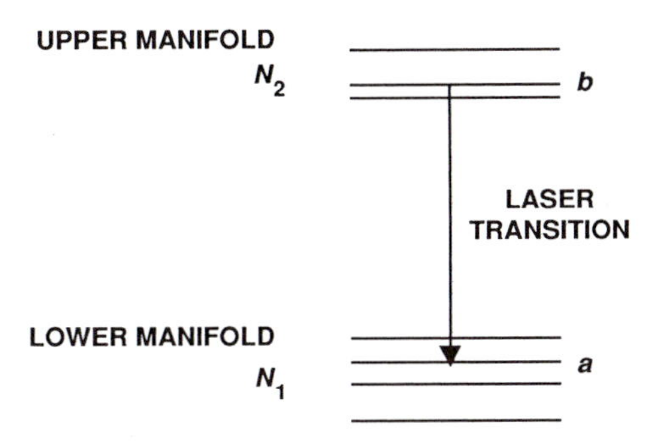

Figure 3. Idealized upper and lower manifolds for a quasi-three-level laser. The upper-laser level is denoted by b and the lower by a. The population densities of the two manifolds are N_2 and N_1 for the upper and lower manifolds respectively.

Quasi-Thermal Equilibrium and Laser Gain

Consider the energy level diagram in Fig. 3 in which there is an upper manifold and a lower manifold that are each split by the crystal field. Quasi-thermal equilibrium implies that the ratios of the populations in each crystal field splitting within a manifold is constant for all times. The fraction of the population withing the level a of the lower manifold is given by

$$f_a = \frac{\exp(-E_{\ell a} / kT)}{Z_\ell} \tag{2a}$$

where

$$Z_\ell = \sum_{i=1}^{m} \exp(-E_{\ell i} / kT) \tag{2b}$$

Table 1. Boltzmann occupation factors for quasi-three-level transitions in YAG at 300 K

Ion	Transition	Wavelength (μm)	f
Nd^{3+}	$^4F_{3/2}$ - $^4I_{9/2}$	0.946	0.123
Ho^{3+}	$^5I_7$ - $^5I_8$	2.097	0.17
Yb^{3+}	$^2F_{5/2}$ - $^2F_{7/2}$	1.03	0.066

f_a is the fractional population within the level a, $E_{\ell a}$ is the energy of the level a, kT is the thermal energy, and Z_ℓ is the partition function of the lower manifold. $E_{\ell i}$ are the energies of the levels in the lower manifold which has of m levels. The fractional population in level b of the upper manifold f_b can be similarly found by calculating the partition function over the levels of the upper manifold. Degeneracy has been ignored in this calculation because for many cases in rare-earth ions, the degeneracies of the levels are the same.

The condition for laser gain is that there be a population inversion, i. e., that the population density in the upper level be greater than that for the lower level. Thus for laser gain on the transition between levels b and a

$$\Delta N = f_b N_2 - f_a N_1 > 0 \tag{3a}$$

which implies

$$f_b N_2 > f_a N_1 \tag{3b}$$

where N_2 and N_1 are the total population densities of the upper and lower manifolds, respectively. If we assume that

$$N_1 + N_2 = N_t \tag{4}$$

where N_t is the total dopant concentration, then the fraction of the population that needs to be in the upper manifold to attain population inversion β_{min} is given by

$$\beta_{min} = \frac{f}{1+f} \tag{5}$$

where f is equal to f_a/f_b and $\beta = N_2/N_t$. Table 1 lists some values of f for quasi-three-level laser transitions in YAG at 300 K. Clearly, for these transitions only a relatively small fraction of the population needs to be in the upper manifold in order to attain population inversion as opposed to a three-level laser like ruby in which a large fraction of the population must be excited to the upper-laser level for inversion.

THREE-LEVEL, FOUR-LEVEL, AND QUASI-THREE-LEVEL LASER MODELS

Laser Rate Equations

Lasers are often described by idealized three and four-level laser models. Here, we constrast these models with a quasi-three-level laser model and show why quasi-three-level lasers have become much more important with the advent of diode-laser pumping.

The energy level diagrams and transition rates for three-level, four-level, and quasi-three-level laser models are shown in Fig. 4. The key difference between these idealized three and four-level laser models and the quasi-three-level models is that the crystal field splittings are taken into account in the latter case. For these three models, the rate equations describing the change in population inversion density with time can be written as

$$\frac{\mathrm{d}\Delta N}{\mathrm{d}t} = 2W_{13}N_1 - \frac{N_t + \Delta N}{\tau_2} - 4W_{21}\Delta N \qquad \text{(three level)}, \tag{6a}$$

$$\frac{\mathrm{d}\Delta N}{\mathrm{d}t} = W_{14}N_1 - \frac{\Delta N}{\tau_3} - 2W_{31}\Delta N \qquad \text{(four level)}, \tag{6b}$$

and

$$\frac{\mathrm{d}\Delta N}{\mathrm{d}t} = f_b'(f_b + f_a)W_{14}N_1 - \frac{f_a N_t + \Delta N}{\tau} - 2(f_b + f_a)W_{31}\Delta N \qquad \text{(quasi-three-level)} \tag{6c}$$

under the assumption of spatially uniform pumping and extraction. Here, ΔN is defined as

$$\Delta N = \begin{cases} N_2 - N_1 & \text{(three level)} \\ N_3 - N_2 = N_3 & \text{(four level)} \\ f_b N_2 - f_a N_1 & \text{(quasi-three level).} \end{cases} \tag{7}$$

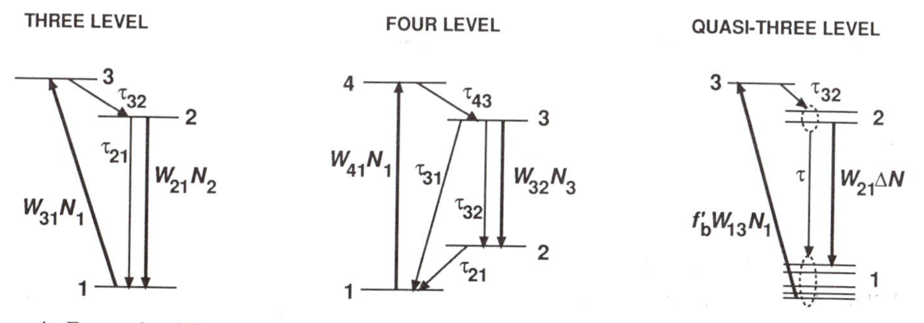

Figure 4. Energy-level diagrams for idealized laser models. From left to right: three-level laser, four-level laser, and quasi-three-level laser.

N_1, N_2, and N_3 are population densities in manifolds 1, 2, and 3, and N_t is the total dopant density. The rate constants W_{xy} are given by

$$W_{xy} = \frac{\sigma I_{xy}}{h\nu_{xy}} \quad . \tag{8}$$

The subscript xy denotes a transition between manifold x and manifold y, σ is the spectroscopic cross section of the transition, and I_{xy} is the intensity of the light at photon energy $h\nu_{xy}$. f_b' is the Boltzmann occupation factor for the initial state of the pumping transition in the ground-state manifold. The spontaneous emission rates $1/\tau_x$ are the total spontaneous relaxation rates out of the upper manifold. It has been assumed that relaxation from the pump level to the upper laser level is infinitely fast.

The first term of the right-hand sides of eq. (6) represents pumping, the second term is spontaneous emission, and the third term is stimulated emission. In eq. (6) the factors of 2 in the stimulated emission terms arise because the laser is assumed to have a standing-wave cavity, thus the population sees the circulating intensity twice per round-trip. There is another factor of two in the pumping term and stimulated emission terms describing the three-level laser because a single absorbed pump photon or a single stimulated photon causes the population inversion to change by two. It is possible to see by inspection of eq. (6) that a three-level laser is intrinsically less efficient than a four-level laser. The stimulated-emission rate is given in all cases by $2W_{xy}\Delta N$; the spontaneous-emission rate in the four-level laser case is proportional to ΔN while in the three-level laser case is proportional to $N_t + \Delta N$ which is a much larger quantity. The quasi-three level case scales between the four-level and three-level lasers because the spontaneous-emission rate is proportional to $f_a N_t + \Delta N$; as f_a approaches 0, the efficiency approaches that of a four-level laser.

The other rate equation required to describe these lasers is the change in intracavity circulating intensity with time. This rate equation is the same for each of the models and is given by

$$\frac{dI_{xy}}{dt} = cW_{xy}\Delta N h\nu_{xy} - \frac{I_{xy}}{\tau_c} \tag{9}$$

where c is the speed of light, and τ_c is the cavity lifetime given by

$$\tau_c = \frac{-2L}{c \ln R} \tag{10}$$

L is the optical path length of the cavity, and R is the product of the power reflectivity of the cavity mirrors. The subscripts xy denote the laser transition. The first term on the right-hand side of eq. (9) is the growth in intracavity circulating intensity due to stimulated emission, and the second term is the decrease in circulating intensity due to cavity losses. It is assumed here that the gain medium occupies the entire cavity length; this assumption has no impact on the calculated efficiency or threshold.

Laser Threshold

In the limit of steady-state pumping in which the time derivatives in eqs. (6) and (9) are 0, the threshold population inversion density ΔN_{th} can be found from eq. (9) in the limit where I_{xy} goes to 0. ΔN_{th} is the same for all three models and is given by

$$\Delta N_{th} = \frac{-\ln R}{2\sigma L} \quad . \tag{11}$$

This is the typical solution derived for the threshold population inversion density if the condition for threshold is that the round-trip gain be equal to the round-trip loss.

The total threshold pump power absorbed by the gain element is given by the product of the pumping density at threshold, the pump photon energy, and the volume of the gain element. The pumping density at threshold can be found by substituting eq. (11) into eq. (6) and then solving eq. (6) for $W_{1z}N_1$ where z is either 3 or 4. Thus the absorbed powers at threshold are given by

$$P_{th} = \frac{1}{4\sigma\tau}(2\sigma N_t L - \ln R)h\nu_p A \qquad \text{(three level),} \tag{12a}$$

$$P_{th} = \frac{1}{2\sigma\tau}(-\ln R)h\nu_p A \qquad \text{(four level),} \tag{12b}$$

and

$$P_{th} = \frac{1}{2(f_a + f_b)\sigma\tau}(2f_a\sigma N_t L - \ln R)h\nu_p A \qquad \text{(quasi-three level).} \tag{12c}$$

A is the cross sectional area of the gain element. From eq. (12), we can understand the scaling laws of the threshold power and understand the reason that quasi-three level lasers have become more important given the improvements in diode laser pump sources.

In the three-level laser, there are two terms that contribute to threshold power. The first term is the power required to reach $\Delta N = 0$ while the second term is the additional amount of power required to overcome the round-trip cavity losses. For cw three-level lasers, this second term is negligible compared with the first term, consequently, the threshold pump power is proportional to the volume of the gain element. In the four-level laser, the only term the contributes to threshold is the need to overcome the cavity loss; the threshold power is proportional to the cross-sectional area of the gain element. Finally, in the quasi-three-level laser, there are two terms that contribute to the threshold power; these terms represent the same contributions as in the three-level laser. The difference here is that even for low cavity loss, the term that represents the need to reach population inversion is not necessarily dominant since this power is proportional to f_a; in the limit of f_a goes to 0, this term vanishes. As before, the contribution to threshold power by the need to invert the population scales proportionally with the volume of the gain element while the power required to overcome intracavity loss scales with gain element cross-sectional area.

The interpretation that can be drawn from these threshold power scaling laws is that the gain element should be made small in order to minimize absorbed threshold power. In the case of a four-level laser it is sufficient to make the gain element cross section small, but in the case of quasi-three-level and three-level lasers, both the cross sectional area and gain element length need to be minimized. Diode lasers have much higher spatial and spectral brightness than lamps which means that their output can be focused into a much smaller volume than a lamp. In fact, it is possible to achieve an order of magnitude greater pump density rates with a low-power diode laser than it is with a high-power lamp. Consequently, with diode pumping small gain elements can be used and operation well above threshold can be achieved even in quasi-three-level lasers. The essential engineering trades for quasi-three-level lasers can be found in eq. (12c). In the first term of eq. (12c) which is the power required to reach the condition $\Delta N = 0$, there are four multiplicative factors; these are $f_a \sigma N_t L$. To reduce the size of this term, their product needs to be made small. N_t cannot be made small because as N_t is reduced to 0, none of the pump radiation is absorbed. σ is actually irrelevant because it is cancelled by a σ outside the bracket. That leaves f_a and L. In lamp-pumped lasers, L needed to be long to match the length of lamps, thus the engineering trade that was performed to achieve good performance was to reduce f_a by operating below room temperature. In diode-pumped lasers, L can be made small and consequently f_a can be larger thus allowing operation of many quasi-three level lasers at room temperature. The term $2f_a \sigma N_t L$ in eq. (12c) has a simple interpretation; it is equal to the round-trip absorption at the laser wavelength for the gain element in thermal equilibrium. Thus the round-trip absorption at the laser wavelength due to lower state population is often referred to as the reabsorption loss.

Quasi-Three-Level Laser Models Including Spatial Dependence

Up to this point uniform pumping and extraction has been assumed; however, in real lasers this is not necessarily a good assumption. Rate equation models have been derived for end-pumped configurations of four-level[22,23] and quasi-three-level[4,24] lasers in which Gaussian pump and cavity mode intensity profiles (i. e., TEM_{00} modes) have been assumed. The threshold power for a quasi-three-level laser is given by[11]

$$P_{th} = \frac{\pi h \nu_p}{4\sigma\tau f}\left(\omega_p^2 + \omega_\ell^2\right)\left[2\sigma f_a N_t L + (1-R)\right] \tag{13}$$

Here ω_p and ω_ℓ are the Gaussian beam radii for the pump beam and laser cavity mode respectively. This is essentially the same as eq. (12c) except for the constants that account for the spatial variation. It has also been assumed that the cavity is low loss thus $-\ln R$ is approximately equal to $1-R$.

The rate equations (6) and (9) have been solved in the cw limit with spatial variation.[4,24] The results can be simply expressed in terms of three parameters. These are

$$p = \frac{\omega_p}{\omega_\ell} \tag{14}$$

which is the ratio of the pump spot size to the laser cavity mode size and

$$d = \frac{2 f_a \sigma N_t L}{1-R} \tag{15}$$

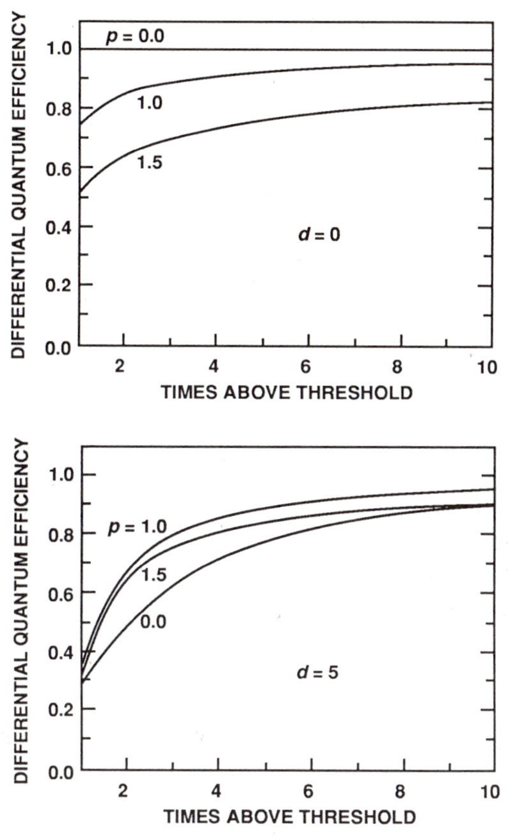

Figure 5. Calculation of differential quantum efficiency including spatial variation as a function of number of times above threshold. (top) $d = 0$, four-level laser. (bottom) $d = 5$. Adapted from Ref. 24

which is the ratio of the round-trip absorption loss in equilibrium induced by the lower state population at the laser wavelength to the round-trip cavity loss. The third parameter is the ratio of the pump power to the threshold pump power.

The differential quantum efficiency η is plotted in Fig. 5 for two different values of d.[12] This differential quantum efficiency is the change in the number of stimulated output photons per unit change in the number of pump photons. The plot for $d = 0$ is that for a four-level laser and agrees with previous calculations for four-level lasers. The differential quantum efficiency is highest for the smallest pump spot size. For larger values of d the differential quantum efficiency is reduced compared with a four-level laser, particularly near threshold; however, once the laser is a few times above threshold, η approaches that for a four-level laser. Thus quasi-three-level lasers can be nearly as efficient as four-level lasers as long as they are operated sufficiently far above threshold. As opposed to a four-level laser, the smallest pump spot size in quasi-three-level lasers does not alway give the highest η. For example, for $d = 5$, a spot size ratio of $p = 1$ gives higher η than $p = 0$ at the same number of times above threshold. However, this does not mean that $p = 1$ gives the highest overall efficiency because for the same value of ω_ℓ, the threshold power for $p = 1$ is two times larger than for $p = 0$.

A QUASI-THREE-LEVEL LASER EXAMPLE: Yb:YAG

Room temperature, quasi-three-level laser operation has been demonstrated now on several transitions in rare-earth ions with diode laser pumping. One example is the $^2F_{5/2}$ - $^2F_{7/2}$ transition in Yb^{3+} near 1 μm. A room-temperature, diode-pumped laser has been demonstrated in Yb:YAG.[16] This laser is an ideal quasi-three-level laser, and because of its simple energy level diagram, there is only one excited-state manifold, thus there is no possibility of parasitic effects such as excited-state absorption, upconversion, or concentration quenching. Yb^{3+} laser systems, particularly Yb:YAG, also offers advantages relative to Nd:YAG for diode-laser pumping. These advantages include a wider absorption band for reduced need of thermal management of the diodes, longer upper-state lifetime for improved energy storage relative to Nd:YAG, and lower heat generation in the laser gain element per unit pump power for reduced thermo-optic distortion.

The energy level diagram of Yb:YAG is shown in Fig. 6 and the room temperature absorption and fluorescence spectra are shown in Fig. 7. The main gain line at 1.03 μm is on a transition between the lowest-lying crystal-field splitting of the upper manifold and a level at 612 cm^{-1} in the ground-state manifold. There are two absorption lines at 0.968 and 0.943 μm with essentially the same absorption coefficient. The feature at 0.943 μm is preferred for diode-pumping because of its large absorption bandwidth. We have measured the spectroscopic gain cross section at 1.03 μm to be 2.6 x 10^{-20} cm^2 and the fluorescence lifetime to be 1.16 ms in low doped material.

We have performed cw laser experiments at room temperature with both $Ti:Al_2O_3$ pumping and InGaAs-diode laser pumping. In the $Ti:Al_2O_3$-pumped laser, the gain element was a 6.5 at. % doped Yb:YAG that was 0.16-cm long. This piece of YAG was polished plane-convex with a 5-cm radius of curvature on the convex surface. The cavity mirrors were coated directly onto the polished surfaces; the coating on the flat side was a high reflector at 1.03 μm and on the convex side was 97.5% reflecting at 1.03 μm. The InGaAs-diode-pumped gain element was 15 at. % doped Yb:YAG also polished plano-convex with a length of 0.11 cm and a radius of curvature of 15 cm. Again the cavity mirrors were coated directly onto the polished surfaces with a high reflector at 1.03 μm on the flat side and a 95% reflector on the convex side. The radii of curvature on these monolithic Yb:YAG gain elements lead to TEM_{00} mode radii of ~45 μm in both gain elements.

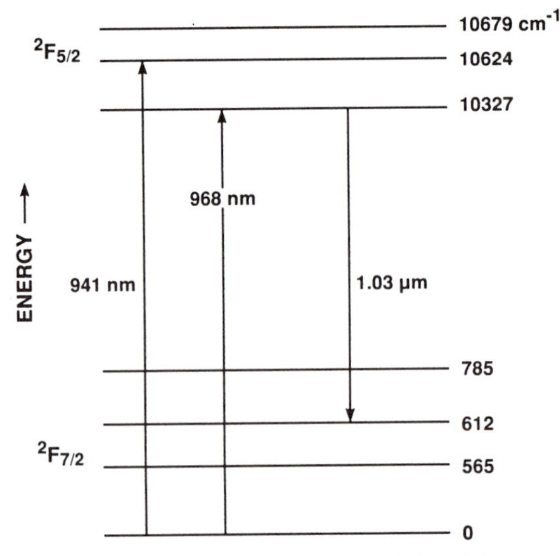

Figure 6. Energy-level diagram of Yb:YAG.

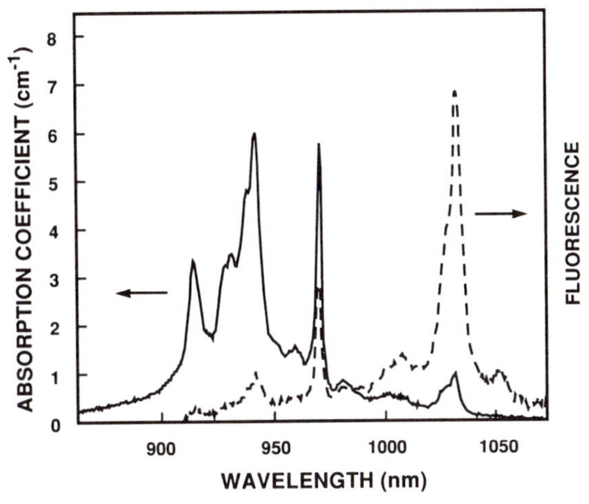

Figure 7. Absorption and fluorescence spectra of 6.5% doped Yb:YAG at 300 K.

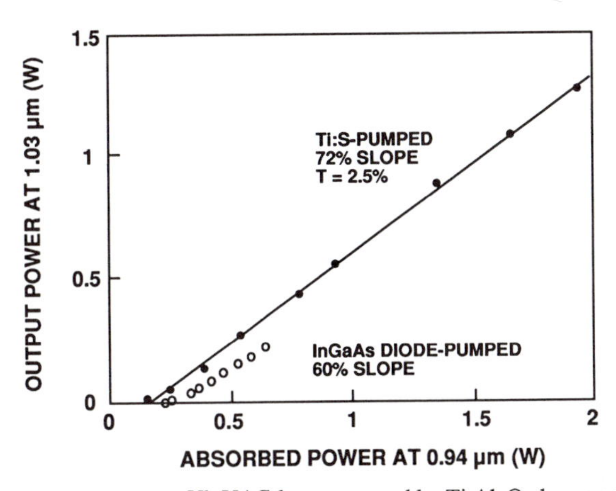

Figure 8. Results from room-temperature Yb:YAG lasers pumped by $Ti:Al_2O_3$ lasers or InGaAs diode lasers.

The laser results are shown in Fig. 8. With $Ti:Al_2O_3$-pumping a power slope efficiency of 72% is obtained by fitting the four highest power points to a straight line. This power slope efficiency is higher than any 0.81-µm-pumped, 1.06-µm Nd^{3+} laser demonstrated to date to our knowledge. This shows that quasi-three-level lasers can indeed be efficient. This power slope efficiency corresponds to a differential quantum efficiency of 0.79. The InGaAs-diode-pumped laser has a power slope efficiency of 60% corresponding to a differential quantum efficiency of 0.66. This power slope efficiency is comparable to that obtained in diode-pumped Nd:YAG lasers.

The pump beam sizes were nominally matched to the TEM_{00} mode radii in both cases so that $p \sim 1$. We can also calculate d for both these cases. For the 1.03 µm transition in Yb:YAG at room temperature, $f_a = 0.046$, and 6.5 at.% and 15 at. % doping are equivalent to N_t of 9.0×10^{20} cm^{-3} and 2.1×10^{21} cm^{-3} respectively. For the 6.5% doped sample, this leads to $d \sim 13$ and for the 15% doped sample, d is ~11. The differential quantum efficiencies are in agreement with their expected values given the large values of d and the number of times above threshold. The slightly lower differential quantum efficiency for the diode-pumped laser is primarily due to the fact that it is not being operated as far above threshold.

The advantages of Yb^{3+} gain media, particularly the long upper-state lifetimes, the low thermal loading, and the efficient cw operation at room temperature have prompted additional investigations into Yb^{3+} lasers. Q-switched[25] and mode-locked[26] Yb:YAG lasers have been demonstrated and other host materials are under active investigation.[27] Yb^{3+}-doped solid-state lasers appear to have a bright future.

SUMMARY

Efficient, room-temperature laser performance has been demonstrated in diode-pumped solid-state lasers on transitions that performed only poorly at room temperature using lamp-pumping. A common thread in many of these lasers is that the lower-laser level is in the ground-state manifold, although it is not the ground state itself, thus leading to increased threshold powers at room temperature. We have shown that a key to low threshold operation in these lasers is to minimize the volume of the gain element. Laser pump sources have the characteristic of high spatial and spectral brightness which permits high pump densities in small volumes and correspondingly low threshold powers. Models and experiments show that these quasi-three-level lasers can be efficient, nearly as efficient as a four-level laser, as long as the laser is operated sufficiently far above threshold. The development of quasi-three-level solid-state lasers opens new wavelengths and applications for solid-state laser technology.

ACKNOWLEDGEMENTS

This work was supported by the Department of the Air Force.

REFERENCES

1. R. L. Byer, Diode pumped solid-state lasers, *Science* 239:742 (1988).
2. T. Y. Fan and R. L. Byer, Diode laser pumped solid-state lasers, *IEEE J. Quantum Electron.* 24:895 (1988).
3. W. Streifer, D. R. Scifres, G. L. Harnagel, D. F. Welch, J. Berger, and M. Sakamoto, Advances in diode laser pumps, *IEEE J. Quantum Electron.* 24:883 (1988).
4. T. Y. Fan and R. L. Byer, Modeling and cw operation of a quasi-three-level 946 nm Nd:YAG laser, *IEEE J. Quantum Electron.* QE-23:605 (1987).
5. T.Y. Fan and R.L. Byer, Continuous wave operation of a room temperature, diode laser-pumped, 946 nm Nd:YAG laser, *Opt. Lett.* 12:809 (1987).
6. W. P. Risk and W. Lenth, Room-temperature, continuous-wave, 946-nm Nd:YAG laser pumped by laser-diode arrays and intracavity frequency doubling to 473 nm, *Opt. Lett.* 12:993, (1987).
7. T.Y. Fan, G. Huber, R.L. Byer, and P. Mitzscherlich, Spectroscopy and diode laser pumped operation of Tm, Ho:YAG, *IEEE J. Quantum Electron.* QE-24:924 (1988).
8. T.Y. Fan, G. Huber, R.L. Byer, and P. Mitzscherlich, Continuous wave operation at 2.1 μm of a diode laser pumped, Tm-sensitized Ho:YAG laser at 300 K, *Opt. Lett.* 12:678 (1987).

9. H. Hemmati, 2.07 μm cw diode-laser-pumped Tm, Ho:YLF room temperature laser, *Opt. Lett.* 14:435 (1989).
10. B. T. McGuckin and R. T. Menzies, Efficient cw diode-pumped Tm, Ho:YLF laser with tunability near 2.067 μm, *IEEE. J. Quantum Electron.* 28:1025 (1992).
11. G. J. Kintz, R. Allen, and L. Esterowitz, Continuous-wave Laser emision at 2.02 μm from diode-pumped Tm^{3+}:YAG at room temperature, in "Technical Digest, Conference on Lasers and Electro-optics," Optical Society of America, Washington, D. C. (1988).
12. P. J. M. Suni and S. W. Henderson, 1-mJ/pulse Tm:YAG laser pumped by a 3-W diode laser, *Opt. Lett.* 16:817 (1991).
13. T. S. Kubo and T. J. Kane, Diode-pumped lasers at five eye-safe wavelengths, *IEEE J. Quantum Electron.* 28:1033 (1992).
14. J. A. Hutchinson and T. H. Allik, Diode array-pumped Er, Yb:phosphate glass laser, *Appl. Phys. Lett.* 60:1424 (1992).
15. T.Y. Fan, Diode-pumped solid-state lasers, *Lincoln Lab J.* 3:413 (1990).
16. P. Lacovara, H.K. Choi, C.A. Wang, R.L. Aggarwal, and T.Y. Fan, "Room temperature diode-pumped Yb:YAG laser," *Opt. Lett.* 16:1089 (1991).
17. G. H. Dieke and H. M. Crosswhite, The spectra of the doubly and triply ionized rare earths, *Appl. Opt.* 2:675 (1963).
18. D. S. Hamilton, Trivalent cerium doped crystals as tunable laser systems: two bad apples, *in*: "Tunable Solid State Lasers," P. Hammerling, A. B. Budgor, and A. Pinto, eds., Springer-Verlag, Berlin (1985).
19. J. M. F. van Dijk and M. F. H. Schuurmans, On the nonradiative and radiative decay rates and a modified exponential energy gap law for 4f-4f transitions in rare-earth ions, *J. Chem. Phys.* 78:5317 (1983).
20. T. T. Basiev, A. Yu. Dergachev, Yu. V. Orlovskii, S. Georgescu, and A. Lupei, Nonradiative multiphonon relaxation and energy transfer from the strongly quenched high-lying levels of Nd^{3+} in laser crystals, in "Tunable Solid-State Lasers," vol. 5 of the OSA Proceedings Series, M. L. Shand and H. P. Jenssen, eds., Optical Society of America, Washington, D. C. (1989).
21. D. W. Hall, M. J. Weber, and R. T. Brundage, Fluorescence line narrowing in neodymium laser glasses, *J. Appl. Phys.* 55:2642 (1984).
22. D. G. Hall, R. J. Smith, and R. R. Rice, Pump size effects in Nd:YAG lasers, *Appl. Opt.* 19:3041 (1980).
23. P. F. Moulton, An investigation of the Co:MgF_2 laser system, *IEEE J. Quantum Electron.*QE21:1582 (1985).
24. W. P. Risk, Modeling of longitudinally pumped solid-state lasers exhibiting reabsorption losses, *J. Opt. Soc. Am. B* 5:1412 (1988).
25. P. Lacovara, T. Y. Fan, S. Klunk, and G. Henein, Q-switched Yb:YAG lasers, *in* "Conference on Lasers and Electro-Optics, 1992," Vol. 12, OSA Technical Digest Series, Optical Society of America, Washington, D. C. (1992).
26. S. R. Henion and P. A. Schulz, Yb:YAG laser: mode-locking and high-power operation, *in* "Conference on Lasers and Electro-Optics, 1992," Vol. 12, OSA Technical Digest Series, Optical Society of America, Washington, D. C. (1992).
27. L. D. DeLoach, S. A. Payne, L. K. Smith, W. L. Kway, L. L. Chase, and W. F. Krupke, Spectral properties of Yb^{3+}-doped crystals for laser applications, in "OSA Proceedings on Advanced Solid-State Lasers, L. L. Chase and A. A. Pinto, eds., Optical Society of America, Washington, D. C. (1992).

Er,Tm,Ho:YLF LASERS IN ASTIGMATICALLY COMPENSATED RESONATORS

P. Minguzzi[1], A. Di Lieto[2], H.P. Jenssen[3], M. Tonelli[1]

[1]Dipartimento di Fisica dell'Università
Piazza Torricelli 2, 56100 Pisa, Italy
[2]Scuola Normale Superiore
Piazza dei Cavalieri 7, 56100 Pisa, Italy
[3]Center for Materials Science and Engineering
Massachussets Institute of Technology
Cambridge, Massachussets 02139

INTRODUCTION

The Holmium emission near 2 μm has long been studied for the purpose of developing efficient solid state lasers. Several authors[1-15] reported laser action for different host crystals, both in pulsed and in CW regime, and for different operating temperatures.These sources are attractive for applications requiring atmospheric propagation because the emission wavelengths are eye-safe: thus they can be useful for optical communications, coherent laser radar, and detection of pollutants. There also may be medical applications because liquid water has a strong absorption peak in this spectral region and the wavelength tuning can be a practical mean of adjusting the absorption depth.

Our initial motivation was to improve the operation of a CW Ho:$LiYF_4$ laser and eventually to develop a device useful for high resolution spectroscopy of atoms and molecules.

Laser action takes place between the $^5I_7$ and $^5I_8$ manifolds of Ho^{3+} ion. The upper $^5I_7$ laser level is efficiently populated by energy transfer from Er^{3+} and Tm^{3+} sensitizers. In this well proven technique the $LiYF_4$ (YLF) host crystal is co-doped with a high concentration of other rare earths ions (Er, Tm) which have a good spectral matching with the pump sources and improve the efficiency through internal processes of energy conversion.

Usually for a high-efficiency performance a selective pumping in bands only slightly higher in energy than the lasing state is desirable, because less pump energy is wasted into heat. Here however the internal complex processes allow the conversion of a single visible pump photon into two or even three infrared laser photons, so the Ho:YLF laser is potentially a very high quantum-efficiency device.

Solid State Lasers: New Developments and Applications
Edited by M. Inguscio and R. Wallenstein, Plenum Press, New York, 1993

In our experiments we used in turn a red Kr-ion laser, a blue-green Ar-ion laser , and a diode laser as the pump source, in a longitudinal pumping scheme. The results for each of these cases will be described in the following Sections.

The Ho:YLF lower laser level within the $^5I_8$ manifold is approximately 300 cm^{-1} above the ground state of the ion. At room temperature this energy is about 1.5 k_BT, so the thermal population of the terminal level is appreciably large making the CW lasing threshold fairly high. Therefore at room temperature it is not possible to disregard reabsorption losses, and the Ho:YLF laser is essentially a three-level laser system. If the crystal temperature is decreased this problem becomes less important, and at 77 K (liquid nitrogen) a description as a four-level laser system is realistic. At intermediate temperatures a quasi-three-level modeling[16] can be applied to understand the laser behaviour. Our initial experiments were performed at 77 K, while, more recently, we studied the laser performance as a function of the crystal temperature, and we could achieve a satisfactory operation above 200 K.

The modeling of longitudinally pumped lasers exhibiting reabsorption losses has been discussed in the literature[16,17]; from these studies one can obtain suggestions about the best pumping geometries and the optimum crystal length. A good recipe appears to be the following: a strong focusing of the pump laser gives a low threshold, while a short length of the active crystal reduces the adverse effect of reabsorption losses.

Among the different resonators compatible with the above requirements we preferred a three-mirror astigmatically-compensated configuration[18], similar to that used for dye lasers or color centre lasers. The disadvantages of a complicated structure are balanced by the versatility of the system which allows the use of the same optical components for different pumping wavelengths and can easily accomodate tuning elements for the control of the emitted frequency. The focusing of pump power can be very strong yielding a low threshold for CW lasing even in the presence of reabsorption losses.

The use of very high pumping density on Er,Tm,Ho co-doped YLF crystals has led us to discover that upconversion effects can be very important in the operation of the laser. These processes are energy transfers between two ions in excited states, in which one ion gives up energy to the other causing the second ion to be excited to a higher energy level. Upconversion effects become important at high concentration of excited ions, that is at very strong pumping rates, and constitute alternative paths to depopulate the upper laser level. Therefore they may represent a limiting factor in the attainment of high efficiency at strong pumping. In these operating conditions the modeling described in Refs. 16,17 does not apply very well, because these excited-state effects are disregarded in those analyses.

It is however important to have guidelines for optimizing the laser performance and we followed the approach of studying the fluorescence from states higher in energy than the upper laser level. The fluorescence from the active crystal, placed inside the resonator, is observed as a function of pumping power. A comparison of the fluorescence signal intensity is made for the case of the laser operating above threshold and the case of laser action inhibited by a deliberate large intracavity loss (in the following these cases will be labelled as laser-on and laser-off conditions). By spectral resolution of the fluorescence we found that some emitted wavelengths have an intensity correlated with the on/off condition, while others did not display such a behaviour. This kind of studies can be a useful tool to get information about what it is happening inside the complicated laser medium and some examples will be given below.

EXPERIMENTAL

In this section we describe the experimental details concerning the structure of the resonator, the pumping beam optics, the tuning elements, and the measurement apparatus.

The Folded Resonator

The design of the three-mirror folded resonator started from the assumption that an approximately 2 mm thick $LiYF_4$ crystal is placed at Brewster angle at the waist of the short arm of the cavity. The schematic diagram of the resonator structure is shown in Figure 1. The end mirror M_1 and the folding mirror M_2 have a rather short radius of curvature to insure a small waist of the laser mode, while keeping a practical distance between them, for confortable crystal handling. The chosen values are 5 cm and 8 cm for M_1 and M_2 respectively, and their distance is about 9 cm. The total length of the resonator is approximately 110 cm and the third outcoupling mirror M_3 is flat. Once the mirror geometry and the crystal thickness are fixed, the folding angle is determined by the requirement of astigmatism compensation[18]. For this computation the refractive index of the crystal is assumed to be unaffected by dopants. It turns out that the good folding angle is 17 degrees. With this geometry we were able to successfully operate laser crystals of thickness ranging from 1.5 to 2.5 mm, without appreciable deformation of the output beam.

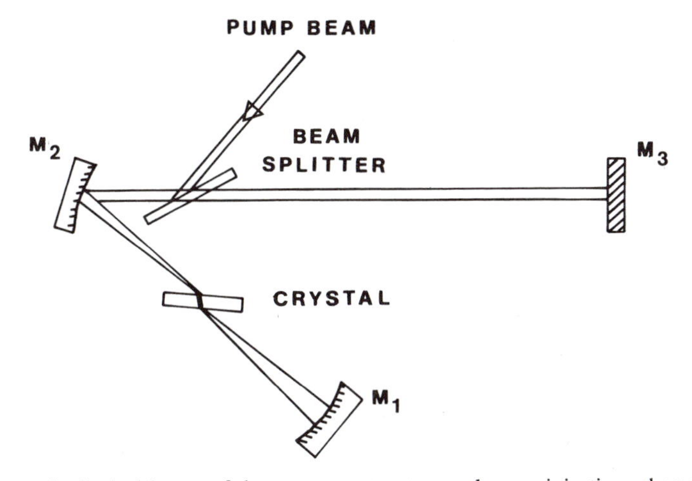

Figure 1. Optical layout of the resonator structure and pump injection scheme.

The Pumping Optics

Usually two different methods have been employed for longitudinal pump injection into a three-mirror folded resonator: either through the M_1 end mirror or by reflection on a beam splitter placed at Brewster angle in the long arm of the cavity. The former method requires a special coating for mirror M_1, with high reflectivity in the infrared and a high transmittance for the pump. Different mirrors will therefore be necessary for each pump pump wavelength and, perhaps, for each tuning range of the laser. The beam-splitter method does not present this drawback because high reflectivity broadband mirrors can be used for M_1 and M_2, and there are additional advantages too: the astigmatism of the pump beam inside the crystal is automatically corrected and the pump power reflected back by M_1 allows a double-pass absorption by the active medium. There are however also some disadvantages, because an uncoated beam splitter requires a pump polarization perpendicular to that of the infrared beam and, even in the case of a high refractive index, it does not reflect all of the pump power onto the crystal.

In our experiments we used a ZnSe beam splitter, which is either uncoated or it has one surface coated to reflect both pump polarizations efficiently and to exhibit a high transmission at laser frequencies. The coated beam splitters allowed us to study the effect of pump polarization on the lasing efficiency: different performance may arise from physical reasons, such as different absorption coefficients or transition rates, but also from the trivial fact that a pump beam polarized orthogonally to the plane of incidence is partially reflected by the crystal (which is uncoated and placed at Brewster angle), while a parallel polarization does not suffer from this drawback.

In our setup the mirrors M_1 and M_2 have a broadband metallic (gold or silver) coating; a collimated pump beam, e.g. in the green, is focused to a 25 μm waist at the crystal with a power density of about 35 kW/cm^2 per watt of incident power.

For easier alignment of the pump beam into the folded resonator, we found it convenient to place a pair of iris diaphragms along the pump path, which are useful for fine alignment and are kept open at full pump power. By following this practice we had a reliable mean of pump laser alignment and we discovered that the pump beam divergence is rather critical for optimum performance. This behaviour is theoretically predicted by the modeling of Ref. 17, which indicates that, for each pump intensity, there is an optimum ratio of the pump waist to the laser waist inside the active crystal. In other words the best beam overlap inside the crystal depends on the pump power. Even before becoming aware of this theoretical analysis we used a pair of lenses for the control of the pump beam divergence and we actually found that this adjustment is important for achieving the best efficiency. We also learned that it is important that any optical component inserted in the pump beam does not degrade the wavefront quality.

The Crystal and the Cryostat

Crystal growth took place in a computer-controlled Czochralski growth system, in a purified Argon atmosphere. The active crystals are rectangular samples of high optical quality and uniform concentration of dopants; they are cut with the **c** axis parallel to the surface and are polished according to good optical pratice. Typical dimensions of the rectangle are 4 ~ 6 mm and the thickness is about 2 mm. The selected concentrations of dopants are different according to the source employed for pumping and to the operating temperature, and they will be detailed in the following Sections.

The crystal is placed at the waist of the short arm of the resonator, usually (but not always) with its **c** axis in the plane of incidence of the laser beam, and therefore parallel to the polarization of the emitted infrared beam.

The crystal is in good thermal contact with a copper holder, which acts as a heat sink. Thin indium foil is used to improve heat dissipation. The crystal holder is in turn clamped to the cold finger of a cryostat, usually employing liquid nitrogen as the coolant. The temperature is sensed by a thermocouple placed close to the crystal. For measurements at different temperatures a controlled heating of the cold finger is possible; in some cases cooling by dry ice was used.

To avoid troublesome condensation of water vapor on the crystal the cryostat chamber is evacuated by a liquid-nitrogen-cooled sorption pump. The vacuum chamber is wide enough to contain the curved mirrors M_1 and M_2, and the ZnSe beam splitter. Bellow couplings allow the refinement of the mirror alignment while the laser is operating. Optical coupling of the infrared beam to the external part of the resonator is obtained through a CaF_2 Brewster window. The chamber has two additional side windows: one is necessary for the injection of the pump laser, the other is useful for viewing the crystal and for observing the emitted fluorescence in a direction orthogonal to the laser beam, while the laser is operating.

A careful design of the supporting structure of resonator is important for a good noise performance of the laser. The cryostat, the mirror holders, and all other optical components are clamped to a thick super-invar plate, to insure a good thermal and mechanical stability.

The Output Beam

The flat mirror M_3 is used as the outcoupler device; the fundamental cavity mode has a waist of about 1 mm at M_3 and the output beam has a clean Gaussian shape with a diffraction limited divergence.

When the tuning range of the laser is studied the mirror M_3 is substituted by a grating in a Littrow mount . In this case the output beam comes from the zero-th order of the grating. A mirror mechanically coupled to the grating rotation allows tuning the laser without steering the output beam; this is a convenient feature because it avoids realignment of the measuring instrumentation. The characteristics of the grating (625 g/mm and blaze angle of 61 degrees) had been selected for high dispersion and good reflectivity near 2 μm, so the output beam cannot be intense. If an independent choice of outcoupling transmission and wavelength selection is desired, a birefringent filter should be preferred. In a preliminary work with a two-plate quartz birefringent selector we found that a reasonable tuning can be achieved with only a 10% insertion decrease of the output power.

The Measuring Instrumentation

The output power of the laser and the pump input power are measured by calibrated thermopile detectors. Three commercial power meters from different companies were used and found to agree within 10% in the worst case. The emitted wavelength is measured by a small IR monochromator; in this application, sensitive detectors are used, such as a PbS photoconductor or a cooled InSb photovoltaic detector. The output beam is chopped and the signal is processed by a lock-in amplifier.

A small fraction of the visible pump power (of the order of 0.1 percent) is transmitted to the output of the laser; while it is generally useful for IR beam tracking, it may sometimes cause trouble for low intensity measurements. In these cases infrared-transmitting filters of known optical density are employed.

For the measurement of high-frequency noise in the intensity of the emitted radiation an InAs fast detector is used; it is operated at room temperature and it has a bandwidth of 10 MHz.

A simple measuring apparatus has been set up for the detection of the crystal fluorescence, observed through the side window of the cryostat. A lens collects the fluorescence and focuses it on the input slit of a monochromator; the scattered pump light is blocked by a suitable absorption filter. A detector is placed at the output slit: infrared emission is observed by PbS or InSb detectors (as described above) while visible emission is detected by photomultipliers with S20 or S1 spectral response. The radiation is chopped before reaching the monochromator and the detector signal is fed into a synchronous amplifier. The data acquisition system operates under the control of a personal computer, equipped with standard interface cards.

RESULTS

As discussed in the Introduction the presence of Er and Tm dopants as sensitizers in the YLF crystal is important for pump absorption and for energy-transfer processes. The schematic diagram of the energy levels of Er, Tm, and Ho ions in YLF, relevant for the laser operation, are displayed in Figure 2. One can see that the first excited manifold of Er ($^4I_{13/2}$) is just slightly higher in energy than the $^3F_4$ Tm multiplet, which in turn is very close to the $^5I_7$ upper laser level of Ho. These remarkable coincidences make the transfer of excitation from Er to Ho very efficient. The energy levels of Er and Tm have further interesting properties which are important for optical pumping by visible lines of ion lasers; these properties will be described below, for each different pumping case.

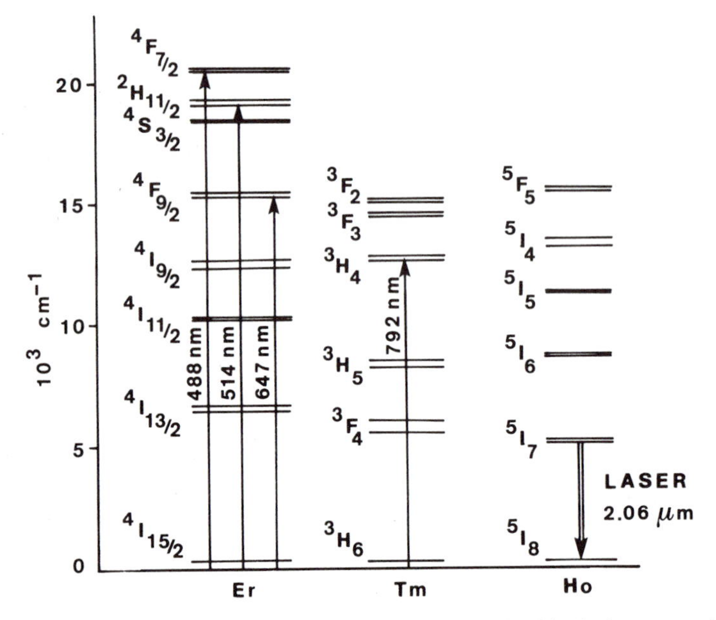

Figure 2. Ionic energy levels of Er, Tm, and Ho in YLF involved in the laser operation.

Kr^+ Laser Pumping

Our first experiments were performed by using the 647 nm red line of a Kr^+ laser as the source of pump power. This line has a frequency coincidence with the $^4I_{5/2}$ - $^4F_{9/2}$ transition of Er^{3+}. The first successful crystal had the following concentrations of dopants: 30% Er, 10% Tm, and 0.1% Ho, which substitute Yttrium in the crystal. When we measured its absorption coefficient at the Kr^+ wavelength, we found that it was much larger for a polarization orthogonal to the **c** axis of the crystal than for parallel polarization: at the temperature of 77 K we observed nearly 60 cm^{-1} for the former and about 5 cm^{-1} for the latter. This result suggested us a configuration with the pump polarization parallel to the **c** axis, otherwise the pumping would not be homogeneous over the crystal thickness of 2.2 mm. For pump injection we used an uncoated beam splitter, so the crystal was placed with its **c** axis perpendicular to the polarization of the infrared beam and therefore parallel to the pump polarization.

Two different outcoupling mirrors M_3 have been tested and the best for high power extraction was the one with 30% transmission. The results for the performance of the laser in this configuration are shown in Figure 3, which displays the emitted power as a function of the incident power when the crystal is at 77 K. For the incident power we mean the power that actually penetrates the front surface of the crystal, i.e. we evaluated it by taking into account all Fresnel reflections at the input window, at the beam splitter, and at the crystal itself. About 90% of this power is absorbed by the crystal.

The maximum output is 220 mW which is obtained with 780 mW of incident power; the slope efficiency is 36% and the threshold is about 180 mW. It can be noted that the quantum slope efficiency is about 120% and the internal quantum efficiency is, of course, larger than this value, even if in these first experiments we had not sufficient information for evaluating it accurately.

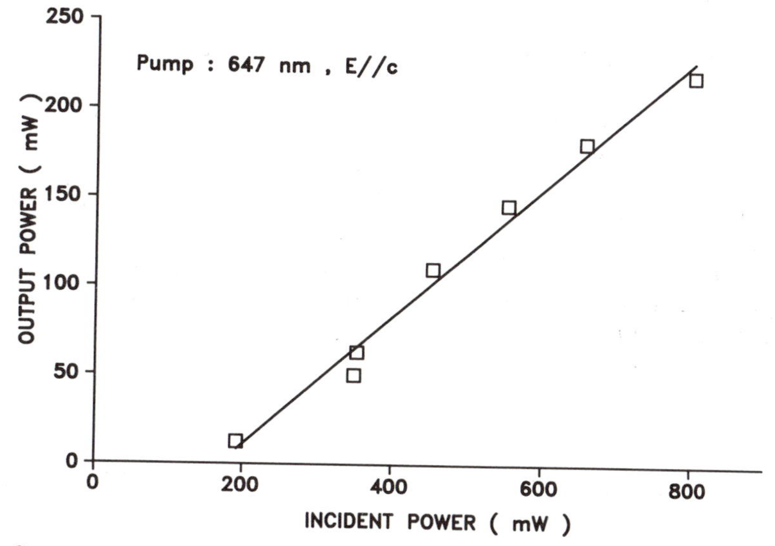

Figure 3. Output power versus incident power for Kr pumping. The crystal temperature is 77 K and the outcoupler transmission is 30%.

The emission wavelength is 1.91 μm, which is surprising because it corresponds to a Tm^{3+} emission from the $^3F_4$ manifold. We tried to understand the reason for this unexpected behaviour by observing the infrared fluorescence of the crystal. We found[19] that, while at low power the Ho emission spectrum is observed, at high power density the Tm^{3+} emission is stronger. The exact reason for this fact is not clear, however probably, at the low Ho concentration of this crystal, upconversion processes deplete the Ho $^5I_7$ level, in favour of Tm $^3F_4$ manifold.

The set of all our experimental results prove that each visible pump photon produces more than one excited Tm ion. A proposed mechanism to explain this larger-than-one internal quantum efficiency is the following (see Figure 2): after excitation to the $^4F_{9/2}$ state the Er ion decays to the $^4I_{13/2}$ level with an almost resonant transfer of a Tm ion from the ground to the $^3H_5$ state. Subsequent relaxation of Tm from $^3H_5$ to $^3F_4$ and energy transfer from Er $^4I_{13/2}$ to Tm $^3F_4$ produce two excited Tm ions in the $^3F_4$ manifold for each pump photon. If this spectroscopic scheme is correct the maximum theoretical quantum efficiency is equal to 2. Upconversion processes which can counterbalance the energy transfer from Tm $^3F_4$ to Ho $^5I_7$ are possible and an example is discussed in Ref. 9.

The tuning range of the laser was also studied, but the grating available at the time of measurements[19] was not optimum for the 2 μm operating region. The observed laser tunability covers the 1.85 - 1.98 μm range. An unusual and interesting feature of the tuning curve is the fact that the longer wavelengths are attainable at a low pump power, which is consistent with the fluorescence measurements and the interpretation that a high power density tends to deplete the $^5I_7$ state of Ho.

As a final remark of this subsection we can state that it is possible to obtain tunable laser action from Tm^{3+} in YLF if a very strong pumping power density by a Kr^+ laser is used.

Ar^+ laser pumping

Further measurements with Kr^+ pump could not be performed because our laser was irreparably damaged. It was later possible to continue the research work by using an Ar^+ laser, capable about 10 W output power in multiline operation. There are two absorption coincidences of Er^{3+} in YLF for the 488 and 514 nm lines, which are the most powerful emissions of typical Ar^+ lasers. The 488 nm line excites the $^4F_{7/2}$ state, while the 514 nm line excites the $^2H_{11/2}$ manifold. From previous work it was known that the absorption coefficient at these wavelengths is smaller than at 647 nm, therefore a larger Er doping was used. A crystal with 72% Er, 7.2% Tm, and 0.3% Ho was employed in a large part of the work; its dimensions are 7 mm long x 3.5 mm wide x 2.5 mm thick. We measured the absorption coefficients of this crystal at the 488 and 514 nm lines for polarization both parallel and orthogonal to the **c** axis. The results for the temperature of 77 K are reported in Table 1; in the same Table we also give the estimates of the fraction of absobed power, based on the assumption of double pass through the crystal. We checked by direct measurement that these estimates are reliable even at the strong power density present at the laser waist.

Table 1. The absorption coefficients (cm^{-1}) and the fraction of absorbed power (%) for the second crystal sample at different wavelengths and polarizations.

wavelength	polarization	
	parallel	perpendicular
488 nm	1.6 cm^{-1} 54%	1.1 cm^{-1} 39%
514 nm	5.4 cm^{-1} 92%	~ 8 cm^{-1} ~ 96%

Since a coated beam splitter with good reflectivity in the blue/green for both polarizations was available, the crystal was placed in the cavity with its **c** axis parallel to the infrared beam polarization. A first set of measurements have been performed with pump polarization perpendicular to the **c** axis (this configuration is possible with an uncoated beam splitter as well). For 77 K operation and an outcoupler transmission of 20% we obtained a 25% slope efficiency with multiline pumping; at 488 nm (single line) the slope efficiency is 20% and at 514 nm is 34%. These efficiencies are evaluated using the incident power, as explained above. The difference between the 488 and 514 nm cases is not unexpected, because the absorption coefficients are rather different (see Table 1). The threshold intensities are always very small, ranging from 6 to 16 mW. These measurements show a clear preference for the 514 nm pumping; however the configuration of multiline pumping has a practical interest

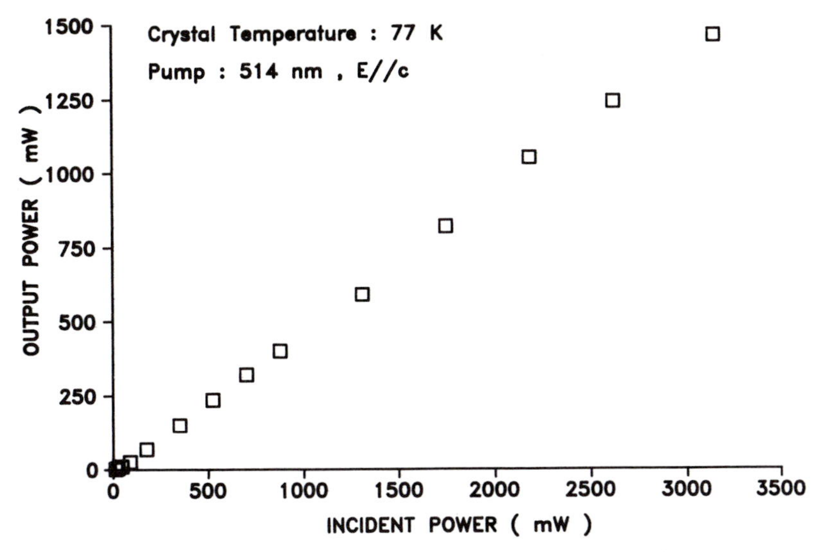

Figure 4. Output power versus incident power for Ar pumping. The outcoupler transmission is 40%.

for the reason that a large output power can be achieved without special components. We could routinely obtain 1.4 W at 2.06 μm with 5.8 W multiline pumping. The maximum output power is limited only by surface damage of the crystal: with the focusing optics described above the damage threshold is at about 6 W impinging on the crystal.

Better performance is obtained when the pump polarization is parallel to the **c** axis. In this case (labelled E//**c** in the figures) a very large efficiency is possible and we tested several different outcoupler transmissions to search for the optimum configuration. The best result is shown in Figure 4: for a 40% outcoupler a slope efficiency of 48% is measured; the maxi-

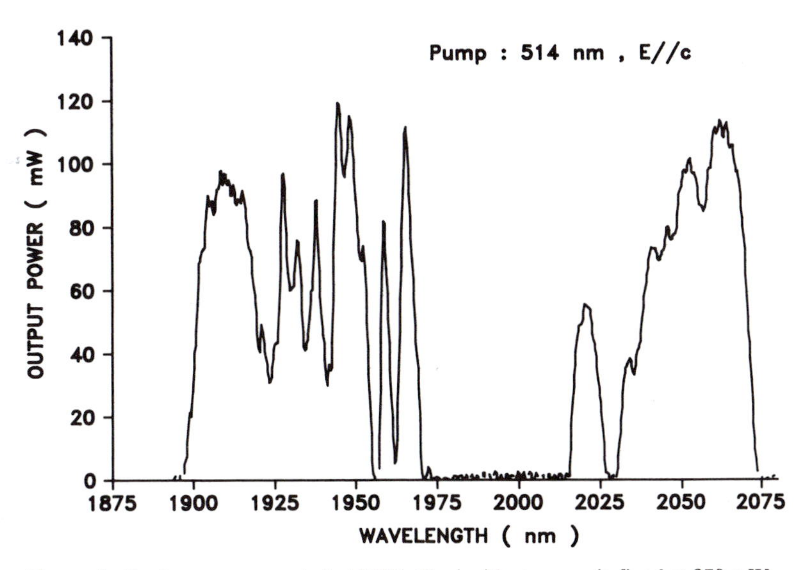

Figure 5. Tuning curve recorded at 77 K. The incident power is fixed at 870 mW.

mum power is 1.46 W with 3.2 W pump at 514 nm. The lasing threshold in this configuration is 16 mW and the emitted wavelength is 2.06 μm. More details about the results obtained with other outcouplers can be found in Ref. 20.

From the knowledge of the fraction η_A of absorbed power and from the measured slope efficiencies at different output transmissions **T** it is possible to evaluate the resonator losses **L** and the internal quantum efficiency η_i of the active medium. By fitting the equation[17]

$$\text{Slope} = \frac{\Delta P_{out}}{\Delta P_{pump}} = \eta_A \frac{h\ \nu_{laser}}{h\ \nu_{pump}} \eta_i \frac{T}{L + T} \quad (1)$$

to the experimental data we found $\eta_i = 2.6 \pm 0.3$ and $\mathbf{L} = 0.12 \pm 0.03$. The large value of **L** is mainly caused by the absorption of the coated beam splitter, which was confirmed by direct measurement, while the high quantum efficiency is really remarkable. The spectroscopic scheme we propose to explain this result is the following: The 514 nm Ar+ line excites the $^2H_{11/2}$ manifold of Er^{3+}; this state decays by radiationless relaxation to the $^4S_{3/2}$ level. Now let us consider the Er cascade from $^4S_{3/2}$ to $^4I_{9/2}$ to $^4I_{13/2}$ to $^4I_{15/2}$: each of these steps is almost resonant with the $^3H_6$ - $^3F_4$ transition of Tm^{3+}. It is therefore acceptable to assume that for each excited Er ion three Tm ions are brought to the $^3F_4$ state; then by energy transfer three Ho ions are excited to the $^5I_7$ upper laser level. This mechanism gives reason of the quantum efficiency close to 3 observed at 77 K, but does not take into account upconversion processes, which are certainly present and decrease the efficiency by depleting the population of the upper laser level.

The tuning range of the laser has been studied with the apparatus described in the Experimental Section. The results are shown in Figure 5: the output wavelength could be tuned from 1.89 μm to 2.07 μm, with the exception of a small gap between 1.97 μm and 2.01 μm. An interesting feature is the presence of emission in the 1.89 - 1.93 μm.range where Tm is probably lasing. By detecting the infrared fluorescence we observed a behaviour similar to that recorded for Kr^+ pumping: an anomalously large signal of Tm fluorescence is measured when the pump power density is strong. The selective action of the grating then causes lasing in this region.

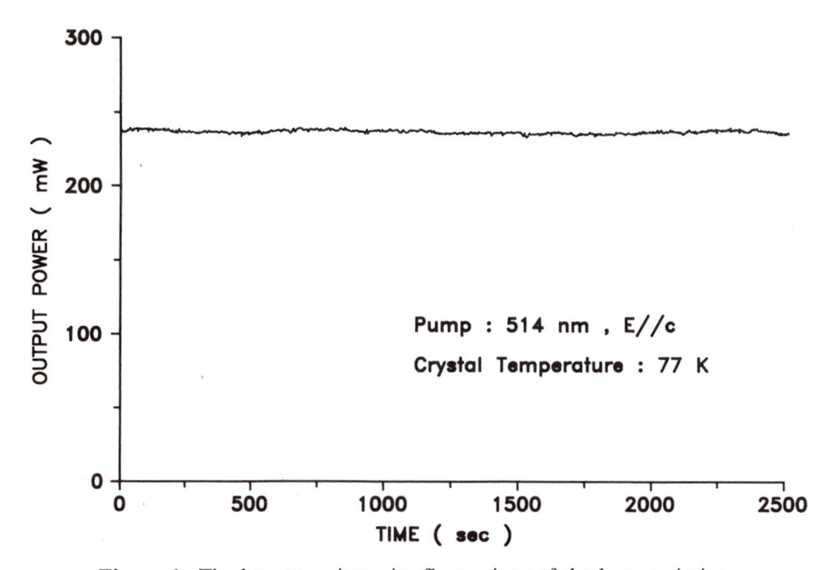

Figure 6. The long term intensity fluctuations of the laser emission.

In order to complete the characterization of the laser operation we measured the noise properties of the emitted power. In Figure 6 we display the long term stability of the output for single line operation (the grating is tuned to 2.03 μm). The fluctuations are smaller than 2% peak-to-peak over a time duration of about 40 minutes. The high frequency noise was also measured by a fast detector connected to a RF spectrum analyzer. The intensity fluctuations are smoothly decreasing for frequencies up to a few hundreds kHz, except for a narrow-band noise peak, whose frequency depends on the pump intensity. Anyway the signal-to-noise ratio is always larger than 10^4 and typically is 10^5, so this high frequency noise can be disregarded in almost every practical application.

All the results described above refer to 77 K operation; we also investigated the laser properties when the crystal is kept at higher temperatures. A practical temperature near 210 K is easily obtained by cooling the cryostat with dry ice. At this temperature appreciable reabsorption losses are present, so a lower efficiency and a smaller output power are expected. We found[20] that at this temperature the best outcoupler transmission is 10% and for pumping at 514 nm with parallel polarization we obtained a slope efficiency of 15%. The threshold power is 60 mW and the maximum output power is 130 mW at 1 W pump: above this pumping rate thermal roll-off was observed. The tuning range at 210 K is limited to the 2.053 - 2.070 μm interval.

In the attempt of collecting information for improving the laser performance we measured the fluorescence from the active crystal, at different wavelengths and temperatures, as a function of the pump power. We found fluorescence lines at 670 and 890 nm whose intensity is significantly correlated with the laser on/off condition[21], i.e. with the population of the Ho^{3+} $^5I_7$ upper laser level. Therefore we think that our fluorescence experimental data support the evidence of the existence of upconversion processes and of their importance for the laser efficiency. However it is not easy to give a reliable interpretation of the fluorescence results and further work on this subject will be necessary. Anyway we were led to think that perhaps an increase of Ho concentration in the host could reduce the adverse upconversion effects at high temperature.

Therefore a crystal with different composition was recently studied for operation at high

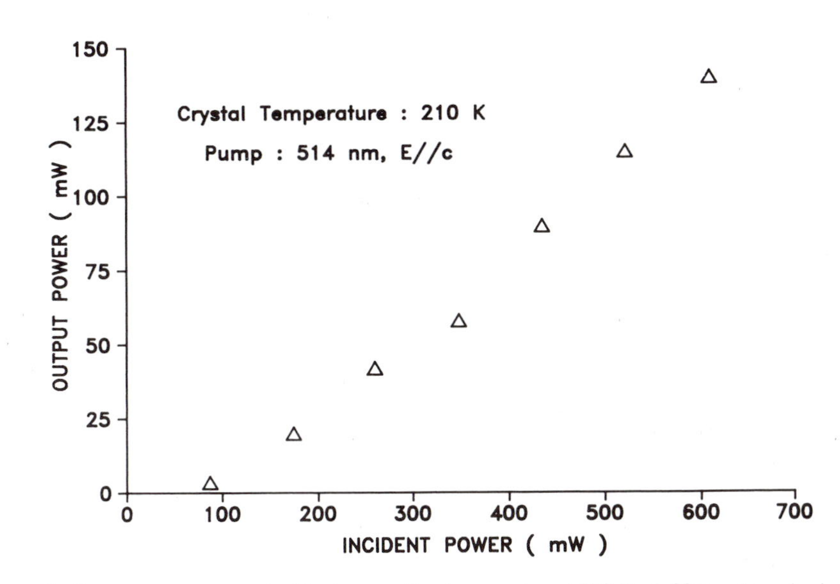

Figure 7. Output power versus incident power for Ar pumping at 210 K. A 30% outcoupler is used.

temperatures. The dopants concentrations are 60% Er, 7.2% Tm, and 1.7% Ho and the crystal thickness is 1.8 mm: i.e. more Ho and less Er in a thinner sample. This crystal is not as efficient as the previous one for operation at 77 K, but at dry ice temperature is significantly superior. At this temperature an absorption coefficient of about 5 cm^{-1} was measured at 514 nm and parallel polarization. At this wavelength the fraction of absorbed power is about 83%, for double pass through the crystal. The best outcoupler transmission was found to be 30% and the results obtained for this configuration are shown in Figure 7. The slope efficiency is 26% and the threshold power is about 80 mW; a 23% slope efficiency was observed with a 20% outcoupler, with a clear indication of large internal losses A maximum output power of nearly 200 mW was obtained at 1.3 W of pump power, however at this pumping rate thermal roll-off was observed. These results show that it is possible to achieve a remarkable efficiency at a more practical temperature than 77 K.

At temperatures above 210 K the configuration with pump polarization perpendicular to the **c** axis can also be interesting for the reason that the absorption coefficient has about the ideal value (~ 6 cm^{-1}) for the strucure of our laser. So we measured the output power in this configuration at different temperatures and an outcoupler transmission of 5%. The results are shown in Figure 8: it is worth noting the remarkable output of 17mW at 259 K.

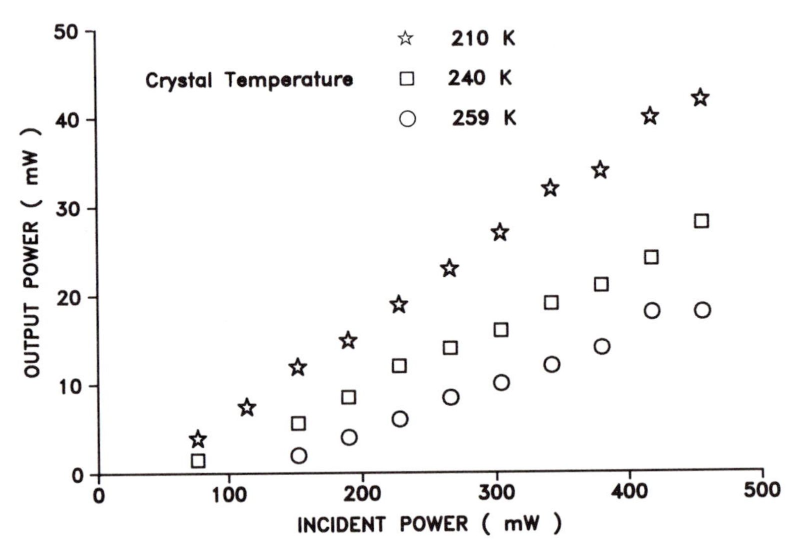

Figure 8. Output power versus incident power for 514 nm orthogonal pumping at high temperatures. The outcoupler transmission is 5%.

It was interesting to find the maximum operating temperature of the laser. For this measurement we selected the high gain configuration of parallel pump polarization, an outcoupler transmission of 5% for low threshold, and a pump power of 350 mW at 514 nm. The output power of the laser is displayed in Figure 9 as a function of crystal temperature. One can see that the cutoff temperature is about 280 K, while at 273 K the laser is still above threshold and a few mwatts are emitted. We think that probably a small improvement is possible, with this crystal, if the losses of the optical components in the resonator (particularly of the beam splitter) can be reduced. We also believe that further work with a crystal of different composition and/or of different thickness can eventually lead to room temperature operation of the laser.

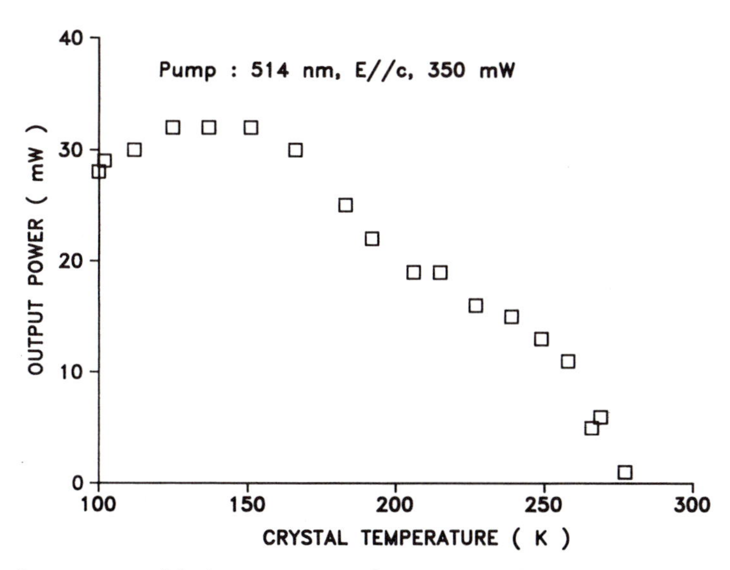

Figure 9. Output power of the laser versus crystal temperature. The outcoupler transmission is 5%.

Diode Laser Pumping

The great development of powerful diode-laser arrays has made these devices attractive sources for compact solid-state laser pumping. Recently we had available one of these arrays, capable of 500 mW output power in the wavelength region near 790 nm. At these frequencies the Tm $^3H_4$ and the Er $^4I_{9/2}$ manifolds can be excited. The diode array was therefore interesting for pumping our YLF crystals, and we started a preliminary study with the (60% Er, 7.2% Tm, 1.7% Ho) sample. Since a suitably coated beam splitter was available, parallel pumping of the crystal was preferred. We measured the absorption coefficient at different wavelengths within the tuning range of the diode and we found a value of about 5 cm^{-1} at 792.6 nm. This was really good for homogeneuos pumping of our 1.8 mm thick sample. The crystal was placed into the resonator with the **c** axis parallel to the infrared and to the pump polarization Some trouble was encountered for obtaining a proper handling of the pump beam, because of the bad rectangular shape of the source, much less confortable than ion lasers. After some efforts and by using astigmatic optical components we succeeded to bring about 83% of the power into a collimated beam that could be well focused onto the crystal by the resonator optics. Several outcoupler transmissions were tested and the best ones were 20% and 30%, which gave almost the same slope efficiency but slightly different thresholds. The results for operation at 77 K with the 30% outcoupler are shown in Figure 10: the threshold is about 40 mW and the slope efficiency is 29%; the maximum output power is 56 mW at 225 mW pump.

The internal quantum efficiency can be roughly estimated from the slopes at different outcouplings: we found $\eta_i = 1.9 \pm 0.4$, with a clear evidence of a mechanism which yields two laser photons for each pump photon. The proposed spectroscopic scheme is the following: after excitation a Tm ion decays from $^3H_4$ to $^3F_4$ and by cross relaxation a second Tm ion is excited from the ground to the $^3F_4$ state; then a final energy tranfer from Tm brings two Ho ions to the upper laser level. An analogous scheme takes place if the ion excited by the pump radiation is Er instead of Tm.

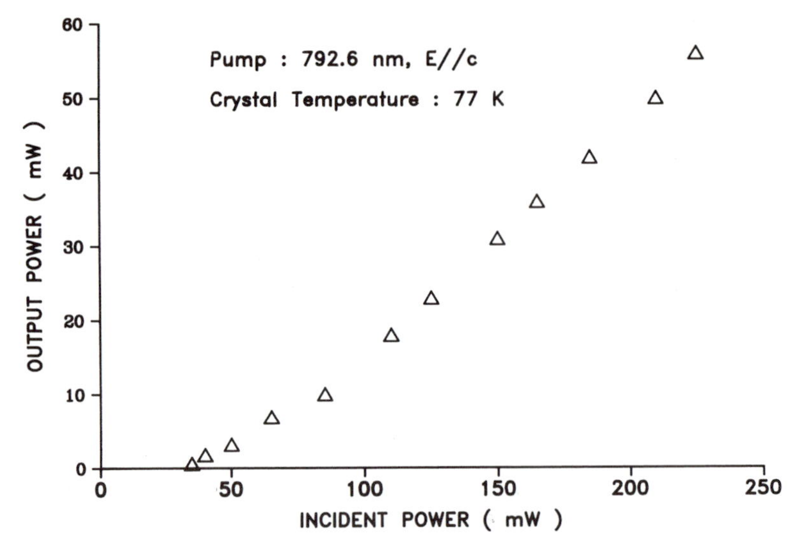

Figure 10. Output power versus incident power for diode-laser pumping. A 30% outcoupler is used.

Laser action was observed also at different pump wavelengths, near to 789.5 nm and 793.7 nm, but the observed performance was inferior.

Using the laser configuration described above we measured the output power as a function of the crystal temperature; Figure 11 displays the results recorded at a fixed pump power of 85 mW. It can be noted that the output drops to zero at a temperature of about 180 K. This result is not surprising for the reason that the outcoupler transmission and the crystal itself are not optimum for high temperature operation. The results of Ref. 10 seem to indicate that Er is not necessary for diode-laser pumping; on the contrary a crystal containing only Tm and Ho as the dopants has a superior performance at high temperature. Work along these guidelines is in progress in our laboratories.

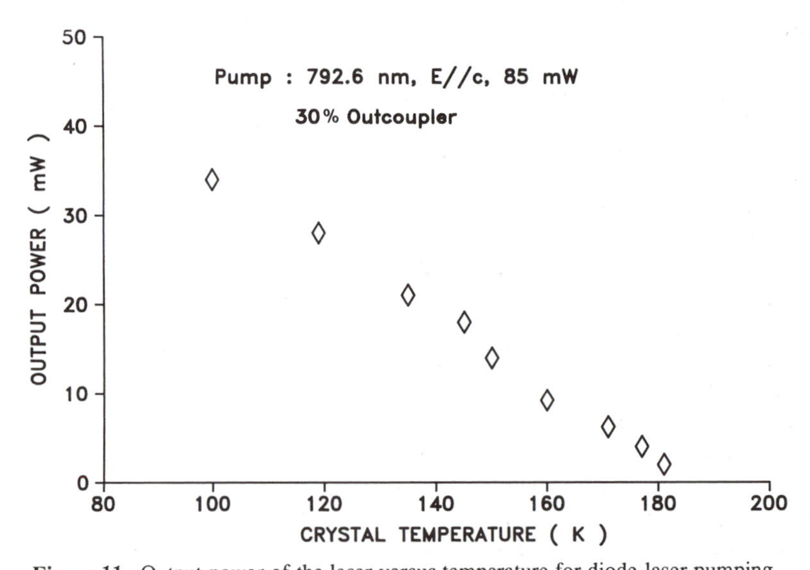

Figure 11. Output power of the laser versus temperature for diode-laser pumping.

CONCLUSIONS

Our work demonstrates that powerful tunable emission can be obtained near 2 μm from CW YLF lasers operated at 77 K. A stable source with a 1 Watt output power is very interesting for experiments of nonlinear high resolution spectroscopy. The complex structure of a three -mirror folded resonator is not a disadvantage in a laboratory environment; on the contrary it is very useful as a versatile workbench for testing new crystals and different pump sources. Research work to obtain efficient room temperature operation and compact diode-laser pumping is in progress

REFERENCES

1. L.F. Johnson, J.E. Geusic, and L.G. Van Uitert, Coherent oscillations from Tm^{3+}, Ho^{3+}, Yb^{3+}, and Er^{3+} ions in Yttrium Aluminium garnet, *Appl. Phys. Lett.* 7:127 (1965)
2. R.H. Hoskins and B.H. Soffer, Energy transfer and CW laser action in Ho^{3+}:Er_2O_3, *IEEE J. Quantum Electron.* QE-2:253 (1966)
3. R.L. Remski, L.T. James, K.H. Gooen, B. DiBartolo, and A. Linz, Pulsed laser action in $LiYF_4$:Er^{3+}, Ho^{3+} at 77 K, *IEEE J. Quantum Electron.* QE-5:214 (1969)
4. E.P. Chicklis, C.S. Naiman, R.C. Folweiler, D.R. Gabbe, H.P. Jenssen, and A. Linz, High efficiency room temperature 2.06 μm laser using sensitized Ho^{3+}:YLF, *Appl. Phys. Lett.* 19:119 (1971)
5. A. Erbil and H.P. Jenssen, Tunable Ho^{3+}:YLF laser at 2.06 μm, *Appl.Opt.* 19:1729 (1980)
6. L. Esterowitz, R. Allen, L. Goldberg, J.F. Weller, M. Storm, and I. Abella, Diode pumped 2 μm Holmium laser, *in*: "Tunable Solid-State Lasers II", A.B. Budgor, L. Esterowitz, and L.G. DeShazer, eds., Springer-Verlag, Berlin(1986)
7. V.A. Smirnov and F.A. Shcherbakov, Rare-earth Scandium Chromium garnets as active media for solid-state lasers, *IEEE J. Quantum Electron.* QE-24:949 (1988)
8. G. Huber, E.W. Duczynski, and K. Petermann, Laser pumping of Ho, Tm, Er-doped garnet lasers at room temperature, *IEEE J. Quantum Electron.* QE-24:920 (1988)
9. T.Y. Fan, G. Huber, R.L. Byer, and P. Mitzscherlich, Spectroscopy and diode laser-pumped operation of Tm,Ho:YAG, *IEEE J. Quantum Electron.* QE-24:949 (1988)
10. H. Hemmati, 2.07-μm diode-laser-pumped Tm,Ho:$YLiF_4$ room-temperature laser, *Opt. Lett.* 14:435 (1989)
11. S.W. Henderson and C.P. Hale, Tunable single-longitudinal-mode diode laser pumped Tm:Ho:YAG laser, *Appl. Opt.* 29:1716 (1990)
12. S.W. Henderson, C.P. Hale, J.R. Magee, M.J. Kavaya, and A.V. Huffakev, Eye-safe coherent laser radar system at 2.1 μm using Tm, Ho:YLF lasers, *Opt.Lett.* 16:773 (1991)
13. B.T. McGuckin and R.T. Menzies, Efficient CW diode-pumped Tm, Ho:YLF laser with tunability near 2.067 μm, *IEEE J. Quantum Electron.* QE-28:1025 (1992)
14. P.A. Budni, M.G. Knights, E.P. Chicklis, and H.P. Jenssen, Performance of a diode-pumped high PRF Tm, Ho:YLF laser, *IEEE J. Quantum Electron.* QE-28:1029 (1992)
15. T.S. Kubo and T.J. Kane, Diode-pumped lasers at five eye-safe wavelengths, *IEEE J. Quantum Electron.* QE-28:1033 (1992)
16. T.Y. Fan and R.L. Byer, Modeling and CW Operation of a quasi-three-level 946 nm Nd:YAG laser, *IEEE J. Quantum Electron.* QE-23:605 (1987)
17. W.P. Risk, Modeling of longitudinally pumped solid-state lasers exhibiting reabsorption losses, *J. Opt. Soc. Am. B* 5:1412 (1988)
18. H.W. Kogelnik, E.P. Ippen, A. Dienes, and C.V. Shank, Astgmatically compensated cavities for CW dye lasers, *IEEE J. Quantum Electron.* QE-8:373 (1972)
19. A. Di Lieto, P. Minguzzi, F. Pozzi, M. Tonelli, and H.P. Jenssen, Er,Tm,Ho:YLF laser for spectroscopic applications, *in*: "OSA Proceedings on Tunable Solid State Lasers", vol 5:227, M.L. Shand and H.P. Jenssen, eds., Optical Society of America, Washington D.C. (1989)
20. A. Di Lieto, A. Neri, P. Minguzzi, F. Pozzi, M. Tonelli, and H.P. Jenssen, Characterization and spectroscopic applications of a high-efficiency Ho:YLF laser, *in*: "OSA Proceedings on Advanced Solid State Lasers", vol 10:213, G. Dube and L. Chase, eds., Optical Society of America, Washington D.C. (1991)
21. A. Di Lieto, A. Neri, P. Minguzzi, F. Pozzi, M. Tonelli, and H.P. Jenssen, Improved operation of a CW YLF laser, *J. Physique IV* C7:407 (1991)

COLOR CENTER LASERS:
STATE OF THE ART AND RECENT DEVELOPMENTS

R. Beigang[1] and Z. Yoon[2]

[1]Fachbereich Physik
Universität Kaiserslautern, FRG
[2]Andong National University
Andong, Korea

INTRODUCTION

Tunable lasers using color centers in alkali halide crystals are well established laser systems in the near infrared spectral region (see e.g. [1] - [6]. The wavelength range from 0,8 µm up to almost 4 µm can, in principle, be covered using different types of color centers in various alkali halides. This is shown in Fig. 1 where the emission bands are displayed. Unfortunately, the performance of some of the laser systems (e.g. F_2^+, $(F_2^+)^*$) is not always ideal with respect to the generation of the desired centers, their stability during laser operation and the shelf lifetime of the crystals. In particular, the wavelength range below 2.2 µm was difficult to be covered with stable and reliable color center crystals.

Recently, a new type of laser active color centers in OH^--doped NaCl crystals was discovered by Pinto et al. [7] and investigated in great detail by Wandt et al. [8,9] and Pollock et al. [10, 11]. The $(F_2^+)_{H^-}$ centers are easy to generate, stable under laser operation and offer the possibility to achieve laser action in the near infrared spectral region. As this wavelength region is of particular importance for fiber optics communications, this new type of color center laser will be described in more detail. First a brief overview about color centers in alkali halides will be given and criteria for the optimum design of color center lasers will be discussed. Finally some applications with certain types of color center lasers are presented.

COLOR CENTERS IN ALKALI HALIDE CRYSTALS

Color centers are point defects in alkali halide crystals. The simplest color center is the F-center (from "Farbe", the German word for color) which consists of an electron bound to a halogen vacancy. The electronic states of the excited electron in the field of the lattice are strongly coupled to lattice vibrations. As a consequence all electronic states are broadened, resulting in broad absorption and emission bands.

The F-center is the basic type of all color centers. Unfortunately it is not suited for laser

Solid State Lasers: New Developments and Applications
Edited by M. Inguscio and R. Wallenstein, Plenum Press, New York, 1993

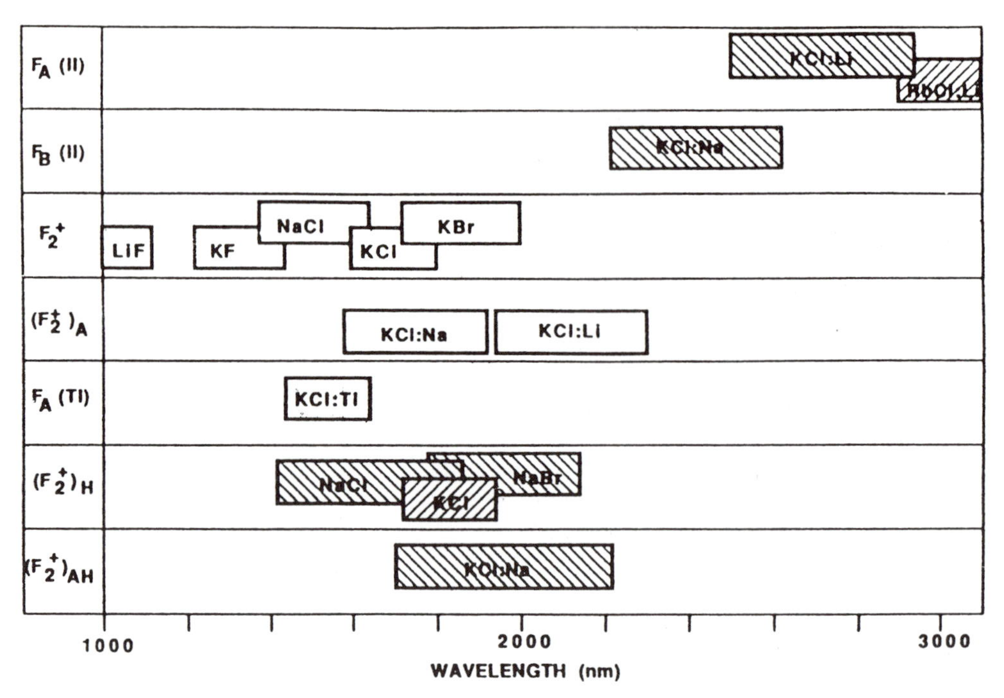

Figure 1. Emission bands of different laser active color centers in various alkali halide crystals in the wavelength range from 1 μm to 3 μm. Stable centers are marked.

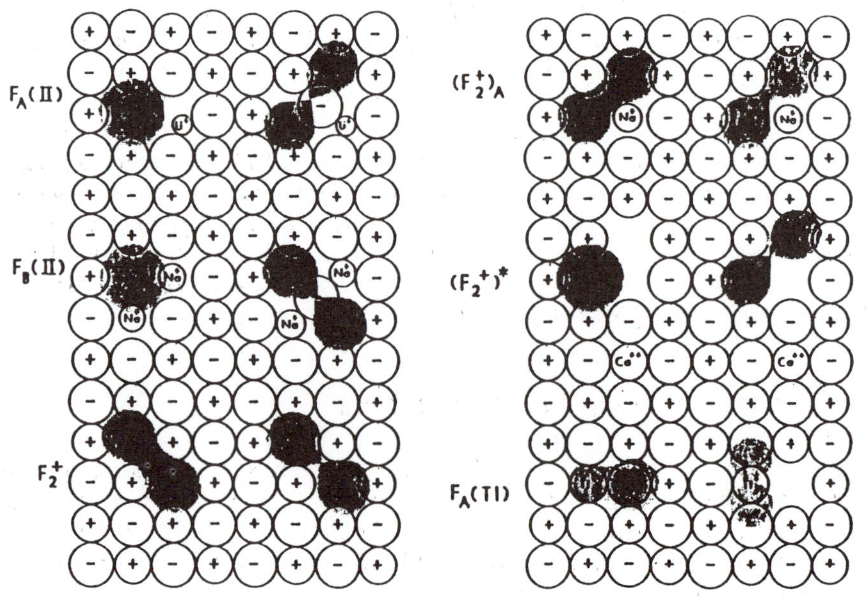

Figure 2. Ionic configuration of various types of F centers. The relaxed excited states are shown in the right part of the figure.

action itself as the oscillator strength for the laser transition is too small. There is, however, a large number of so-called F-aggregate centers which basicly consist of one or more F-centers and substitute ions as nearest neighbors. Typical examples are shown in Fig. 2.
F_A(II) and F_B (II) centers, where one or two nearest kations are substituted by a Li or Na ion, are ideally suited for tunable laser action. If two F-centers are aligned along a 110-direction an F_2-center is formed. The ionized counterpart is called the F_2^+- center. Here, one electron is bond in a double well configuration. More aggregate centers can be formed attaching foreign ions to F_2, or F_2^+-centers, like $(F_2^+)_A$ or $(F_2^+)_H$ centers, which will be discussed in more detail later.

A simplified energy level diagram of an F center is shown in Fig. 3. It represents an almost ideal 4 level scheme. After excitation from the ground state the F-centers relax within 10^{-12} sec

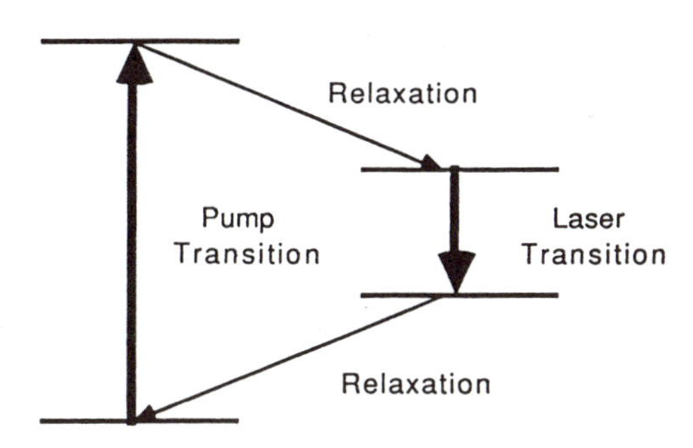

Figure 3. Simplified energy level diagram of an F center.

into the upper laser level. This relaxation may lead to a somewhat different ionic configuration (saddle point configuration). The lifetime in the upper laser level is in the order of 10 to 200 nsec with an oscillator strength for a transition to the lower laser level between 0.3 and almost 1 for F_2^+ centers. The oscillator strength may also be temperature dependent. From the lower laser level (the relaxed ground state) the F-center relaxes back to the ionic ground state again on a time scale of 10^{-12} sec. All emission and absoprtion lines are broad bands due to lattice interaction.

The generation of the desired centers strongly depends on the center under consideration. In general additive coloration or radiation damage will lead to the formation of F-centers, in most cases followed by optical aggregation resulting in F-aggregate centers. Details are given in the literature [see e.g. 4,5,6].

In the following chapters the properties of $(F_2^+)_H$ centers, their generation and the behaviour of laser systems with these centers will be discussed in more detail.

GENERATION OF $(F_2^+)_H$-CENTERS

The production of $(F_2^+)_H$- centers consists of three distinct steps:

1. NaCl crystals are grown with 5 x 10^{-4} mol% NaOH in the melt.

2. The crystals are additively colored in Na vapor at a pressure of approximately 30 mbar and a temperature of 650 º C for up to 2 hours. Using slabs not exceeding 2 mm thickness guarantees a uniform coloration of the crystals. After coloration the crystals can be stored at room temperature without deterioration. Before further processing they have to be heated to a temperature of 650 º C for about 3 to 5 minutes and then quenched immediately.

3. Optical aggregation with uv-light leads to the formation of $(F_2^+)_H$- centers. Using the uv-lines of an ion laser at a power level of 200 mW an exposure time of 2-3 minutes is sufficient to produce an optical density up to 0.8 at the pump wavelength in the infrared (λ = 1.064 μm). With a low pressure mercury lamp an exposure time of 15 to 20 minutes is required to obtain the same optical density. The $(F_2^+)_H$- centers are not stable at room temperature and the crystal has to be cooled down to low temperatures after optical aggregation. Therefore, the third step should be performed immediately before laser operation to ensure an optimum center concentration and, as a consequence, laser performance.

The quenching procdure and optical aggregation can be performed several times with the same crystal without losing optical density considerably. However, each quenching process results always in a loss of laser active centers, so that the number of preperation cycles is limited.The quality of the crystal after optical aggregation can be inspected taking an absorption spectrum as described in detail by Wandt et. al. [8]. An optical density at 1.064 μm between 0.5 and 0.8 for a 2 mm thick sample should result in excellent laser performance.

DESIGN OF COLOR CENTER LASERS

The laser configuration usually consists of an astigmatically compensated 3-or 4-mirror resonator as described in detail elsewhere [12-14]. A new improved cooling system for the color center crystal results in a very effective cooling of the crystal which is of particular importance for the $(F_2^+)_H$ centers. In this set up the crystal is mounted on a copper cold finger which is in close contact to the cooling medium (in most cases liquid nitrogen) via a pipe line system. The two spherical mirrors are integral part of the vacuum system which also contains the cooling system with the color center crystal. All other optical components are easily accessible outside the vacuum chamber. The laser was operated broadband cw, in modelocked and in single mode operation in a ring laser cavity. As the configuration for cw broadband operation is described e. g. in [3] the ring laser will be discussed in more detail.The ring cavity is shown in Fig. 4.

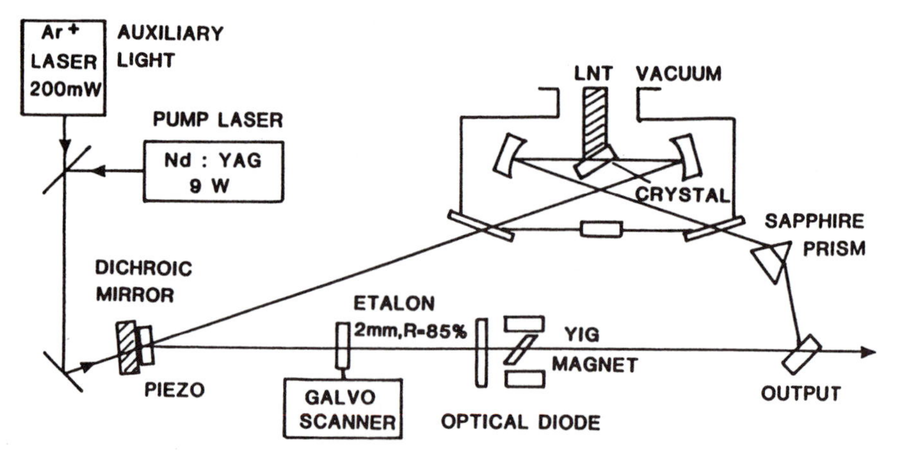

Figure 4. Schematic diagram of an astigmatically compensated 4-mirror ring laserconfiguration (for details see text).

Stable unidirectional operation can be achieved using an optical diode and an internal Fabry Perot etalon (2 mm thickness and 85 % reflectivity). The optical diode consists of a 0.5 mm YIG plate mounted at Brewster's angle in a permanent magnet as the Faraday rotator and a birefringent quartz plate. The rotation of the laser polarization varies between 0.5 degrees at 1.45 μm and 0.2 degrees at 1.75 μm due to the fact that the Verdet constant decreases with increasing wavelength. Spontaneous unidirectional and single mode operation was already achieved with the Fabry Perot etalon alone. However, as the difference in losses for the two possible directions of operation were almost negligible, the directions of the travelling wave was not predetermined. Consequently, each small perturbation of the laser cavity resulted in a change of direction of the laser oscillation. In particular, when the laser was tuned in frequency a highly unstable behavior was the result. Only the use of an optical diode, although at the expense of somewhat higher losses in the laser cavity, resulted in stable unidirectional single mode operation.

Coarse tuning over the entire tuning range was easily accomplished using a Brewster-cut sapphire prism or a birefringent filter.However, due to the high gain of the $(F_2^+)_H$ center laser at the center of the tuning range it was to allow tuning over the whole gain profile. The frequency selective losses of the three stage birefringent filter were still too small to prevent laser oscillation at the gain center when tuned to the wings of the gain profile. In combination with a particularly designed output coupler with increasing reflectivity at the wings of the reflectivity range tuning over more than 200 nm became possible [15]. A grating in Littrow mount can also be used as tuning element as shown in ref. [8] and [11]. In this case the output coupling is taken in zero order from the grating and cannot be varied independently.

In order to obtain continuous tuning in single mode operation one cavity mirror was mounted on a piezo ceramic to change the cavity length. Synchronously the internal Fabry Perot etalon was tilted by use of a galvo drive (see Fig. 4). The maximum tuning range in single mode operation was thus limited by the maximum piezo displacement. In our case a maximum displacement of 18 μm resulted in a tuning range of roughly 4 GHz. The additional losses induced by the titled etalon are even for the maximum detuning below 2 % and thus negligible.

For mode-locked operation the length of a 3-mirror cavity was matched with the length of the mode-locked pump laser. Tuning was accomlished with a two- or one-stage birefrigent filter. As in the case of the ring laser the mode-locked laser could not be tuned over the entire gain curve of the $(F_2^+)_H$ crystal with one set of resonator mirrors. Using output couplers with a particular designed coating only one set of mirrors was required to tune the mode locked laser from 1.49 μm to 1.72 μm with a two stage birefringent filter. The use of a birefringent filter instead of a prism has the advantage that there is no beam walking when the laser is tuned. This can be important for applications where the laser has to be tuned over a rather wide wavelength range and where at the same time an accurate alignment of the output beam is required.

TYPICAL PERFORMANCE OF $(F_2^+)_H$-CENTER LASERS

Cw Operation

In this chapter typical results obtained with a $(F_2^+)_H$ center laser will be discussed as this laser is of particular importance for investigations of optical fibers and fiber optic communications. In addition it represents the first **stable** laser active color center (both during laser operation and shelf lifetime) in the wavelength region between 1.4 μm and 1.9 μm. In all laser experiments reported here the auxiliary ligth which is required to sustain laser action at higher power levels was provided by an Ar^+-ion laser, which was aligned collinearly with the Nd:YAG pump laser.

The output power of the color center laser strongly depends on the intensitiy of the auxiliary

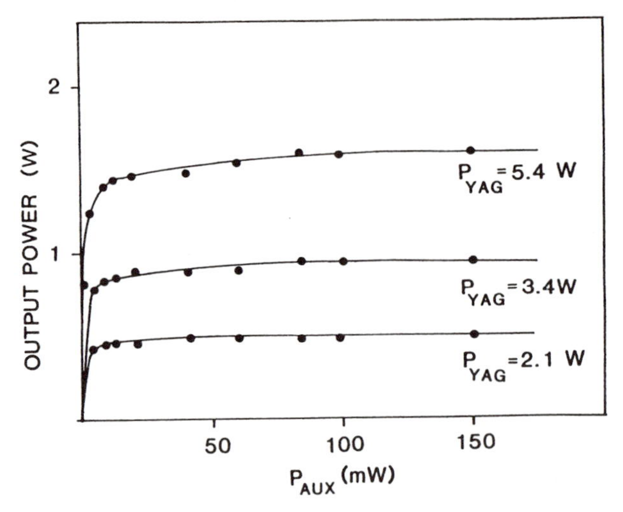

Figure 5. Output power as a function of auxiliary light.

laser and shows a saturation behavior. About 20 mW of auxiliary light were sufficient to obtain more than 90 % of the maximum output power of the color center laser. The power level where saturation occurs increases with increasing temperature of the crystal and increasing pump power. The increasing temperature obviously leads to an enhanced reorientation of the $(F_2^+)_H$-centers which can be counterbalanced by additional auxiliary light, as demonstrated experimentally. As a consequence it is essential to have an efficient cooling system for the color center crystal in order to guarantee optimum laser performance with minimum auxiliary ligth intensity. We have obtained 4 W output power at a wavelength of 1.58 µm with 15 W pump power and 80 mW auxiliary light in true cw operation without any fading of the output power or deterioriation of the crystal. Up to a pump power of 12 W no temperature effect was observed and the power dependence was still linear as shown in Fig. 6.

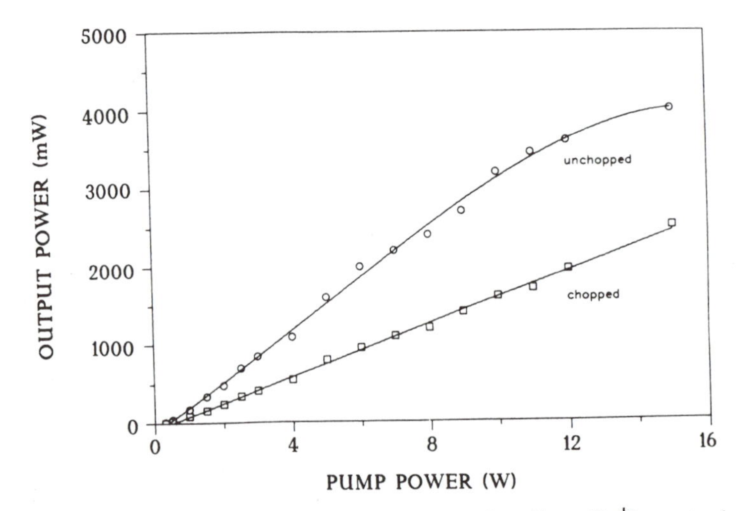

Figure 6. Output power as a function of pump power for a linear $(F_2^+)_H$ center laser.

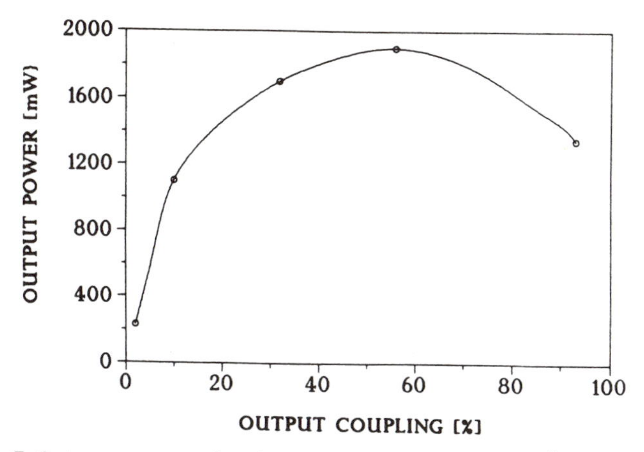

Figure 7. Output power as a function of output coupling for a $(F_2^+)_H$ center laser.

We found experimentally that the output power did not strongly depend on the output coupling. A variation of the output coupling between 10 % and 80 % did not result in a considerable change of the output power (see Fig. 7). This behavior is characteristic for a high gain laser medium. Even with a simple uncoated quartz plate as end mirror an output power of more than 1 Watt was achieved demonstrating the high gain of the system.

A typical tuning curve using a prism as tuning element is shown in Fig. 8. The tuning range is mainly limited by the reflectivity of the dielectric resonator mirrors, which extended from 1.4.μm to 1.75 μm. Although the transmission of the output coupler varied over the entire tuning range from 30 % at the maximum of the gain curve to about 50 % at 1.7 μm and 1.4 μm, respectively, the output power is nearly constant over a wide wavelength range. This is again in agreement with the above mentioned behavior.

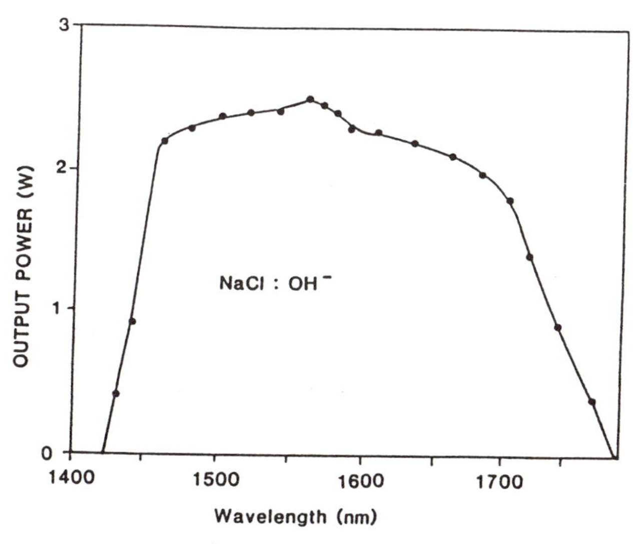

Figure 8. Tuning curve of a linear $(F_2^+)_H$ center laser. A sapphire prism was used as tuning element.

In single mode operation with the ring laser cavity as described before a maximum output power of 2.2. W was achieved. The optical diode and the internal Fabry Perot etalon provided stable unidirectional operation over the entire tuning range. Using a prism as coarse tuning element the laser was tunable between 1.45 μm and 1.78 μm. For longer wavelengths the single mode operation became instable due to the reduced rotation of the polarization caused by the decreased Verdet constant of the Faraday rotator. With a thicker YIG plate stable single mode operation should be obtainable over the whole tuning range.

The linewidth of the free running color center laser was investigated by means of a stable external Fabry Perot etalon with a free spectral range of 2 GHz. Amplitude fluctuations in the linear part of the transmission curve were analyzed with a computer controlled multichannel analyzer. The linewidth of the NaCl:OH^- $(F_2^+)_H$ color center ring laser was measured to be smaller than 2 MHz. This is an upper limit of the actual linewidth as amplitude fluctuations which also contribute to the measured frequency fluctuations were not separated. The main reason for the observed linewidth is the amplitude instability of our pump laser as pump power fluctuations directly result in frequency fluctuations of the color center laser. Using an amplitued stabilization of the pump laser the linewidth should be considerably reduced as in the case of the other color center lasers [16, 17].

Mode Locked Operation

In mode-locked operation we have obtained stable mode-locking performance with pulses as short as 4 ps. The pulse length was measured using an autocorrelation technique with a $LiJO_3$ crystal as the frequency doubling crystal. A typical result of a pulse width measurement is shown in Fig. 9. The measured pulse width of 8 ps corresponds to an actual length of the pulse of approximately 5ps when a $(sech)^2$ pulse shape is assumed. The frequency spectrum as

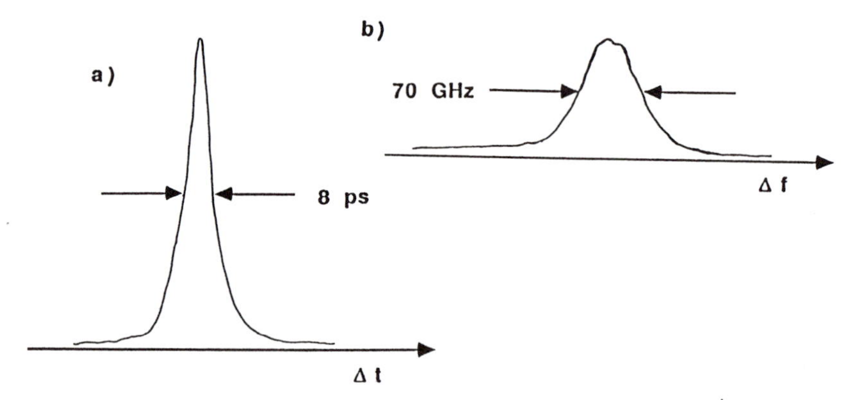

Figure 9. a) Autocorrelation trace as obtained from a mode-locked $(F_2^+)_H$ color center laser. b) Frequency spectrum of the pulse measured in a).

displayed in the left part of Fig. 8 indicates that the pulses are Fourier transform limited. Typical output powers of 1.5 W were obtained with pump power levels of 6 W at a repetition rate of 82 MHz. The length of the pump pulses was in the order of 150 ps. The conditions for mode-locked operation were easily adjusted and due to a stable invar resonator construction the laser was operated over a long period of time without pulse broadening or instabilities over the entire wavelength range from 1.5 μm to 1.7 μm. Using a ring cavity and a Nd:YLF pump laser with pump pulse lengths of about 50 ps the color center laser produced tunable pulses between 1 and 2 ps [18].

The advent of new mode locking techniques like additive pulse mode locking (APM) or Kerr lens mode locking (KLM) (see e. g. [19, 20, 21]) resulted in the generation of sub picosecond pulses with a minimum pulse length of 60 fs [22].

APPLICATIONS

The characteristic properties of color center lasers as discussed here make those laser systems very attractive for a variety of applications. In basic research the spectroscopy of atoms and molecules is certainly the main field of applications taking advatage of the narrow bandwidth and wide tuning range in both cw and pulsed operation.The small linewidth in combination with high single mode output power in cw operation is essential for nonlinear spectroscopy. Due to the emission range in the near infrared spectral region molecular species are preferable objects of investigations. In combination with other excitation schemes, however, also highly excited atomic states can be studied. With pulsed systems time resolved experiments are possible in atoms and molecules.

In addition to spectroscopy, applications in laser chemistry seem feasible where the color center laser is used to initiate a chemical reaction by selective excitation of ro-vibrational levels of molecules. For this kind of experiments where high energy is required a flashlamp pumped system is most appropriate as it delivers energies per pulse in the mJ-range. For environmental research tunable infrared lasers are also ideal light sources. Most species to be detected in the atmosphere contain compounds which absorb strongly in the near infrared (e. g. CH, HF, etc.). As all these species have characteristic absorption bands the color center laser can be tuned to the specific absorption band so that particular pollutions can be detected without interfering with other atmospheric molecules.

Another field where the color center was already applied successfully is solid state physics and the investigation of optical fibers. The wavelength range between 0.8 μm and 3.5 μm is of particular importance for the investigation of narrow bandgap semiconductors and the propagation of light in fiber optic communications. The availability of tunable laser systems with picosecond pulses is the main key for investigations in these areas. Using a mode-locked F_2^+ center laser in KF around 1.3 μm the distortionless propagation of picosecond pulses in kilometer-length optical fibers was demonstrated (see e. g. [3] and references therein). Another intersesting phenomena connected with the nonlinearity of the index of refraction is the propagation of solitons. These pulses either do not change shape or have shapes that periodically vary with propagation along the fiber. This periodic pulse shape variation with propagation was fisrt demonstrated by Mollenauer et al. [3 and references therein] using a mode-locked F_2^+ center laser in NaCl and a 700 m long single mode silica glass fiber.

Finally it should be mentioned that color center lasers have also been used as part of an optical frequency synthesis chain as a link between the infrared HeNe laser and the visible.

SUMMARY

Color center lasers are efficient powerful laser systems covering the near infrared spectral region joining directly the emission range of tunable dye lasers. They are widely tunable and can be operated in continuous wave, pulsed or mode-locked operation. As a consequence, they can be used for a great number of applications in basic and applied research. In particular, $(F_2^+)_H$-and similar centers are stable and efficient laser centers for the wavelength range between 1.4 μm and 2.2 μm.

REFERENCES

[1] L. F. Mollenauer and D. H. Olson, Appl.Phys. Lett.**24** (1974) 386

[2] H. Welling, G. Litfin and R. Beigang,Springer Series in Optical Sciences, **7:** Laser Spectroscopy III (1977) p. 370

[3] L. F. Mollenauer, Methods of Experimantal Physics **15B** (1979) p. 1
L. F. Mollenauer, CRC Handbook of Laser Science and Technology, Vol. 1, ed. M. J. Weber, Boca Raton, Fl. (1982)

[4] C. R. Pollock, J. Luminescence **35** (1986) 65

[5] W. Gellermann, K. P. Koch and F. Lüty, Laser Focus **4** (1982) 71

[6] I. Schneider and L. L. Marquardt, Optics Lett. **6** (1981) 627

[7] J. F. Pinto, E. Georgiou and C. R. Pollock, Optics Lett. **11** (1986) 519

[8] D. Wandt, W. Gellermann, F. Lüty and H. Welling, J. Appl. Phys. **61** (1987) 864

[9] D. Wandt and W. Gellermann, Optics Comm. **61** (1987) 405

[10] E. Georgiou, J. F. Pinto and C. R. Pollock, Phys. Rev **B35** (1987) 7636

[11] K. R. German and C. R. Pollock, Optics Lett. **12** (1987) 474

[12] H. W. Kogelnik, E. P. Ippen, A. Dienes and C. V. Shank, J. Quant. Electr. **QE-8** (1972) 373

[13] W. W. Rigrod, Bell System Tech. Journal **5** (1965) 907

[14] C. R. Pollock, F. K. Tittel and G. Litfin, Proc. Int. Conf. on Lasers 80 (1983) p. 123

[15] G. Phillips, P. Hinske, W. Demtröder, K. Möllmann and R. Beigang,Appl. Phys. **B47**, 127 (1988)

[16] J. L. Hall, T. Baer, L. Hollberg and G. Robinson, Springer Series in Optical Sciences: Laser Spectroscopy V (1981) p. 15

[17] R. Beigang, G. Litfin and H. Welling, Optics Comm. **22** (1977) 269

[18] T. Kurobori, A. Nebel, R. Beigang, and H. Welling, Opt. Commun. **73**, 365 (1989)

[19] C. P. Yakymyshyn, J. F. Pinto, and C. R. Pollock, Opt. Lett. **12**, 621 (1989)

[20] G. Sucha, Opt. Lett. **16**, 922 (1991)

[21] R. S. Grant and W. Sibbett, Opt. Commun. **86**, 177 (1991)

[22] F. M. Mitschke and L. F. Mollenauer, Opt. Lett. **12**, 407 (1987)

FIBRE LASERS

David C. Hanna

Optoelectronics Research Centre
University of Southampton
Southampton, S09 5NH
United Kingdom

INTRODUCTION

The first fibre laser, described by Snitzer[1], was also the first glass laser. The glass fibre, doped with trivalent neodymium ions (Nd^{3+}) was transversely pumped by a flash lamp. The fibre geometry was chosen to exploit the brightness enhancement of pump light in the core after passing through the clear cladding, and also to provide a structure which was more forgiving of the optical inhomogeneities of the doped glass[2]. These early fibres clearly demonstrated the capability for very high gain, with 47dB measured[3] at 1.063μm, even if the transverse pumping geometry was of very low efficiency.

The next major development was the demonstration that longitudinal pumping, using a laser as the pump source and end-launching the pump light into the doped core, gave greatly reduced threshold, and high efficiency[4]. In this way cw operation was easily obtained and it was recognised that pumping of Nd-doped fibre using laser diodes operating at $\sim 0.8\mu$m would provide a simple and practical form of fibre laser.

The third major development, coming 12 years later, and which initiated an enormous growth of interest in fibre lasers and amplifiers, was the demonstration of lasing and amplification in rare-earth-doped monomode silica fibres[5,6]. The erbium-doped fibre amplifier (EDFA) has been the main reason for this expositon of interest. Its success is based on the fact that erbium-doped silica fibre offers an amplifying transition[7] at $\sim 1.5\mu$m, matching perfectly to the third telecommunications window, in a fibre which is compatible with telecom fibre, and which can be readily pumped by a diode laser to give high gain (in excess of 30dB, i.e. an intensity gain of 1000). These, and other attractive features of EDFAs have revolutionised the field of optical fibre communications.

While the EDFA has received the greatest share of attention, there is also a growing appreciation that fibre lasers and amplifiers have a great deal more to offer. For example, as a result of developments in fibres based on heavy metal fluoride glasses, there are now seen to be very exciting prospects for compact, visible, sources in the form of blue, green and red fibre lasers pumped by infrared diode lasers[8-13]. There is also a growing realisation that while fibre lasers may have very low threshold pump powers, this does not imply that they are necessarily low power devices and indeed it has been demonstrated that powers in excess of 1 watt are achievable from diode-pumped fibre lasers[14]. The broadened lines of rare-earth ions in glasses means that significant tuning ranges are available, in some cases[15] broader than the range typically offered by a dye laser, and yet despite this broad linewidth it has been possible to demonstrate very narrow linewidth operation (a few kHz) in single-frequency fibre lasers[16]. By contrast, on the other hand, the high gain can allow "mirrorless lasing", resulting in a broad, mode-less continuum as a result of amplified spontaneous emission (ASE)[17]. The broad gain profile has also been exploited in numerous demonstrations of self-mode-locked operation of fibre lasers where soliton pulse propagation behaviour is clearly implicated[18,19].

Thus the range of operating conditions, for fibre lasers, in power, wavelength, spectral and temporal behaviour, provides a very rich field for investigation and exploitation. In this brief review we therefore concentrate on basic features of fibre lasers and discuss particular lasers mainly as a means of illustrating general ideas. A list of laser transitions that have so far operated as fibre lasers is also provided with appropriate references. In fact the number of publications on fibre lasers is growing very rapidly and the list provided in this review is only a very small fraction of the total. Many further references, and an extensive review of fibre lasers and amplifiers can be found in the book edited by France[20].

BACKGROUND

The most striking feature of fibre lasers and amplifiers is the very high gain that can be achieved for very modest pump power, when the active core has the typical dimensions of a monomode fibre (few microns diameter) and the fibre is pumped by end-launching of the pump into the core. A figure of ~10dB gain per milliwatt of absorbed pump power has been reported for the EDFA[21]. Another striking feature is the very low pump power needed to saturate the pump transition, typically a few milliwatts. Even with this very low power, most of the population can be excited out of the ground level so that a three-level laser takes on the character of a four-level laser, i.e. having an essentially empty lower laser level, when a few milliwatss of pump are used. The ease with which a strong inversion can be produced has enabled laser operation to be achieved on many transitions which had not previously been operated in bulk form, including transitions with very low quantum efficiency. We begin therefore by giving a simple analysis which leads to an expression for the gain per unit of absorbed power.

We assume the fibre to have a step-index-profile, with refractive indices n_{co}, n_{cl} in the core and cladding respectively and a core diameter of 2a. It is customary to define a quantity, referred to as the V-value, which, for light of wavelength, λ, (in vacuum), is given by[22]

$$V = \frac{2\pi a}{\lambda} (n^2_{co} - n^2_{cl})^{1/2} \tag{1}$$

The quantity $(n^2_{co} - n^2_{cl})^{1/2}$ is known as the Numerical Aperture (NA) and defines the maximum acceptance angle θ_m for the end-launched light incident on the core via

$$\sin\theta_m = NA \tag{2}$$

For the fibre to be monomode at the wavelength λ (ie for the LP_{11} mode to be cut off, leaving only the LP_{01} able to propagate), V must be less than 2.4. With typical values inserted in Equation (1), such as $\lambda = 1\mu m$ and $NA = 0.15$, then for $V = 2.4$, one has a $\sim 2.5\mu m$. Since the small core area leads to the high gain it is beneficial to minimise a. This can be done by using as large an NA as possible, although this will generally be limited by the maximum available index difference. One can also operate with a V-value somewhat smaller than 2.4, hence a smaller value of a, but for much smaller V-values the problems of bending loss can become significant[23].

We now consider a 4-level laser, with energy levels 1-4 as shown in Figure 1, and population densities N_1, N_2 etc respectively. The upper laser level, 3, has a lifetime τ, the lower laser level, 4, is assumed empty, and the emission cross-section for the laser transition 3-4 is σ_l. Pumping is from level 1 to level 2, with subsequent decay to level 3 and we shall assume a pumping quantum efficiency ϕ_p, this being the fraction of absorbed pump photons that yield an ion in level 3.

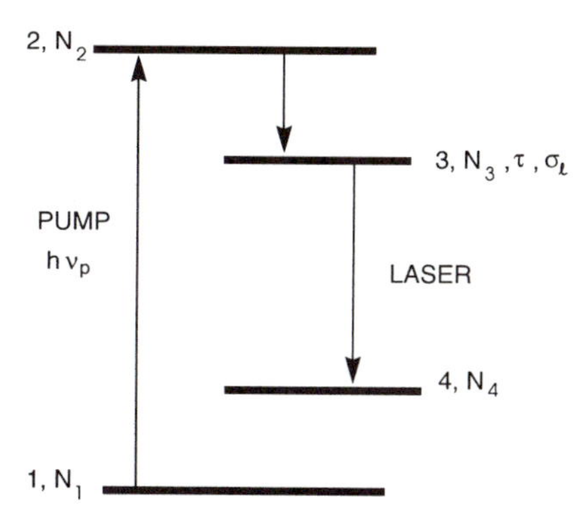

Figure 1. Schematic 4-level laser. For a true 3-level laser, the lower laser level, 4, and the ground level, 1, are the same.

Suppose that a continuous pump power P is absorbed in the fibre core and that the laser is below threshold, so that stimulated emission does not contribute to de-excitation from level 3. Then the total rate of excitation of population into level 3 is $P\phi_p/h\nu_p$ and this must equal the rate of de-excitation, ie

$$P\phi_p/h\nu_p = \frac{\int_0^l N_3(z)\, dz.\, A_{core}}{\tau} \tag{3}$$

where A_{core} is the core area, and l is the fibre length

The gain (ie gain exponent) at frequency ν is given by $\sigma_l(\nu) \int_0^l N_3(z)\, dz$, ie

$$\text{gain} = \frac{\sigma_l(\nu)\tau\, P\phi_p}{h\nu_p\, A_{core}} \tag{4}$$

This can be re-expressed, by making the following substitutions:

$$\tau = \phi_r \tau_r \tag{5}$$

where τ_r is the radiative lifetime corresponding to the laser transition, ϕ_r is the radiative quantum efficiency, (ie the fraction of ions in the upper level that decay via radiation on the laser transition) and[24]

$$\tau_r = \frac{\lambda_l^2}{8\pi n^2 \int \sigma_l(\nu) d\nu} \tag{6}$$

so that the gain, expressed in dB/W becomes

$$\text{gain(dB/W)} = \frac{4.3\ \sigma_l(\nu)}{\int \sigma_l(\nu) d\nu} \quad \frac{\phi(NA)^2}{2n^2V^2 h\nu_p} \tag{7}$$

where $\phi = \phi_r \phi_p$

Taking the case of a Lorenztian line, for which[24]

$$\sigma_l(\nu) \;/\; \int \sigma_l(\nu) d\nu = 2/\ \pi\ \Delta\nu\ \text{(FWHM)} \tag{8}$$

and assuming $V = 2.4$ yields

$$\text{gain (dB/mW)} = (NA)^2\ \phi\ /(10^4 \Delta\nu h\nu_p) \tag{9}$$

For a three level laser one can derive an appropriate analytical expression very similar to (9), ie

$$\text{gain (dB/mW)} \sim \frac{(I_p/I_{sat} - 1)}{(I_p/I_{sat} + 1)}\ (NA)^2\ \phi\ /(10^4 \Delta\nu h\nu_p) \tag{10}$$

where I_{sat} is the saturation intensity for the pump transition, i.e. $I_{sat} = h\nu_p/\sigma_A\tau$ where σ_A is the absorption cross-section.

From Equation (10) it is seen that for $I_p >> I_{sat}$ the gain expression for the three level laser is essentially the same as for the four-level laser, as expected on the basis that most of the ground state population is excited so that the lower laser level is essentially empty.

In practice many laser transitions in fibres involve a final laser level which is an excited Stark level of the ground manifold, which is not completely empty, so that one deals in practice with a system that is intermediate between 3- and 4-level ("quasi-3-level" or "quasi-4-level").

It is instructive to look at the typical magnitude of gains indicated by Equations (9) and (10). In the case of Nd-doped fibre, pumped at $\sim 0.8\mu m$, taking the pumping quantum efficiency to be ~ 1 and for lasing at $\sim 1\mu m$ the radiative quantum efficiency into the $1\mu m$ transition to be ~ 0.5 (emission also occurs at $1.3\mu m$ and $\sim 0.94\mu m$ [25-27]). The linewidth is typically 3THz. With an NA of 0.15 this implies a gain of 1.5dB/mW.

A somewhat larger figure has been achieved in Er-doped silica, where $\phi \sim 1$, the pump wavelength can be longer (eg 0.98μm), and where a large NA has been deliberately used to enhance the gain, giving up to 11dB/mW [21].

These very high gains have a number of important consequences. First it means that diode lasers can be used as pumps and readily produce gains of 20dB or more. Thus one has available a very practical means of pumping. It also means that in many situations high losses can be tolerated. This can give greater freedom of design in a fibre laser resonator where the inclusion of a large number of lossy elements need not preclude efficient low threshold operation. With such high gains it is easy to achieve oscillation from the bare, uncoated fibre ends. This oscillation can be suppressed by such means as immersing the fibre ends in index-matching liquid, use of antireflection coatings, or of deliberately angled ends. Even with complete suppression of feedback, at sufficiently high gains, typically 30-40dB per single pass, the process of amplified spontaneous emission (ASE)[17] will take place with high efficiency and limit the achievable gain. This corresponds to a situation where photons spontaneously emitted at one end of the fibre are amplified on a single pass to an intensity that significantly depletes the gain. Such conditions can be readily achieved with diode-laser pumping.

The ability to produce high gain for modest pump power opens up the possibility of achieving laser oscillation on transitions of low quantum efficiency[17]. Even quantum efficiencies as low as 10^{-3} can be successfully exploited since this would still imply gains of several dB/watt. With care over the cavity design, losses can be kept down to a fraction of a dB so that thresholds of $\sim$100mW are not unreasonable for these low quantum efficiency transitions. It should also be noted that a low quantum efficiency need not imply a low slope efficiency for the laser. While the threshold varies as ϕ^{-1} (see Equation (9)), the slope efficiency varies as ϕ_p and does not depend on the radiative quantum efficiency ϕ_r. So, if the low quantum efficiency ϕ arises from a low ϕ_r, but $\phi_p \sim 1$, then although the threshold is increased by the low ϕ_r, the slope efficiency can still approach unity since, once threshold is exceeded, each absorbed pump photon can yield a laser photon. In fact the slope efficiency $\sim \phi_p(\nu_l/\nu_p)T/(L+T)$ where T is the output mirror transmission and L represents all other losses per round trip.

Another benefit of the high gain is that it can allow oscillation to be tuned into the far wings of the laser transition since enough gain can still be accessed in these regions. Generally the tuning extent is in fact limited, not by any inherent insufficiency of gain in the wings, but by excessive gain at line centre where ASE will set in and limit the gain. Substantial tuning ranges have been achieved in Yb [5,28] and Tm [29,30], 1.01-1.17μm and 1.65-2.05μm, respectively, before ASE limitations set in. By careful resonator design, to minimise the losses at the desired wavelength in the wings, it should be possible to achieve even wider tuning ranges on these and other transitions. With three-level lasers the effects of ASE at line centre can be countered by using a longer length of fibre so that at the distant end (from the pump) the pump intensity has dropped to a level which does not give sufficient inversion for net gain at line centre. In the long wavelength wing where the emission approaches a four level character, involving termal Stark levels which are above the ground level, gain is achieved for a smaller population in the upper laser level. Thus line centre ASE can be suppressed while gain is accessible in the long wavelength wing.

The last part of the above discussion indicates that in practice, for many of the broadband transitions of rare-earth dopants, in glass, it is not appropriate to refer to the laser level scheme as simply a "3-level laser" or a "4-level laser" since in practice the character of the emission can vary from essentially pure 3-level to pure 4-level, (ie empty laser level) as the emission wavelength goes from line centre to the long wavelength limit. In practice, for the purpose of calculating gain, all that is needed are the absorption and emission cross-sections $\sigma_a(\nu_l)$, $\sigma_e(\nu_l)$, and the population densities N_3

and N_1 ($=N_4$, where we have assumed level 4 and level 1 are the same manifold, see Figure 1). The gain exponent is then

$$\int_o^l (N_3(z)\sigma_e(\nu_l) - N_1(z)\sigma_a(\nu_l))dz$$

The emission cross-section can be calculated from the observed fluorescence emission, combined with the known radiative decay rates[31]. $N_3(z)$ and $N_1(z)$ are calculated from the following equations (11)-(13):

$$dI_p(z)/dz = -\alpha_p(I_p,z)I_p(z) \tag{11}$$

where $\alpha_p(I_p,z)$ is the pump absorption coefficient at location z. α_p is given by

$$\alpha_p(I_p,z) = \frac{\alpha_p(o,z)}{1 + I_p(z)/I_s} \tag{12}$$

where the pump saturation intensity I_s is given by

$$I_s = h\nu_p/(\sigma_e(\nu_p) + \sigma_a(\nu_p))\tau \tag{13}$$

With (13) and (12) substituted into (11) one can solve for $I_p(z)$, thus from (13) obtain α_p as a function of z, and then by using the relation

$$\begin{aligned}\alpha_p(z) &= N_1(z)\sigma_a(\nu_p) - N_3(z)\sigma_e(\nu_p)\\ &= N_1(z)\sigma_a(\nu_p) - (N_0-N_1(z))\sigma_e(\nu_p)\end{aligned} \tag{14}$$

$N_1(z)$ can be calculated, hence $N_3(z) = N_0-N_1(z)$ is found and finally the gain exponent can be calculated from

$$\int_o^l (N_3(z)\sigma_e(\nu_l) - N_1(z)\sigma_a(\nu_l))dz$$

Here we have described a general procedure for calculating gain from a given length of fibre, where the pump wavelength may in general overlap with the laser emission transition, so that pump photons cause both excitation and de-excitation. This degree of generality is necessary in circumstances where so-called "in-band" pumping is used, as for example in the EDFA where a pump at $\sim 1.48\mu m$ can be used[32] to produce gain at $\sim 1.53\mu m$ so that the pump is essentially in the short wavelength wing of the pump band but also has a significant overlap with the short wavelenth wing of the emission band. Another important example of such in-band pumping is the $0.8\mu m$ transition in Tm, where pumping at $\sim 0.78\mu m$ produces gain at $\sim 0.81\mu m$ [33]. Where the pump band is well separated from the emission band the analysis is of course simpler and one notes for example that the saturation intensity given in Equation (13) then reduces to the familiar form $I_s = h\nu_p/\sigma_a(\nu_p)\tau$, as quoted after Equation (10).

The saturation power, P_s, is simply I_sA_{core}, and it is interesting to note how low a value the saturation power can have typically. In erbium-doped silica with

$\sigma_a \sim 4x10^{-25}m^2$ for 980nm, and $\tau \sim 10^{-2}s$, one has $I_s \sim 5x10^7 w/m^2$, and for a 6μm diameter, P_s is $\sim$1.5mW. Thus, strong saturation by the pump is a typical situation in practice. This fact emphasises the potential of fibre lasers for sequential pumping to higher levels since with powers of a few milliwatts it is possible to essentially empty the ground level and put all the population in an excited level from which further excitation is possible.

Since the pump can strongly saturate the absorption the pump will propagate much further into the fibre than suggested simply by the extinction length calculated from the small-signal pump absorption coefficient. In fact, a useful approximate result is that if the pump power at the input is x times the saturation power, it will only drop to the saturation power after travelling a distance of $\sim$x times the small signal extinction length. So, for a three level laser, pumped by x times the saturation power, the gain would increase with fibre length, up to x extinction lengths, and thereafter decrease as the medium would no longer be inverted at greater distances along the fibre. One sees that 3-level lasers therefore have an optimum length in terms of maximum gain and that this length is pump-power dependent, increasing with increasing pump power. For four-level lasers, on the other hand, assuming no loss, there would be no optimum length, gain being greatest for infinite length. In practice, of course, finite losses imply an optimum length even for four-level systems. Since the character of the emission can vary from 3- to 4-level across an emission band, so the optimum length for maximum gain is a function, not only of pump intensity but also of emission wavelength.

As a final word on this background material it should be added that the procedure for calculating gain as described here is an approximate one, susceptible to straight-forward numerical calculation. In practice, for a detailed and realistic modelling of actual gain it may also be necessary to include a number of other features, such as the fact that the dopant may not be uniformly distributed across the core. Also the fact that the pump and signal modes are not top-hat functions needs to be included in the form of overlap integrals[34], and the model also needs to include the fact that the saturation behaviour now has a transverse dependence. Finally at high gains the effect of ASE[35] must also be included since it affects the gain, by saturation, and also influences the noise behaviour of the amplifier.

A Survey of Fibre Laser Transitions

Figure 2 shows the energy levels of the tri-valent rare-earth ions and at a first glance suggests a very wide choice of potential laser transitions.

In practice many of the levels have very short lifetimes, so they are not good candidates for laser levels. The main mechanism responsible for shortening level lifetimes is decay to a lower level via multiphonon emission, where conservation of energy requires that the total energy of the created phonons equals the energy difference, ΔE, between the initial and final levels of the ion. This nonradiative process has a decay rate W_{nr} that obeys a relation[36],

$$W_{nr} = B\exp\left[-\alpha(\Delta E - 2h\nu)\right] \qquad (15)$$

where $h\nu$ is the highest phonon energy in the phonon spectrum of the host. The equation shows that the decay rate in a given host (fixed α, B, $h\nu$) decreases orientially with increasing energy gap ΔE and the multi-phonon decay rate is lower in materials for which the maximum phonon energy is lower. This feature has given considerable

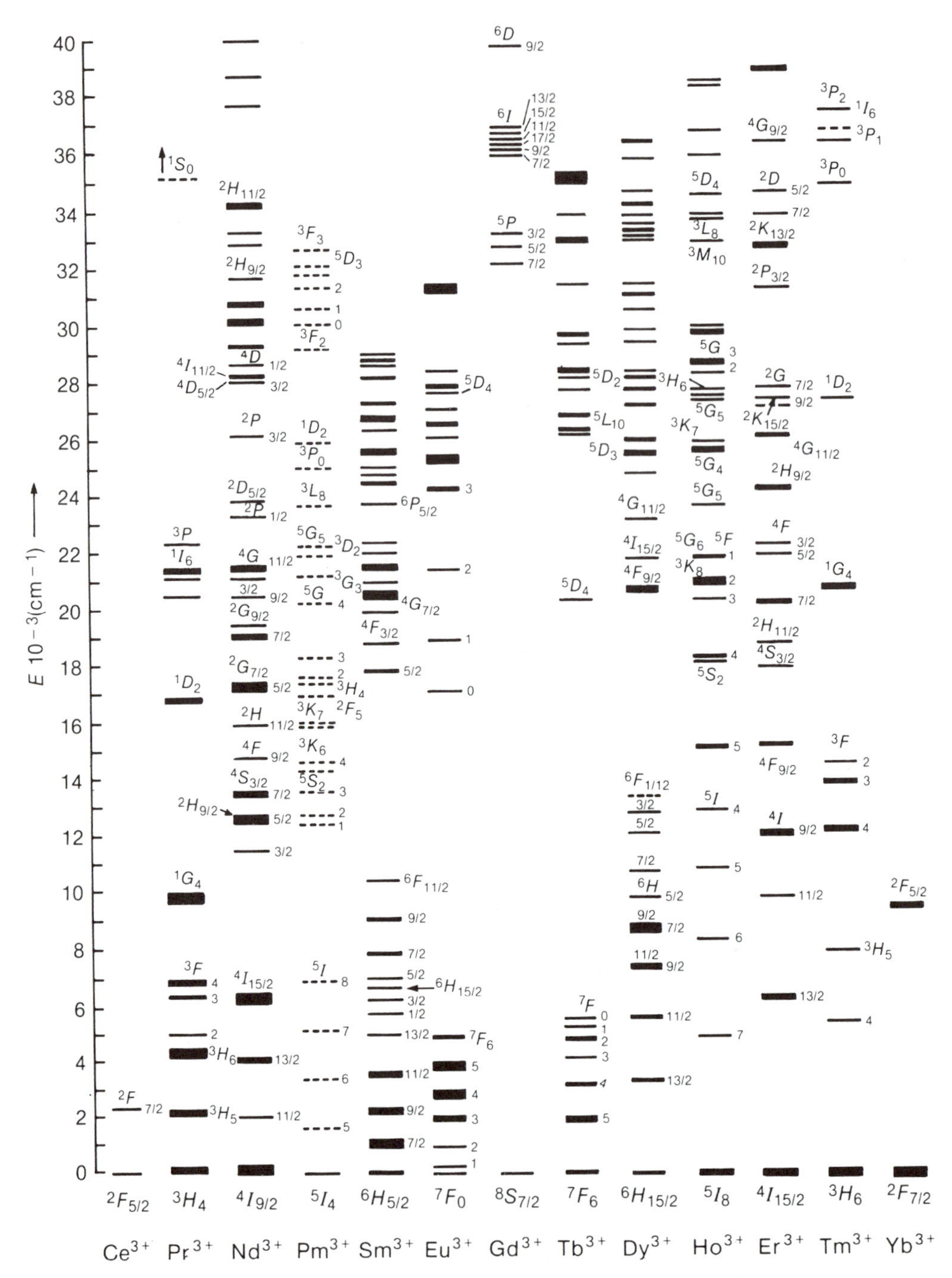

Figure 2. Energy levels of tri-valent rare-earth ions (after ref 70).

prominence in recent years to glasses with lower phonon energies, such as the heavy-metal fluoride glasses[37,38], of which the Zirconium Fluoride glass known as ZBLAN[37] has been most extensively used. Particular interest in the rare-earth doped ZBLAN fibres has arisen for two reasons. The first relates to the search for a fibre amplifier suitable for the important telecom window at $\sim 1.3\mu m$, and the second relates to their potential for visible upconversion lasers. A third reason, which is gaining prominence, is the capability for longer wavelength infrared laser operation (to $\sim 3.5\mu m$ so far[39]). The lower phonon energy is then necessary for two reasons first to prevent multiphonon decay across the small energy gap of the infrared transition, and secondly to provide good transparency in the infrared by virtue of the infrared absorption edge being shifted to longer wavelengths in low-phonon-energy materials.

The search for a suitable fibre amplifier for the $1.3\mu m$ window has proved much less straightforward than for the $1.5\mu m$ window. A transition at $1.3\mu m$ in Nd^{3+} ($^4F_{3/2}$ - $^4I_{13/2}$) was found to suffer from strong excited state absorption (ESA) at the $1.3\mu m$ signal wavelength[26], this absorption being particularly strong in a silica host. While the ESA was found to be weaker in Nd:ZBLAN[40], it was still troublesome and in any case the rather weak $1.3\mu m$ transition suffered from the problem of there being a much higher gain (with the associated problem of ASE) on the $1\mu m$ transition which shares the same initial level. The search for a alternative $1.3\mu m$ transition has now centred on the $^1G_4$ - $^3H_5$ transition in Pr^{3+}. However the $^1G_4$ level has other levels below it, much closer than $^3H_5$, (the $^3F_4$ level is only $\sim 3000cm^{-1}$ below $^1G_4$), so that multiphonon decay out of $^1G_4$ precludes the use of silica host. The decay rate of Pr^{3+} $^1G_4$ in ZBLAN is much reduced (although still significantly shortened by multiphonon decay) and appears to offer a viable $1.3\mu m$ amplifier. In fact, a very rough guide to the effect of multiphonon decay would indicate that an energy gap of $\sim 4500cm^{-1}$ in silica, and $\sim 3000cm^{-1}$ in ZBLAN are the minimum required to avoid rapid multiphonon decay.

The role of multiphonon decay is also of very great importance to upconversion lasers, where the aim is to pump an ion to a high energy level by sequential absorption of two or more photons, followed by laser emission of a single more energetic photon. For this to succeed the intermediate level(s) and the upper laser level must not suffer from an excessive rate of multiphonon decay. This again puts hosts with low phonon energies at an advantage and it is notable that all of the upconversion fibre lasers to date (in Tm[13,41], Ho[8], Er[11,12], Pr[9,10]) have all made use of a ZBLAN host. The use of ZBLAN has greatly extended the range and number of laser transitions and there are undoubtedly very many more laser transitions yet to be demonstrated in ZBLAN fibre. The success of ZBLAN has also prompted the search for alternative low phonon energy glasses which may lend themselves to simple preparation and simpler fibre fabrication than is the case for ZBLAN. The table below gives a list of fibre laser transitions observed to date, with representative, but not exhaustive, references, where detailed performance data can be found. Where a range of wavelengths is indicated this corresponds to the tuning range that has been achieved.

From the wealth of material represented in the table we shall single out Tm fibre lasers for further discussion since Tm offers a good example for illustrating a number of general points and for showing the rich variety of laser characteristics available in a fibre geometry. The relevant energy levels of Tm^{3+} are shown in Figure 3.

The first demonstration of laser action in Tm^{3+}-doped fibre was in silica fibre[61], on the $^3H_4$ - $^3H_6$ transition at $\sim 1.9\mu m$, by pumping at $\sim 0.8\mu m$ into the $^3F_4$ level. Rapid multiphonon decay from this level leads to efficient population of $^3H_4$. Low threshold, efficient operation has been achieved on this transition, including diode-pumped operation[30] and widely tunable operation[29,30]. The $^3H_4$ level lifetime is

Table 1. Fibre laser transitions

Wavelength (μm)	Element	Transition	Ref
3.4	Er	$^4F_{9/2}$ - $^4I_{9/2}$	39
2.9 (2.83-2.95)	Ho	$^5I_6$ - $^5I_7$	42
2.75 (2.70-2.83)	Er	$^4I_{11/2}$ - $^4I_{13/2}$	43
2.3 (2.25-2.5)	Tm	$^3F_4$ - $^3H_5$	44
2.04	Ho	$^5I_7$ - $^5I_8$	45,46
1.9 (1.65-2.05)	Tm	$^3H_4$ - $^3H_6$	29,30,47
1.72	Er	$^4S_{3/2}$ - $^4I_{9/2}$	48
1.66	Er	$^2H_{11/2}$ - $^4I_{9/2}$	48
1.55 (1.52-1.62)	Er	$^4I_{13/2}$ - $^4I_{15/2}$	20
1.47	Tm	$^3F_4$ - $^3H_4$	47,49
1.38	Ho	$^5S_2$, $^5F_4$ - $^5I_5$	45
1.34	Nd	$^4F_{3/2}$ - $^4I_{13/2}$	40,27
1.31	Pr	$^1G_4$ - $^3H_5$	50-52
1.2	Ho	$^5I_6$ - $^5I_8$	53
1.08	Pr	$^1D_2$ - $^3F_{3/4}$	54
1.06 (1.05-1.14)	Nd	$^4F_{3/2}$ - $^4I_{11/2}$	20
1.04 (1.01-1.17)	Yb	$^2F_{5/2}$ - $^2F_{7/2}$	15,28
0.98	Er	$^4I_{11/2}$ - $^4I_{15/2}$	55
0.975	Yb	$^2F_{5/2}$ - $^2F_{7/2}$	15
0.94 (0.9-0.95)	Nd	$^4F_{3/2}$ - $^4I_{9/2}$	26,56
0.91	Pr	$^3P_0$ - $^1G_4$	57
0.88	Pr	$^3P_1$ - $^1G_4$	57

Table 1. (cont'd)

Wavelength (μm)	Element	Transition	Ref
0.88	Pr	$^1D_2$ - $^3H_6$,$^3F_2$	54
0.85	Er	$^4S_{3/2}$ - $^4I_{13/2}$	58
0.8	Tm	$^3F_4$ - $^3H_6$	33
0.715	Pr	$^3P_0$ - $^3F_4$	57,9,60
0.695	Pr	$^3P_1$ - $^3F_4$	57,9,60
0.651	Sm	$^4G_{5/2}$ - $^6H_{9/2}$	59
0.635	Pr	$^3P_0$ - $^3F_2$	57,9,10,60
0.61	Pr	$^3P_0$ - $^3H_6$	57,9,60
0.55	Ho	$^5S_2$ - $^5I_8$	8
0.546	Er	$^4S_{3/2}$ - $^4I_{15/2}$	11,12
0.520	Pr	$^3P_0$ - $^3H_5$	9,60
0.491	Pr	$^3P_0$ - $^3H_4$	9
0.48	Tm	$^1G_4$ - $^3H_6$	41
0.455	Tm	$^1D_2$ - $^3H_4$	41

nevertheless somewhat shortened by nonradiative multiphonon decay, from an expected radiative lifetime of ~6 sec to a few hundred microseconds. This raises the threshold, but as indicated earlier does not affect the slope efficiency and in practice slope efficiencies well in excess of the radiative quantum efficiency of $^3H_4$ have been observed. In fact the possibility exists for pumping quantum efficiencies in excess of 100% and approaching 200%, by energy transfer between Tm ions, where a Tm ion in $^3F_4$ shares its energy with a neighbouring unexcited Tm ion (in $^3H_6$), so that both end up in $^3H_4$. For this to occur efficiently however the Tm - Tm transfer rate must exceed the multiphonon decay rate $^3F_4$ - $^3H_5$. This situation would be favoured in a host of low phonon energy such as ZBLAN, and also requires a high dopant concentration.

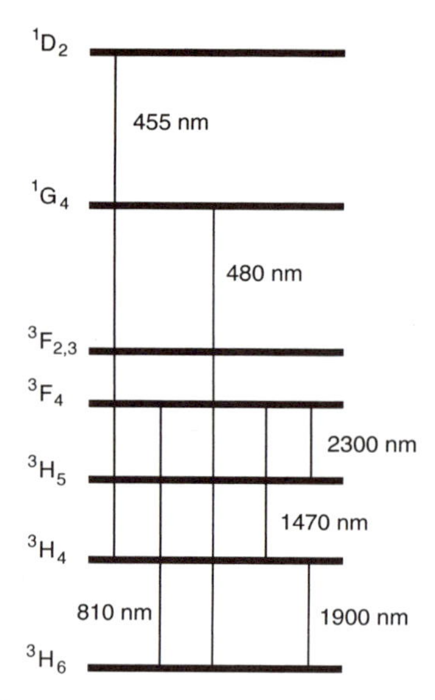

Figure 3. Energy levels of Tm^{3+} with demonstrated fibre laser transitions and their wavelengths indicated by vertical lines.

Tm doped ZBLAN shows many more laser transitions than in silica as a result of the reduced decay rate of the excited levels[47,62]. From $^3F_4$, laser transitions at 2300nm ($^3F_4$-$^3H_5$), 1470nm ($^3F_4$--$^3H_4$) and 810nm ($^3F_4$-$^3H_6$) are all observed. In fact in ZBLAN the $^3F_4$-$^3H_6$ branching ratio ($\sim$0.88) is dominant, so that when pumping the $^3F_4$ level the pumping quantum efficiency for the $^3H_4$ level is now much reduced compared to the case in silica. Possible ways of increasing the pumping efficiency of $^3H_4$ include the use of high doping levels to allow Tm - Tm energy transfer as described above, or enforcement of lasing on the 2300nm and/or 1470nm transitions[47,63]. The lifetime of $^3H_4$ is essentially equal to the radiative lifetime of ($\sim$ 6 msec) as multiphonon decay from $^3H_4$ level is suppressed in ZBLAN. In fact, for optimum operation of the $^3H_4$-$^3H_6$ laser (minimum threshold) one ideally requires a host glass with maximum phonon energy intermediate between those of ZBLAN and silica. In this way nonradiative decay can efficiently channel population from $^3F_4$ to $^3H_4$, while the $^3H_4$ lifetime remains at essentially the pure radiative value. This outcome has indeed been achieved in Tm-doped lead germanate glass fibre[64], this glass composition having been deliberately chosen for its intermediate phonon energy. As expected this has shown a lower threshold for 1900nm lasing than either the ZBLAN or silica fibre.

Another way of pumping the $^3H_4$ level and avoiding problems of inefficient branching out of the $^3F_4$ level is to pump into the $^3H_5$ level. However peak absorption is at an inconvenient pump wavelength of $\sim$ 1.2μm for which sources are not readily available. Nevertheless the absorption in the high energy wing of the $^3H_6$ - $^3H_5$ is sufficient in Tm silica to allow pumping with a NdYAG laser at 1.064μm[65]. In this way high power operation of a Tm fibre laser has been achieved (> 1W) at 1.9μm[66] by virtue of the fact that a copious amount of pump power was available with a good beam quality that allowed efficient launching into the core.

While pumping with the 1.064μm source, a considerable amount of blue side-light emission could be seen, corresponding to upconversion pumping, via three absorbed photons; $^3H_6$ - $^3H_5$, followed by relaxation to $^3H_4$, then $^3H_4$ - $^3F_{2,3}$, followed by relaxation to $^3F_4$, and finally $^3F_4$ to $^1G_4$ from which emission at 480nm, $^1G_4$ - $^3H_6$ was observed. A detailed study[67] of this upconversion pumping led to the conclusion that it would be greatly enhanced in ZBLAN fibre, which therefore presented the preferred route to upconversion lasing. The first fibre upconversion laser was in fact demonstrated[41] in Tm:ZBLAN, but using two red pump photons rather than infrared. Very recently however Tm:ZBLAN has been shown[13] to have excellent upconversion laser performance when infrared pumped via the three photon route described above, but using longer wavelength pump radiation ($\sim 1.12\mu m$) to enhance the absorption cross section on the pumping stages.

This completes a brief survey of the variety of lasing processes that can be observed in Tm doped fibres. It is clear that there is a need for such more detailed characterisation and that there is great scope for further optimisation of all of the laser transitions reported. Particular interest is attached to the blue upconversion laser, since with its already demonstrated efficiency of $\sim 30\%$ (from pump photons to output), it appears to be a particular attractive candidate for the much sought after compact blue light source.

Concluding Remarks

In this review we have given emphasis to the wide spectral coverage that is now possible via fibre lasers. Here the prospect is that ultimately most of the spectral range from $\sim 0.4\mu m$ to $\sim 4\mu m$ should be continously covered with diode-pumped fibre lasers. Applications of such lasers may make various demands on the lasers' performance, such as single-frequency operation, mode-locked operation, Q-switched operation, tunable operation, high power operation etc etc. All of these modes of operation have been demonstrated in fibre lasers, and while the demonstrations have perhaps been confined so far to particular laser transitions, the techniques involved are generally capable of being applied very widely, so that these laser performance characteristics should ultimately be available across the whole spectral range indicated above.

Meanwhile rapid progress is being made on the development of fibre components, such as fibre Fabry-Perot etalons, highly reflective Bragg gratings, written directly into the fibre core, fibre couplers, isolators etc, so that we can look forward to fibre lasers with a high degree of temporal and spectral control with the entire laser in the form of a rugged, compact, all-fibre geometry.

References

1. E. Snitzer. *Phys. Rev. Lett.*, 7:444 (1961).
2. E. Snitzer. Perspective and overview, *in*: "Optical Fibre Lasers and Amplifiers", P.W. France, ed., Blackie, Glasgow and London (1991).
3. B. Ross and E. Snitzer. *IEEE J. Quant. Electron.*, QE-6:361 (1972).
4. J. Stone and C. Burrus. *Appl. Phys. Lett.*, 23:388 (1973).
5. S.B. Poole, D.N. Payne and M.E. Fermann. *Electron. Lett.*, 21:737 (1985).
6. R.J. Mears, L. Reekie, S.B. Poole and D.N. Payne. *Electron. Lett.*, 21:738 (1985).
7. R.J. Mears, L. Reekie, S.B. Poole and D.N. Payne. *Electron. Lett.*, 22:159 (1986).
8. J.Y. Allain, M. Monerie and H. Poignant. *Electron. Lett.*, 26:261 (1990).
9. R.G. Smart, D.C. Hanna, A.C. Tropper, S.T. Davey, S.F. Carter, D. Szebesta. *Electron. Lett.*, 27:1307 (1991).
10. J.Y. Allain, M. Monerie and H. Poignant. *Electron. Lett.*, 27:1156 (1991).
11. T.J. Whitley, C.A. Millar, R. Wyatt, M.C. Brierley and D. Szebesta. *Electron. Lett.*, 27:1785 (1991).

12. J.Y. Allain, M. Monerie and H. Poignant. *Electron. Lett.*, 28:111 (1992).
13. S.G. Grubb, K.W. Bennett, R.S. Cannon, W.F. Humer. Postdeadline Paper CPD18, Conference on Lasers and Electro-Optics, Anaheim, Ca (1992).
14. J.D. Minelly, E.R.Taylor, K.P.Jedrzejewski, J.Wang and D.N.Payne. Paper CWE6, Conference on Lasers and Electro-Optics, Anaheim, Ca. (1992).
15. D.C. Hanna, R.M. Percival, I.R. Perry, R.G. Smart, P.J. Suni and A.C. Tropper. *J. Mod. Opt.* 37:517 (1990).
16. J.L. Zyskind, J.W. Sulhoff, Y. Sun, J. Stone, L.W. Stulz, G.T. Harvey, D.J. DiGiovanni, H.M. Presby, A. Piccirilli, U. Koren and R.M. Jopson. *Electron. Lett.*, 27:2148 (1991).
17. I.N. Duling, W.K. Burns and L. Goldberg. *Opt. Lett.,* 15:33 (1990).
18. N. Langford and A.I. Ferguson. Q-switched and mode-locked fibre lasers, *in* "Optical Fibre Lasers and Amplifiers", P.W. France, ed., Blackie, Glasgow and London (1991).
19. D.J. Richardson, A.B. Grudinin and D.N. Payne. *Electron. Lett.*, 28:778 (1992).
20. P.W. France, ed. "Optical Fibre Lasers and Amplifiers", Blackie, Glasgow and London (1991).
21. M. Shimuzu, M. Yamada, M. Horiguchi, T. Takeshita and M. Okayasau. *Electron. Lett.*, 26:1642 (1990).
22. A.W. Snyder and J.D. Love. "Optical Waveguide Theory", Chapman and Hall, London and New York (1983).
23. E.G. Neumann. "Single-Mode Fibres", Vol. 57, Springer Series in Optical Sciences, Springer-Verlag (1988).
24. O. Svelto. "Principles of Lasers", 3rd Edition, p. 33, Plenum, New York and London (1989).
25. W.L. Barnes, P.R. Morkel and J.E. Townsend. *Optics Commun.*, 82:282 (1991).
26. I.P. Alcock, A.I. Ferguson, D.C. Hanna and A.C. Tropper. *Optics Commun.*, 58:405 (1986).
27. S.G. Grubb, W.L. Barnes, E.R. Taylor and D.N. Payne. *Electron. Lett.*, 26:121 (1990).
28. V.P. Gapontsev, I.E. Samartsev, A.A. Zayats and R.R. Loryan. Paper WC1, OSA Topical Meeting : Advanced Solid State Lasers, Hilton Head, South Carolina (1991).
29. D.C. Hanna, R.M. Percival, R.G. Smart and A.C. Tropper. *Optics Commun.* 75:283 (1990).
30. W.L. Barnes and J.E. Townsend. *Electron. Letts.*, 26:746 (1990).
31. O. Svelto. "Principles of Lasers", 3rd Edition, p.52, Plenum, New York and London (1989).
32. J.R. Armitage. *IEEE J. Quantum Electron.*, 26:423 (1990).
33. J.N. Carter, R.G. Smart, D.C. Hanna and A.C. Tropper. *Electron. Letts.* 26:1760 (1990).
34. M.J.F. Digonnet. *IEEE J. Quantum Electron.* 26:1788 (1990).
35. C.R. Giles and E. Desurvire. *J. Lightwave Tech.*, 9:271 (1991).
36. J.M.F. van Dijk and M.F. Schurmans. *J. Chem. Phys.* 78:5317 (1983).
37. P.W. France and M.C. Brierley. Fluoride fibre lasers and amplifier, *in* "Optical Fibre Lasers and Amplifiers", P.W. France, ed., Blackie, Glasgow and London, (1991).
38. P.W. France, M.G. Drexhage, J.M. Parker, M.W. Moore, S.F. Carter and J.V. Wright. "Fluoride Glass Optical Fibres", Blackie, Glasgow and London (1989).
39. H. Tobben. *Electron. Lett.* 28:1361 (1992).
40. J.E. Pedersen and M.C. Brierley. *Electron. Lett.* 6:819 (1990).

41. J.Y. Allain, M. Monerie and H. Poignant. *Electron. Lett.* 26:166 (1990).
42. L. Wetenkamp. *Electron. Lett.* 26:883 (1990).
43. M.C. Brierley and P.W. France. *Electron. Lett.* 24:935 (1988).
44. R.M. Percival, S.F. Carter, D. Szebesta, S.T. Davey and W. Stallord. *Electron. Lett.* 27:1912 (1991).
45. M.C. Brierley, P.W. France and C.A. Millar. *Electron. Lett.* 24:539 (1988).
46. D.C. Hanna, R.M. Percival, R.G. Smart, J.E. Townsend and A.C. Tropper. *Electron. Lett.* 25:593 (1989).
47. R.G. Smart, J.N. Carter, A.C. Tropper and D.C. Hanna. *Opt. Commun.* 82:563 (1991).
48. R.G. Smart, J.N. Carter, D.C. Hanna and A.C. Tropper. *Electron. Lett.* 26:649 (1990).
49. T. Komukai, T. Yamamoto, T. Sugawa and Y. Miyajima. *Electron. Lett.* 28:830 (1992).
50. Y. Oshishi, T. Kanamori, T. Kitagawa, S. Takahashi, E. Snitzer and G.H. Sigel. *Optics Lett.* 16:1747 (1991).
51. R. Lobbett, R. Wyatt, R. Eardley, T.J. Whitley, P. Smyth, D. Szebesta, S.F. Carter, S.T. Davey, C.A. Millar and M.C. Brierley. *Electron. Lett.* 27:1472 (1991).
52. Y. Miyajima, T. Sugawa and Y. Fukasaku. *Electron. Lett.* 27:1706 (1991).
53. H. Tobben and L. Wetenkamp, paper MD2, OSA Topical Meeting:Advanced Solid State Lasers, Sante Fe N.M, 1992.
54. R.M. Percival, M.W. Phillips, D.C. Hanna and A.C. Tropper. *IEEE J. Quantum Electron.* 25:2119 (1989).
55. J.Y. Allain, M. Monerie and H. Poignant. *Electron. Lett.* 25:318 (1989).
56. I.P. Alcock, A.I. Ferguson, D.C. Hanna and A.C. Tropper. *Optics Lett.* 11:709 (1986).
57. J.Y. Allain, M. Monerie and H. Poignant. *Electron. Lett.* 27:189 (1991).
58. T.J. Whitley, C.A. Millar, M.C. Brierley and S.F. Carter. *Electron. Lett.* 27:185 (1991).
59. M.C. Farries, P.R. Morkel and J.E. Townsend. *Electron. Lett.* 24:709 (1988).
60. R.G. Smart, J.N. Carter, A.C. Tropper, D.C. Hanna, S.T. Davey, S.F. Carter, D. Szebesta. *Optics Commun.* 86:337 (1991).
61. D.C. Hanna, I.M. Jauncey, R.M. Percival, I.R. Perry, R.G. Smart, P.J. Suni, J.E. Townsend and A.C. Tropper. *Electron. Lett.* 24:1222 (1988).
62. J.Y. Allain, M. Monerie and H. Poignant. *Electron. Lett.* 25:1660 (1989).
63. R.M. Percival, D. Szebesta and S.T. Davey. *Electron. Lett.* 28:671 (1992).
64. J.R. Lincoln, C.J. Mackechnie, J. Wang, W.S. Brocklesby, R.S. Deol, A. Pearson, D.C. Hanna and D.N. Payne. *Electron. Lett.*, 28:1021 (1992).
65. D.C. Hanna, M.J. McCarthy, I.R. Perry and P.J. Suni. *Electron. Lett.* 25:1365 (1989).
66. D.C. Hanna, I.R. Perry, J.R. Lincoln and J.E. Townsend. *Optics Commun.* 80:52 (1990).
67. D.C. Hanna, R.M. Percival, I.R. Perry, R.G. Smart, J.E. Townsend and A.C. Tropper. *Optics Commun.*, 80:147 (1990).
68. R.G. Smart, J.N. Carter, A.C. Tropper, D.C. Hanna, S.F. Carter, D. Szebesta. *Electron. Lett.* 27:1123 (1991).
69. J.N. Carter, R.G. Smart, A.C. Tropper, D.C. Hanna, S.F. Carter, D. Szebesta. *J. Lightwave Tech.* 9:1548 (1991).
70. G.H. Dieck and H.M. Crosswhite. *Appl. Opt.*, 2:675 (1963).

THE ROLE OF SOLID STATE LASERS IN SOFT-X-RAY PROJECTION LITHOGRAPHY

W.T. Silfvast, M.C. Richardson, H. Bender, A. Hanzo, V. Yanovsky,
F. Jin, J. Thorpe

Center for Research in Electro-Optics and Lasers (CREOL)
University of Central Florida
Orlando, FL 32826

1. FUTURE LITHOGRAPHY REOUIREMENTS

The ever increasing demand for more sophisticated electronic devices and higher-speed data processing have placed ever-increasing pressure on microchip manufacturers to produce more sophisticated microprocessing chips. These chips must not only have the capabilities of higher processing rates but also have more features on each chip. Manufacturers accomplish this by shrinking the size of each element on the chip and also by increasing the chip size. The greatest need associated with the reduction of the feature size on an individual chip is in the production of memory chips. By the turn of the century, minimum feature sizes of less than 0.2μm will be required in order to manufacture a one gigabit D-RAM. Such small feature sizes will most likely be beyond the capabilities of conventional optical lithography even with the use of phase-shift masks. The only likely alternatives at this point, for high wafer throughput manufacturing facilities, would be x-ray proximity lithography, projection e-beam lithography or soft-x-ray projection lithography (SXPL). This paper will briefly summarize the advantages of SXPL and specifically address the issues relating to the use of solid state lasers and the associated laser produced-plasma sources for this type of lithography.

2. ADVANTAGES OF SOFT-X-RAY PROJECTION LITHOGRAPHY

There are two primary advantages in using an imaging or projection type of lithography. First, it is a parallel type of process that causes patterns to be produced over

Solid State Lasers: New Developments and Applications
Edited by M. Inguscio and R. Wallenstein, Plenum Press, New York, 1993

an entire chip area simultaneously, thereby significantly reducing the time required to expose the entire wafer. Second, the imaging process allows the mask pattern to be much larger than the pattern on the wafer due to the reduction capability of the imaging optics. This can significantly reduce the cost of the mask since larger mask features are much easier and less costly to fabricate than smaller features. This lower cost estimate is based upon the assumption that the decreased cost of the larger features would far outweigh the cost of a larger mask, which is probably reasonable since the total number of information pixels would be the same in each case. SXPL takes advantage of these imaging capabilities by offering the possibility of producing very small feature sizes (0.1μm) while maintaining a large depth of focus (±1.0μm). The large depth of focus is very desirable to be able to provide adequate leeway in the manufacturing process.

By considering the fundamental effects of diffraction, the above requirements for minimum feature size and depth of focus can be shown to place strict limitations on the illumination wavelength and the numerical aperture of the imaging optics. A simple analysis[1], using the Rayleigh criterion, indicates that the illumination wavelength must be 200Å or less and the numerical aperture of the imaging optic must be less than or equal to 0.1. Fortunately these wavelength and numerical aperture requirements can be achieved by taking advantage of two relatively recent technological advances. The first is the development of highly reflective multilayer optical coatings in the soft-x-ray spectral region[2] in the wavelength region from approximately 50-300Å. Multilayer reflection coatings of up to 63% have recently been achieved[3] at a wavelength of 130Å using coatings consisting of alternate layers of molybdenum and silicon. Thus the most desirable wavelength for SXPL is presently at or near 130Å since a lithography system would most likely involve 7 or more reflecting surfaces and the optical throughput is therefore very sensitive to mirror reflectivity.

The second technological advance is the development of large field diffraction limited reflective imaging optics[4] with numerical apertures that satisfy the above requirements. Recently designed optics that will provide such imaging are advanced designs of the ring field type[5]. They produce an arc shaped diffraction-limited image of approximately 2.5cm length and 0.1cm width (approximately $0.25cm^2$) on the wafer at a 5:1 reduction from mask to wafer. Such optics are axially symmetric, distortion corrected, telecentric at the wafer and near telecentric at the mask.

3. POSSIBLE SOURCES FOR SOFT-X-RAY PROJECTION LITHOGRAPHY

Possible sources that have been considered for SXPL include synchrotrons, laser-produced plasmas (referred to here as laser plasmas) and discharge plasma sources. Of these, only synchrotrons and laser plasmas have currently been demonstrated to have the capabilities necessary for SXPL.

Synchrotrons are reasonably efficient, their energy can be concentrated in a relatively narrow wavelength spectrum, they produce a continuous beam, and they are relatively reliable. Their disadvantages include very high cost, high spatial coherence and potentially large down time if a fault occurs. From the standpoint of high cost, most likely a number of lithographic machines would have to be attached to a single synchrotron source, thereby creating a particularly undesirable disruption when a breakdown occurs. Some of the first SXPL experiments were carried out at the NSLS synchrotron using 130Å light in which features of less than 0.1μm were obtained in photoresist[6].

Laser plasmas have advantages of relatively low cost, compactness and potentially high reliability. The disadvantages include low efficiency, the necessity to control the debris from the plasma, and the fact that the technology has not yet been completely developed for this source. Successful demonstration of SXPL using a laser-produced plasma source has recently been made yielding 0.1μm features[7].

4. EMISSION CHARACTERISTICS OF LASER-PLASMA SOURCES

A laser plasma soft-x-ray source consists of a dense high temperature plasma of highly ionized ions and associated electrons emitting a radiation spectrum that is characteristic of matter at such a high temperature. The plasma is produced by focusing a short-pulse laser (typically 10^{-8}sec duration) onto a solid target at an intensity of the order of 10^{11}-10^{12}Watts/cm^2. The focused laser energy is absorbed by the target material, thereby heating the material and eventually stripping the electrons from the material. As the plasma is heated, more and more electrons are removed from each atom of the plasma and the atomic ions move to higher ionization stages. As they move to higher stages, they access higher energy levels and the radiative decay from these levels produces short wavelength radiation.

Due to the small absorption depth of the laser radiation incident upon the target material (typically 100-250Å), only a very thin layer of material is initially heated and ionized. As it is heated, the material expands and the density is reduced, producing a density gradient of plasma atoms and ions decreasing with distance from the solid surface. The density of solid material is of the order of 5×10^{22}atoms/cm^3. The plasma density at which absorption of the laser energy is most efficient (by inverse Brehmstrahlung) is just under 10^{21}/cm^3 for a wavelength of 1.06μm and varies with the inverse square of the wavelength. For shorter wavelength lasers, the energy absorption and plasma heating would occur at a location much nearer the solid material than that for the 1.06μm laser. For example if a laser operating at a wavelength of 248nm (KrF excimer laser) is used to produce the plasma, the absorption would occur at a plasma density of just under 1.6×10^{22}/cm^3 which is nearly that of solid density. This makes shorter wavelengths much less attractive for plasma sources in this wavelength region since a higher thermal conductivity to the substrate would reduce the radiation efficiency and the debris problem could be greater due to more direct heating of the adjacent substrate material

The brightness and wavelength of the plasma emission are associated with both the specific target material and the laser intensity. A typical example is shown in Fig. 1 for a solid tin (Sn) target[8]. It can be seen that the emission occurs over a broad spectrum with a number of sharp features superimposed upon the broad background.

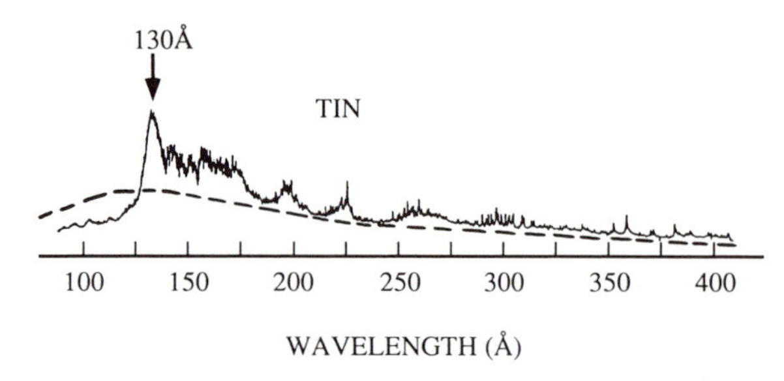

Figure 1. Tin (Sn) plasma emission spectrum overlaid with the shape of a blackbody spectral distribution of temperature T=225,000°K (20eV).

The broad emission of such a plasma is found to be sensitive to the laser intensity and can be characterized by a temperature associated with that of a blackbody radiator. Such a blackbody spectral distribution, of temperature 225,000°K (20eV), is shown superimposed (as a dashed curve) upon the Sn spectrum of Fig. 1 and was chosen to have a maximum emission at 130Å, the optimum wavelength for the Mo:Si multilayer reflectors and the wavelength currently contemplated for SXPL. The sharp emission peak at 130Å, enhanced above the blackbody spectrum, makes Sn an attractive target material for a laser plasma source for SXPL.

Assuming that the plasma radiates as a blackbody, the maximum energy E emitted from such a plasma of temperature T(°K), emitting blackbody radiation of wavelength λ(cm), from a surface area πr^2(cm^2), over a spectral bandwidth $\Delta\lambda$(Å), for a duration Δt(sec), can be expressed as

$$E(\lambda,T)=\frac{3.75\times10^{-2}}{\lambda^5[e^{(1.44/\lambda T)}-1]}\Delta\lambda\pi r^2\Delta t \quad \text{(Joules)} \qquad (1)$$

For a 200μm diameter plasma emitting at 130Å over a bandwidth of 3Å (the maximum bandwidth that would pass through a SXPL imaging system) and emitting for a duration of 10^{-8}sec (the duration of a typical laser plasma), the emitted energy would be 2.3×10^{-4} Joules for a single pulse from a 20eV laser plasma. For the case of a Sn plasma, additional emission above that of the blackbody emission could arise from the sharp spectral features associated with Sn as indicated in Fig. 1.

The radiation efficiency η of a blackbody radiator at a specific wavelength λ and temperature T can be obtained by dividing the energy E emitted at that wavelength and temperature by the total energy emitted by the blackbody radiator at all wavelengths (at that temperature) as obtained from the Stefan Boltzmann law ($I=\sigma T^4$ where $\sigma=0.567\times10^{-11}$W/cm^2 and T is in °K). For a wavelength of 130Å and a bandwidth of 3Å, η is given by the expression

$$\eta=\frac{4.57\times10^{21}}{[e^{(1.1\times10^6/T)}-1]T^4} \qquad (2)$$

where T is in °K. A plot of η vs. plasma temperature, shown in Fig. 2, goes through a maximum of η=1.49% at T=280,000°K, which is a slightly higher temperature than that which produces a maximum emission at 130Å.

An approximate relationship of the plasma temperature T to the incident laser intensity I_l can be obtained by equating the absorbed laser energy incident upon the target to the energy loss due to thermal conduction by the plasma electrons. This leads to an approximate relationship of $T=CI_l^{4/9}$ where C is a constant[9]. Although the exact value of the exponent and the value of C have varied with different experimental studies, the expression for T is a reasonably good approximation for the laser intensity region under

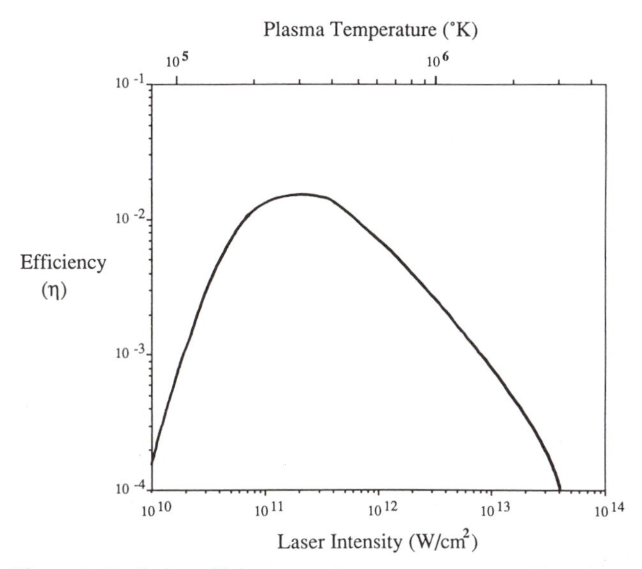

Figure 2. Radiation efficiency vs. plasma temperature and laser intensity for a blackbody radiator emitting within a 3Å bandwidth at 130Å.

consideration. From an experimental result in which the actual plasma temperature was deduced for a similar laser intensity and plasma size[10], the value of C can be determined to be approximately 2.8 for a plasma temperature given in °K and a laser intensity given in Watts/cm^2. Thus using this expression $T=2.8I_l^{4/9}$ in Eq. 2, the blackbody radiation efficiency η can be plotted in terms of laser intensity I_l instead of plasma temperature T and is also shown in Fig. 2. From this graph, it can be seen that the maximum radiation efficiency would be expected to occur at a laser intensity of just under 2×10^{11} W/cm^2.

Two separate measurements of conversion efficiency of laser plasmas have recently been made under conditions thought to be appropriate for laser plasma sources for SXPL. The conversion efficiencies obtained from these measurements are the total plasma radiation flux emitted from the target region in the given spectral bandwidth divided by the incident laser energy on target and are therefore not the same as the radiative efficiency η. Measurements made at Lawrence Livermore National Laboratories were made with a frequency doubled YAG laser at 0.53μm with a pulse duration of 7.5nsec[11]. Those made at Sandia National Laboratories were made with a KrF laser at 0.248μm that had a 3.2nsec pulse duration[12]. A summary of these conversion efficiency measurements are shown in Table I. The 1% measured with the 0.53μm laser at LLNL was made over a 3Å bandwidth and thus is the most similar to the conditions used for the estimate of η from Eq. 2.

Table 1. Experimental parameters and respective measured conversion efficiencies at LLNL and SNL.

Experimental Conditions	Lawrence Livermore National Laboratory	Sandia National Laboratory
λ(laser)	0.53μm	0.248μm
I(laser)	$2x10^{11}W/cm^2$	$3x10^{11}W/cm^2$
Pulse Energy	0.2J	1.1J
Bandwidth	3Å	6Å
Pulse Duration	7.5nsec	32nsec
Plasma Diameter	130μm	150μm
Target Material	Tin(Sn)	Gold(Au)
Measured Conversion Efficiency	1%	0.8%

As stated previously, the conversion efficiencies of Table I are (Energy Radiated at 130Å/Laser Energy In) or $(E_{130Å}/E^{l}_{in})$. Thus $\eta=(E_{130Å}/\beta E^{l}_{in})$ where β is the fraction of the incident laser energy converted to radiation. If we assume that approximately 40% of the energy incident on the target is converted to radiation (β=0.4), which is not unreasonable[13], then the LLNL value of η would be 2.5% which is about 70% above the maximum predicted value of 1.49% in Eq. 2. The additional 70% of the radiation above the predicted value is most likely due to the structured features that are superimposed upon the continuum in Fig. 1. This simple analysis suggests that the only way to increase the conversion efficiency would be to increase the structured emission features within the desired emission bandwidth, since increasing the blackbody temperature would only decrease the radiation efficiency η according to Eq. 2. These structured features are most likely associated with line radiation within one of the highly ionized stages[14] of Sn.

Based upon the preceding analysis and experimental results, a preliminary set of conditions can be summarized for producing optimum laser plasma emission for a SXPL system. These would include a laser with an intensity of $1\text{-}5x10^{11}W/cm^2$ with a pulse duration of the order of 10nsec irradiating a Sn target. Shorter time durations would decrease the total radiation flux. Longer durations would significantly increase the plasma debris. It is expected that refinements relating to laser pulse duration and shape, optimum laser wavelength, target material composition and thickness, and plasma diameter, will no doubt produce an enhancement above the preliminary values of conversion efficiency discussed in this paper.

5. PARTICULATE EMISSION FROM LASER-PLASMA SOURCES

When a plasma is formed at the surface of a material by focusing a laser on that material, a portion of that material ablates from the surface and becomes a plasma of highly ionized ions and electrons as previously described. The energy of the focused laser that is absorbed by the plasma is channeled in two directions. First it accelerates the electrons which in turn collide with and thereby accelerate the surrounding ions such that both the electrons and ions acquire very high velocities or effectively a very high temperature (225,000°K in this case). Second, the ions, via collisions with electrons, are stripped of many of their electrons to evolve towards very high ionization stages from which soft-x-ray radiation occurs. Thus during the early stages of the plasma evolution, the highly stripped ions are the emitting species that provide the soft-x-ray source, which in this case would be used as an illuminating source for SXPL. At the later stages of the plasma development, after the laser heating pulse has ceased, these ions are streaming outward from the target region with the velocities they have acquired from the plasma formation. Many of these ions recombine with the surrounding plasma electrons thereby reducing the stage of ionization. At some point in time after the ions have left the target region, they have been reduced to neutral atoms of that species but they are still moving with high velocities. Also during the early stages the plasma electrons conduct heat into the remaining solid target material thereby heating and vaporizing that material causing a "puff" of vapor to evolve from the target material. Sometimes large particulate or clusters of that material (sometimes referred to as hot rocks) are also ejected from the surface region.

The particulate emission can thereby be classified into three categories. The first is high velocity atomic ions from the plasma. The second is neutral atoms vaporized from the heated target. The third is clusters, ejected from the target material. All three of these types of particulates have the potential to reach the illuminating optics and thereby damage the optical surfaces. In addition, the presence of the atoms and ions can lead to absorption of the soft-x-ray flux before it reaches the optics and mask. It is thus important to measure the amounts and velocities of these particulates and to develop techniques to intercept or deflect them to prevent them from reaching those optics.

6. MEASUREMENTS OF PARTICULATE EMISSION

Cluster Emission

In order to characterize the cluster or "hot rock" emission from a solid target a stroboscopic photographic technique was employed. This technique utilized a 511/578nm copper vapor laser (CVL)[15] operating at a pulse repetition frequency of 14kHz with pulse energy of 3mJ and pulse duration of 50ns. The 25mm diameter CVL beam was situated parallel and directly in front of a rotating solid target material (Sn, Au, Fe). The plasma was created by focusing the output of a Nd:YAG laser at 1.06μm and 500mJ per pulse onto the target in a focal spot of 140μm diameter. This beam was incident at 45° to give near normal emission from the target surface and reduce incident laser light and plasma interaction. The ejected particles are illuminated by the CVL light approximately every 74μs, thus creating a slow motion effect. This is seen as a sequence of bright dots created

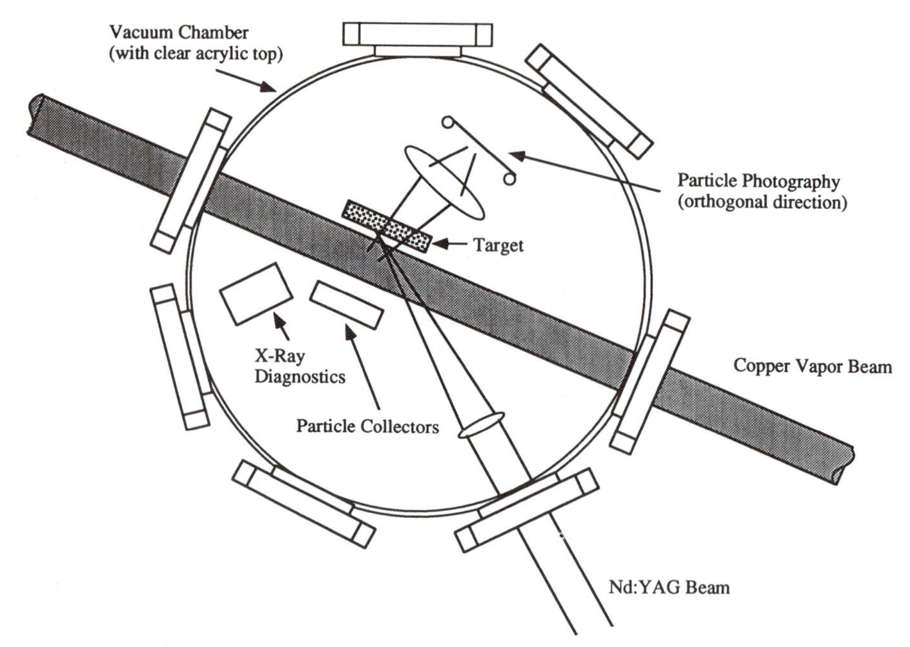

Figure 3. Schematic diagram of setup used to photograph particles. Clusters are illuminated by the copper vapor laser and imaged at 1:1.

by light scattering from the particles. This effect was photographed at 1:1 with a film camera situated above and normal to the plane containing the Nd:YAG and CVL beams. This configuration is shown in Fig. 3.

Using this technique, several types of information were obtained by analyzing the trajectories as recorded on film. By measuring the cluster size and separation it is possible to determine mass and velocity. In addition, angular distribution and total mass may be estimated. In experiments on clean target surfaces, minimal large cluster emission (greater than 50μm) was observed. It was found that large cluster emission increased when progressive laser shots were made on the same spot, thus creating a crater in the target. In preliminary experiments performed using such a cratered surface on Sn targets, particle sizes were observed to range from less than 10μm to over 200μm. Velocities were found to range from 200 cm/sec to 2500cm/sec using this technique.

Atomic Particle Flux

The combined atomic and ionic flux were measured using a cylindrically curved acetate strip, or a planar silicon wafer, as a collecting plate in front of the plasma as shown in Fig. 4. These plates would then indicate the angular distribution of the flux as well as the total amount of flux. Optical transmission through the acetate as well as profilometer measurements were then used to determine the amount of flux per pulse. This amount measured for a Sn planar target is shown in Fig. 5 as a function of angle from the target normal for conditions of vacuum and also for conditions of .05 Torr of helium as a background gas. It can be seen that the helium significantly reduces the flux reaching the acetate as well as randomizing its distribution. These data were taken for a fresh or new target for each laser pulse. Targets of gold and iron produced less flux per shot than for tin targets but similar angular distributions.

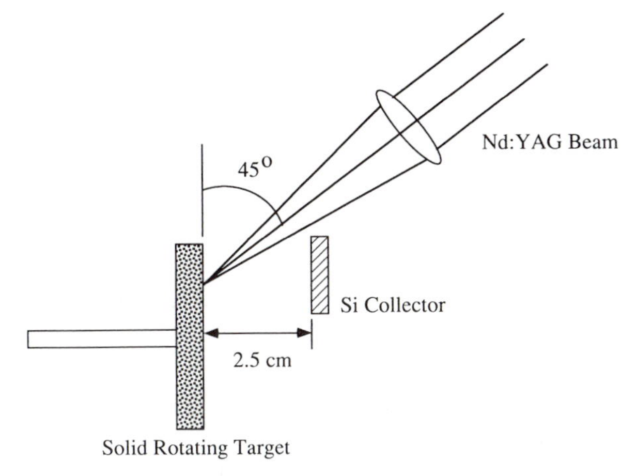

Figure 4. Technique indicating acetate collection of atomic and ionic flux.

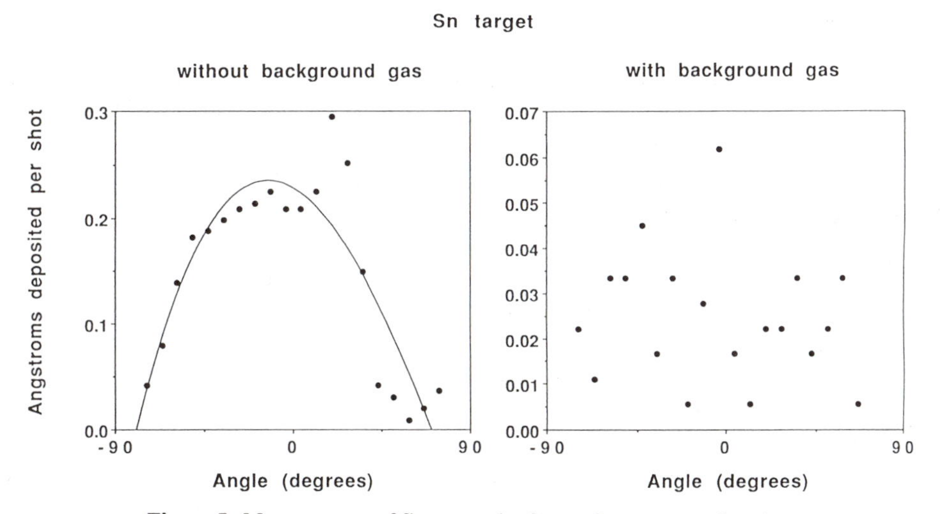

Figure 5. Measurements of flux per pulse from a Sn target as a function of angle under vacuum and with 0.05 Torr of helium as a background gas.

Table II summarizes some of our measurements as well as our best estimates of those quantities we are still in the process of measuring. These quantities include both the number of particles per laser pulse of each type and the velocity of those particles. These quantities will be useful in the next section in analyzing how to prevent the particulates from reaching the collecting optics for a lithography system.

Table 2. Quantity and velocity estimates of the three types of particulate emission per laser pulse.

	QUANTITY/PULSE	VELOCITY
NEUTRALS	up to 5×10^{14}-5×10^{15} (maximum)	up to 5×10^{4} cm/sec (estimated)
IONS	up to 5×10^{14} (probably much less)	up to 10^{7} cm/sec (estimated)
CLUSTERS	several	200-2500 cm/sec

7. POSSIBLE TECHNIOUES FOR INTERDICTING THE PARTICULATE EMISSION

The possible methods of preventing the particulate emission from reaching the collecting optics include mechanical interruption, gaseous interruption, deflection with magnetic fields (charged particles), and possibly laser interdiction. Of these, the first two appear to be the most likely possibilities. We will discuss the interdiction of each of the types of debris of Table II separately since they may require different techniques.

Atoms and Ions

The atomic and ionic particle flux/pulse combined with the required pulse repetition rate of 1kHz (as shown in the next section) leads to a generation of from 5×10^{17} to 5×10^{18} atoms and ions/sec to be removed or approximately 5×10^{-5} to 5×10^{-4} gm/sec. This amounts to a substantial quantity of mass that must not only be blocked from reaching the collecting optic but also must be removed from the target region since it represents an accumulation of up to one gram/hour of debris. For example, it could be potentially deleterious to the operation of a mechanical interruption device by causing significant mass buildup on moving parts. The use of a flowing buffer gas, probably helium, is the most likely solution to the problem of atomic mass removal. It has been shown previously that the use of a small amount of buffer gas (0.1 Torr helium) reduces the amount of debris reaching a collecting plate 1.5 cm from an iron target by a factor of 10 when using a 248nm KrF excimer laser to produce the plasma[15]. Results in our laboratory are similar in that we have shown a reduction of a factor of 7 for a tin target at a distance of 3.0cm with a helium pressure of 0.05 Torr and using a 1.06μm YAG laser to form the plasma. Significant debris reduction at this pressure is fortunate since a gas pressure any higher would absorb the 130Å radiation before it reaches the illuminating optics. For example the absorption cross section for helium at 130Å is 0.016/cm-Torr and for hydrogen is 0.002/cm-Torr[17]. Other gases have higher absorption coefficients than this and would therefore probably be less useful.

The relevant factor in the removal of the atomic flux by a buffer gas is that the gas impart enough momentum to the atomic and ionic flux, in the transverse direction from the target normal, to change the direction of the flux moving toward the collecting optic. In order to do this, the momentum of the buffer gas moving transversely must be at least as great as the momentum of the atomic flux. The momentum would then be effectively transferred since the mean free path between collisions of the buffer gas with the atomic neutral flux is of the order of 0.1cm at these pressures. Since the density of the buffer gas is of the order of $5x10^{15}/cm^3$ and the gas could be flowing at velocities of up to 10^5cm/sec (for helium), the momentum would be of the order of $4x10^{-3}$gm-cm/sec per cm^3 of helium. This could be compared to the momentum of the atomic flux of up to $5x10^{-2}$ gm-cm/sec using the maximum velocity data summarized in Table II for neutral atoms. Since this momentum is somewhat higher than the transverse helium momentum, the neutral flux must be reduced by an order of magnitude, as also suggested in Table II, or a greater path length over which the helium could interact must be used. Thus a rapidly flowing transverse gas, exiting from a nozzle or multiple nozzles would probably be the best solution for the removal of this type of particulate emission.

High-Velocity Ions

The removal of the high velocity ions would be more difficult since the momentum/particle is much higher. Fortunately the quantity of these particles is expected to be significantly less than for the atomic flux moving at lower velocities. Small quantities of these ions therefore could also be taken care of by a rapidly flowing buffer gas since the total momentum of the ions would be much less than the transverse momentum of the buffer gas. Measurements are being made of this flux and its velocity to be able to more accurately determine the parameters necessary for interdiction.

Clusters

Clusters (small chunks of material of various sizes) pose a serious problem. These clusters can range in size from hundreds of angstroms to hundreds of microns. As pointed out in the section above, it is fortunate that for flat (fresh) target surfaces the cluster emission is much smaller than for targets that have been subjected to multiple shots and thus become pitted. It is therefore anticipated that a target arrangement (most likely a continuous strip of thin target material, possibly coated on a thin inert backing material) that provides a fresh target for every pulse will be used for an SXPL source. The cluster emission from such a target will therefore be minimized and consequently the problems associated with such clusters in damaging the illuminating optic will be either eliminated or reduced to tolerable levels if interdictive devices are used. The interdiction of clusters will most likely require mechanical means. The use of a thin film tape filter of thickness small enough to transmit the soft-x-ray flux at 130Å (of the order of 0.5μm) would be costly and may not be able to stop the debris from passing on through the film and reaching the optic. In addition, the cost of such a device would most likely be prohibitive. An analysis by LLNL suggests that the cost of each target shot, in this case including the tape filter, could be no more than 10^{-5} dollars/shot[18]. It does not seem likely that the filter could meet this requirement since it would have to have a relatively large surface area in order to be sufficiently far enough away from the plasma source to prevent it from being damaged by the atomic and radiation flux before the clusters reach the filter.

Thus a mechanical device, such as a rotating disc, will probably be required to interdict the clusters. However, using such a disc is not a simple solution to the problem. Assume that a rotating disc with a one cm diameter hole is located one cm from the target and rotating on an axis normal to the target. In order for that device to interdict clusters with velocities up to 2500cm/sec the disc would have to rotate at a speed of 10,000rpm if the distance from the center of the disc to the center of the hole is 2.5cm. Such a speed is not impossible to achieve but would require a motor that is capable of operating in a low pressure environment and not emitting any vapors or other particulates that would lead to absorption of the soft-x-ray flux. It will be shown in the next section that the laser repetition rate might have to be as high as 1,000 pulses per second which would therefore require a rotational rate of up to 60,000rpm for a single hole in the disc. Increasing the size of the disc would reduce the required rate of revolution but the moment of inertia of the disc would increase as the square of the disc radius thereby placing significantly larger demands upon the stability of the disc while rotating. Placing more holes in the disc to reduce the rotational speed requirements may not be an alternative due to the need to intercept the slowest as well as the fastest particles. In order to intercept the 200cm/sec particles as well as the faster particles, only one hole could be located in a disc rotating at 10,000rpm for example.

The possibility of significant mass accumulation, in an uneven distribution, could unbalance the disc to a point where it would not operate at a high speed. Many of the clusters would bounce off of the disc but some would stick and also a significant number of atomic size particles would also accumulate. A gas flow system would be essential to reduce the mass flux arriving at the disc but may not inhibit the accumulation of clusters. Experiments are necessary to determine the mass accumulation and thus determine the lifetime of a disc before replacement is necessary.

8. FLUX REOUIREMENTS AND LASER REPETITION RATE

Estimates of the flux requirements of a laser plasma source for SXPL can be made assuming a production throughput of 60 six inch wafers per hour or an exposure rate of $3cm^2/sec$. If we also assume a resist sensitivity[19] at 130Å of $10mJ/cm^2$, then the power required at the wafer would be

$$3cm^2/sec \times 10mJ/cm^2 = 0.03J/sec = 30mW$$

For a laser energy of 1J/pulse and a 1% conversion efficiency into the 3Å bandwidth around 130Å, 10mJ/pulse of useful soft-x-ray radiation would be available. Assume 10% of the useful soft-x-ray radiation emitted from the source is collected by the first illumination optic. Also assume that a 7 mirror optical system is used (from source to wafer), with mirror reflectivity of 60% per mirror, such that the fraction of the light transmitted through the optical system is $(0.6)^7$=2.3%. This implies that $2.8x10^{-5}$ J/pulse will reach the wafer. In order to meet the requirements of 30mW outlined above, a repetition rate of 1,000Hz or 1kHz would be required for the laser system.

A number of factors could be improved upon to either lower the pulse repetition rate or lower the energy/pulse requirements. The resist sensitivity could be reduced to a range of 1-5mJ/cm^2. This possibility has been suggested in the literature[20] and is not an unreasonable value when it is considered that most materials absorb very highly at 130Å and thus a small amount of energy can yield a very high absorbed energy density in the top surface of the resist.

Mirror reflectivities might be improved above 60% per surface. Such values have in fact been achieved in the laboratory. The highest reflectivity for a Mo:Si mirror is 63% and maximum theoretical reflectivity[21] is approximately 70%. Thus increasing the reflectivity from 60% to 70% for 7 surfaces would reduce the flux loss by nearly a factor of 3 which is a very large energy savings.

Increasing the x-ray conversion efficiency above 1% is also a very likely possibility. An increase of a factor of 2 might be possible based upon the arguments given earlier in this paper, if more energy can be channeled into line emission rather than blackbody emission.

Also increasing the collection efficiency of light emanating from the source might be possible. One problem associated with this is the larger the numerical aperture of the collecting optic, the more difficult it might be to remove the plasma debris. It is much easier to periodically interrupt a small aperture at a rate of 1kHz than a large aperture. The first collecting optic would most likely be located at a distance of between 5 and 50cm from the laser-plasma source. For a small collecting numerical aperture the distance is not as important an issue as it would be for a large numerical aperture optic due to the large cost increase for larger diameter optics. On the other hand, the closer the optic is to the source, the more the optic would be subjected to the debris, since it would have less buffer gas between it and the plasma source. These issues are closely related to the design of the collecting optic and the flux requirements of the source and therefore could vary significantly from one illumination system design to another.

9. POSSIBLE LASER SOURCES

A laser requirement of 1J/pulse with a 10nsec pulse duration operating at a repetition rate of 1kHz is a fairly demanding requirement and is beyond the present state of the art. The closest that present day commercial lasers come to meeting this requirement is a 0.67J/pulse excimer laser[22] operating at 308nm at 300Hz or a 1J/pulse Nd:YAG laser[23] operating at 1.06μm at 30Hz. These lasers are limited in repetition rate primarily by the cooling requirements, in the case of the solid state laser, and by the power input requirements in the case of the excimer laser.

Diode-pumped solid state lasers have been operated at 1.06μm in the laboratory at a 1J/pulse level at a 50Hz repetition rate[24] and also at a 0.1J/pulse level at a 1kHz repetition rate[25]. Diode pumped solid state lasers will most likely be the lasers eventually used for factory lithographic systems. They are compact, very efficient, and offer high reliability. Due to their pumping wavelength they also offer a minimum heat transfer to the laser material during the pumping process. At the present time the cost is quite high for such lasers but has come down significantly in recent years and by the time SXPL becomes a practical process, the cost of laser diodes should be very reasonable thereby keeping the laser cost to a small fraction of the total lithographic system.

10. FUTURE DIRECTIONS

Several areas of investigation are necessary in order to perfect a laser plasma source for SXPL. These include laser source development, characterization of plasma particulates, design and testing of particulate control techniques and development of advanced target designs to improve x-ray conversion efficiency.

As far as laser source development is concerned, the emphasis should be on high repetition rate solid state laser development. Devising techniques for optimally pumping a laser rod or slab efficiently at high repetition rate with laser diodes should take first priority. Excimer lasers will play a role in SXPL in the short term but in the long term they will never be as efficient, or compact or reliable as a solid state laser system.

Continued characterization of plasma particulates will move in the direction of making measurements on thin targets including mass limited targets. Also the velocity of the atomic particle flux will be measured using optical absorption techniques. Efforts will be made to identify sub-micron size clusters and measure their velocities.

Debris control techniques will include interdiction devices such as rapidly rotating discs. Mass removal with flowing gases should also be investigated. Other possible debris deflectors could include magnetic fields and laser interdiction (a laser irradiating the plasma at a direction normal to that of the collecting optic).

Target design should include shaped targets, mass limited targets, multi-material targets, and layered targets. Also laser pulse shape and duration should be investigated in terms of their effect on conversion efficiency.

References

1. D.L. White, J. E. Bjorkholm, J. Bokor, L. Eichner, R.R. Freeman, T.E. Jewell, W.M. Mansfield, A.A. MacDowell, L.H. Szeto, D.W. Taylor, D.M. Tennant, W.K. Waskewicz, D.L. Windt, and O.R. Wood, II, Solid State Technology, 34, 37 (1991).
2. E. Spiller, Chapter 7,*Thin Films for Optical Coatings,* Edited by K. Guenther, (Springer-Verlag Topics in Applied Physics, New York, to be published in 1992-93). ; and T. Barbee, Optical Eng., 25, 898 (1986).
3. S.P. Vernon, et.al., Proceedings of the SPIE, 1547, 39 (1991).
4. L. G. Seppala, LLNL, and T. Jewell, AT&T, private communication.
5. A. Offner, Opt. Eng. 14, 130 (1975).

6. J.E. Bjorkholm, et.al., J. Vac. Sci. Technol., B8, 1509 (1990).
7. G.D. Kubiak, et.al., J. Vac. Sci. Technol., B9, 3184 (1991).
8. J. M. Bridges, C.L. Cromer and T.J. McIlrath, Appl. Opt. 25, 2208 (1986).
9. T.P. Hughes, "Plasmas and Laser Light" (Wiley, New York, 1975), pp. 287-293.
10. O.R. Wood II, W. T. Silfvast, J.J. Macklin, and P.J. Maloney, Opt. Lett. 11, 198 (1986).
11. R.L. Kauffman and D.W. Phillion, OSA Proceedings of the Topical Meeting on Soft-X-Ray Projection Lithography, 12, 68 (1991).
12. P.D. Rockett, et.al., OSA Proceedings of the Topical Meeting on Soft-X-Ray Projection Lithography, 12, 76 (1991).
13. R. Kauffman, LLNL, private communication.
14. C. Bauche-Arnault, J. Bauche, and M. Klapisch, Phys. Rev. A31, 2248 (1985).
15. Supplied by Oxford Lasers
16. M.L. Ginter and T.J. McIlrath, Appl. Opt. 27, 885 (1988).
17. J. Berkowitz, *Photoabsorption, Photoionization, and Photoelectron Spectroscopy,* (Academic Press, New York, 1979). Chapter V.
18. N. Ceglio, et.al., to be published in Journal of X-Ray Science and Technology, 1992.
19. Not an unreasonable value based upon the following studies: W. Mansfield, et.al., OSA Proceedings of theTopical Meeting on Soft-X-Ray Projection Lithography, 12, 129 (1991); and G.D. Kubiak, et.al., J. Vac. Sci. Technol., B8, 1643 (1990).
20. G. N. Taylor, R.S. Hutton, D.L. Windt, and W.M. Mansfield, Proceedings of the SPIE, 1343, 258 (1990).
21. D. Stearns, Lawrence Livermore National Laboratory, private communication.
22. Lambda Physik, Acton, MA, private communication.
23. Spectra Physics, Mountain View, CA, private communication.
24. McDonnell-Douglas, St. Louis, MO, private communication.
25. Lawrence Livermore National Laboratory, Livermore, CA, private communication
26. F. Bijkerk, E. Louis, E.C.I. Turcu, G.J. Tallents, Proc. Microcircuit Engineering (ME91), Rome, September 17-19, 1991.
27. E. Louis, F. Bijkerk, G.E. van Dorssen, I.C.E. Turcu, 21st European Conference on Laser Interaction with Matter, Warsaw, October 21-25, 1991.

SECOND HARMONIC GENERATION AND ITS APPLICATION TO HIGH RESOLUTION SPECTROSCOPY OF ATOMIC HYDROGEN

C. Zimmermann, A. Hemmerich and T.W. Hänsch

Max-Planck-Institut für Quantenoptik, 8046 Garching, Germany
Sektion Physik der Universität München, 8000 München, Germany

ABSTRACT

The advent of high power single mode diode lasers in combination with state of the art techniques of cw second harmonic generation appears to provide a superior tool for high resolution spectroscopy. A multi milliwatt all solid state laser source operating in the UV near 243nm seems to be within reach. It may become a key device in future experiments which aim at the optical comparison of the 1s-2s transition between hydrogen and antihydrogen.

INTRODUCTION

Spectroscopy of atomic hydrogen has a long tradition in atomic physics. Since the times of Balmer the hydrogen atom serves as one of the most important test objects to prove the validity of fundamental theories. Its atomic spectrum has been observed with increasing precision until in the1930´s the experimental determination of the energy levels has reached an accuracy of several ppm.[1] Up to this time the results seemed mostly in good agreement with Dirac´s relativistic theory of the electron which predicted a strict degeneracy between the two states of the n=2 shell with j=1/2 (i. e., $2S_{1/2}$ and $2P_{1/2}$) and it was a challenging question up to which level of accuracy this degeneracy holds. In 1949 Lamb and Retherford were the first who applied the newly developed microwave technology to induce radiofrequency transitions between the 2s and the 2p states in a strong magnetic field. In

Solid State Lasers: New Developments and Applications
Edited by M. Inguscio and R. Wallenstein, Plenum Press, New York, 1993

fact, their experiment revealed an energy shift of the 2S level which could not be explained by Dirac's theory of the hydrogen atom.

The discovery of the Lamb-shift provided the impetus for the development of quantum electro dynamics (QED) - the quantized theory of electromagnetic interaction - by Tomonaga, Schwinger and Feynman. This theory could not only explain the Lamb-shift but also serves as a basic model for all modern quantum field theories. Concerning the hydrogen atom QED predicts energy shifts which are largest for the s-states and scale with the principle quantum number as n^{-3}. For the ground state this shift of about 8 GHz cannot be observed by radio frequency methods because there is no nearby p-level.

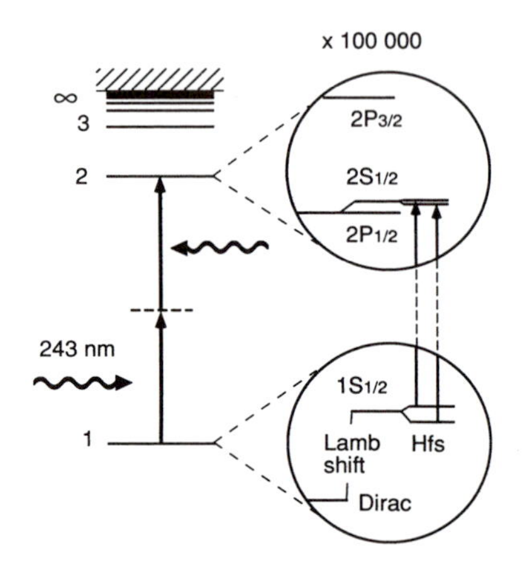

Figure 1. Energy levels of atomic hydrogen.

The first observation of the 1s Lamb-shift became possible after the invention of tunable dye lasers in 1974. Hänsch et al. succeeded in measuring the ground state Lamb-shift by Doppler-free two photon excitation of the 1s-2s transition. The experimental resolution was limited by the bandwidth of the pulsed laser system to about 2 parts in 10^7 and the results were in perfect agreement with theory. Further improvements of the experimental accuracy were closely related to the development of suitable laser sources. About ten years ago it became possible to reliably generate several milliwatts continuous wave radiation near 243nm in a nonlinear crystal.[2] The complicated setup was based on two large frame argon ion lasers, one of which was used to pump a dye laser, and allowed the first continuous wave measurement of the 1s Lamb-shift by Beausolleil et al. in 1986 in

Stanford.[3] Today the 1s Lamb-shift measurement of Weitz et al.[4] has reached the precision of the best radiofrequency measurement of the classical 2s Lamb-shift (1part in 10^5) and challenges the theorists to further refine their calculations. At this level of precision one is usually limited by the natural linewidth of the transitions. Fortunately the 2s state constitutes an important exception because it can not decay to the ground state via an electric dipole transition, which is forbidden by parity. Its lifetime is limited by a two photon decay to about 0.14 s and the corresponding natural linewidth (1.3Hz) is 8 orders of magnitude smaller than for instance the linewidth of the 1s-2p electric dipole transition (100 MHz). This unique situation offers the opportunity for ultimate tests of fundamental physics and challenges the experimentalists to observe the 1s-2s transition with a 1Hz resolution.

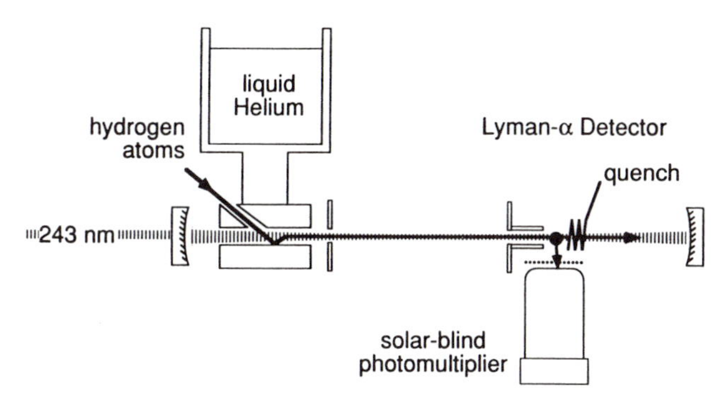

Figure 2. Experimental apparatus for high resolution spectroscopy of the hydrogen 1s-2s two photon transition.

The state of the art of two-photon spectroscopy of the 1s-2s transition[5,6] is shown in Fig.2. In a collinear atomic beam geometry the atomic resonance has been resolved with a linewidth of 12kHz (at 243nm). The atoms thermalize by wall collisions inside a cold nozzle (8K) and form an atomic beam parallel to the axis of an optical resonator. The atoms, which travel along the laser light inside the resonator are excited by Doppler free two photon absorption. At the end of the beam, 20cm downstream, a detector probes the excited atoms. To obtain a decent signal of several hundred counts per second one needs about 1mW incident light near 243nm. For its production we use ß-barium borate, a nonlinear crystal which became commercially available in 1986 and which allows angle tuned frequency doubling at 486nm. The fundamental radiation is taken from a dye laser which is pumped with 5W from a krypton ion laser near 415nm. With 200mW light power at 486nm the yield at 243nm is about 4mW. In contrast to the high performance and the impressing efficiencies of modern solid state lasers as being reported at the present summer school, the conversion efficiency (electric power into desired radiation) of our dye laser based systems holds a negative record of about 10^{-7}. For future precision experiments a more reliable UV-source with better efficiency is certainly desirable. Unfortunately it is not yet possible to replace the dye laser system by a semiconductor laser diode. Recently reported blue diode lasers[7] are still far away from being able to generate 200mW single mode radiation at 486nm. However at 972nm high power laser diodes do exist.[8] The radiation at 243nm may be obtained by frequency quadrupling in two succeeding nonlinear crystals. As an alternative approach, one may start with a laser source at 729nm, double its output in a nonlinear crystal, and then mix it with the fundamental wave in a second crystal.

A diode laser based UV-source would find a most attractive application in spectroscopic studies of antihydrogen as discussed at a workshop in Munich this year.[9] Until recently probably most physicists would have dismissed experiments with artificially produced antiatoms as pure science fiction. However the latest progress in trapping and cooling antiprotons in a Penning trap[10] and new results in spectroscopy of magnetically stored hydrogen[11] show that research is picking up momentum in this direction such that production of antihydrogen as the most simple antiatom seems to be within reach even in this decade. Two experimental questions are of particular interest. Gravitational mass comparison between hydrogen and antihydrogen could provide a stringent test for Einsteins equivalent principle and secondly the validity of CPT-invariance may be studied by comparing the electronic (i.e positronic) energy levels. Spectroscopy of the optical transitions is the obvious tool to detect possible differences between hydrogen and antihydrogen with high accuracy and among those the extremely narrow 1s-2s transition is the most promising candidate. Such experiments are of formidable difficulty and a more reliable UV source will become necessary which is compatible with the rough environment of a high energy laboratory. 10mW at 243nm and a sample of thousand magnetically trapped antihydrogen atoms are probably sufficient to compare the 1s-2s transition frequencies with a relative precision of 10^{-12}.

Other applications of diode laser based light sources in the blue and the UV have been recently discussed for optical memories, biological research, communication, color separation for offset printing, and in cases in which now air cooled argon ion laser are used. It is obvious that for the construction of such coherent UV-sources, techniques of efficient second harmonic generation and frequency mixing in nonlinear crystals will play a key role. This article aims to be a tutorial which summarizes state of the art techniques for efficient continuous wave second harmonic generation which may become important for the realization of an "all solid state" laser source in the ultraviolet.

CONTINUOUS WAVE SECOND HARMONIC GENERATION

General Remarks

If a laser beam is focused into a nonlinear crystal under proper conditions, part of its power will be converted into a coherent light beam of the second harmonic frequency. This effect is known since more than 30 years and a large body of literature exists about its different theoretical and experimental aspects.[12,13] To get an intuitive picture of the underlying physics it is sufficient to apply the Lorentz model of electrons in a crystal: the electrons are viewed as classical particles moving in a potential given by the ion lattice of the crystal. The incident light wave causes oscillations of the electrons and thus creates a polarization wave inside the crystal. To lowest order the potential in which the electrons move depends quadratically on the displacement of the electron from the equilibrium position. For larger displacement higher order terms may come into play. A cubic term for example describes asymmetric distortions of the quadratic potential. It is nonzero only for crystals without inversion symmetry. These crystals are suitable for second harmonic generation (SHG). A Fourier-analysis of the motion of the electron reveals that in potentials with cubic distortion there is a nonzero frequency component at twice the frequency of the incoming light wave. The electrons act like antennas and thus emit radiation which contains Fourier components at the fundamental and the second harmonic frequency. The fundamental component superposes with the original incoming light wave and leads to an overall phase shift of the fundamental light transmitted through the crystal. In other words, this phase shift is responsible for an index of refraction different from 1. The second harmonic component not only forms a laser like beam at twice the fundamental frequency but also acts back on the harmonic polarization wave: the harmonic light generated at one point in the crystal may either enhance or attenuate the harmonic polarization at another point in the crystal depending on the relative time phase between the light and the polarization at this point. To achieve optimum energy flow from the harmonic polarization wave to the harmonic light wave their relative time phase has to be kept constant along the crystal. This is the "phase matching" condition.

Since the time phase of the harmonic polarization wave is determined by the fundamental light, one obtains phase matching if the index of refraction is equal for the fundamental and for the harmonic light wave. To overcome the dispersion in optical crystals one uses birefringent materials. In a negative uniaxial crystal for example the refraction index of the extraordinary beam may be identical with the refraction index of the ordinary beam at twice the wavelength. To achieve phase matching at a given wavelength it is possible to "tune" the crystal either by varying the temperature or by changing the angle between the optical axes of the crystal and the incident laser beam. If this "phase match angle" is different from 90° the birefringence of the crystal leads to the so called "walkoff" effect: inside the crystal the generated harmonic light does not propagate parallel to the fundamental wave. The deviation angle ("walkoff angle") is small but it limits the interaction region between the harmonic polarization wave and the harmonic light wave. Noncritical 90° phase matching thus allows a higher conversion efficiency, but it is not always possible, in particular in crystals which show only a weak temperature dependence of its indices of refraction. A quantitative expression for the harmonic output power, P_{sh}, is derived in Ref.14 using a simple heuristic approach.

$$P_{sh} = \gamma P_f^2$$

$$\gamma = \frac{2\omega^2 d_{eff}^2 k}{\Pi n^3 \varepsilon_o c^2} L \; h(B,\xi,\sigma)$$

with the fundamental power P_f, the conversion factor γ, the nonlinear coefficient d_{eff}, the fundamental angular frequency ω, the fundamental wavevector k_f, the index of refraction n, the crystal length L, and the fundamental constants ε_o and c. The function $h(B,\xi,\sigma)$ contains effects due to focusing of the laser beam, walkoff, and phase matching. Here, ξ is defined as L/b where $b:=k_f*w_o^2$ is the confocal parameter of the laser beam inside the crystal and w_o is the beam waist. The walkoff parameter $B:=\rho/2*(Lk_f)^{1/2}$ (ρ walk off angle) is 0 for 90° phase-matching and may take values up to 20 in cases of large walkoff. The phase matching parameter $\sigma:=1/2*\Delta k*l$ (with $\Delta k=2k_f-k_{sh}$) describes the total phase difference between the harmonic polarization wave and the harmonic light wave which is accumulated along the crystal due to a deviation of the refraction index at fundamental frequency from that at the harmonic frequency. Usually $h(B,\xi,\sigma)$ has to be evaluated numerically from an integral expression, but for practical purpose one may look it up in Ref.(14) where the results are plotted in Fig.2-6. For optimum focusing and phase matching $h(B,\xi_{opt},\sigma_{opt})=h_{mm}(B)$ only depends on the walkoff parameter B. The values of ξ for which h is maximal vary between 2.84 for B=0 and 1.34 for B>6. As a rule of thumb the confocal parameter b should be more or less the length of the crystal. Approximate expressions of $h_{mm}(B)$ are possible for B>2, $h(b)\approx 0.714/B$ and B<0.2, $h(B)\approx 1.068-1.36B^2$. The maximum value for h=1.068 in the case of B=0 (noncritical phase matching). Note that for B=0 the harmonic output power increases linearly with the crystal length. In the presence of walkoff and B>2 the gain is only proportional to $L^{1/2}$. Typical conversion efficiencies P_{sh}/P_f at a fundamental power of 500mW vary between 10ppm for angle tuned crystals and 0.1% for temperature tuned crystals. This is by far too small to generate sufficient amounts of light for most spectroscopic applications but there are techniques which may increase the conversion efficiency by several orders of magnitude.

Second Harmonic Generation in Passive Enhancement Cavities

The most common method for second harmonic generation with enhanced conversion efficiency utilizes the fact, that the harmonic output power depends quadratically on the fundamental light power. If the crystal is placed inside a passive optical cavity the circulating power may be enhanced by one or two orders of magnitude. The enhancement factor not only depends on the linear cavity losses like scattering and absorption but also on losses due to power conversion into harmonic light. This type of loss increases with the fundamental power in the cavity and saturates the enhancement factor at high circulating power. In addition the coupling of the incident light into the cavity strongly depends on the transmission of the input coupler which, for perfect impedance matching, should be adjusted to the cavity round trip losses. The optimum input coupler transmission

$$T_{opt} = L/2 + \sqrt{(L/2)^2 + \gamma P_f}$$

for a given incident fundamental power P_f, and linear losses L where γ is the single pass conversion factor (P_{sh}/P_f^2) (i. e. the conversion in absence of a cavity).[15] The fundamental power enhancement factor is simply the inverse of T_{opt} and the harmonic output obeys:

$$P_{sh} = \gamma \left(\frac{P_f}{L/2 + \sqrt{(L/2)^2 + \gamma P_f}} \right)^2$$

High quality crystals and cavity optic (L=1%) and efficient crystals (γ=0.005W^{-1}) may lead to a total conversion efficiencies of up to 60% even at relatively moderate fundamental input powers of a few hundred miliwatts. The resonator may be constructed from discrete elements but better results have been obtained with crystals which themselves form an optical cavity.

In Fig. 3 and Fig. 4 we sketch two typical examples. The first (Fig.3) is the doubler which we use for hydrogen spectroscopy. A ß-barium-borate crystal is placed inside a passive ring resonator consists of two curved and two plan mirrors and has two foci. The small focus with a confocal parameter of 4mm is located inside the crystal and matches the optimum focus according to theory. The incident laser beam is coupled to the large focus half way between the two plan mirrors because "mode matching"[16] is less difficult between beams with large foci. One of the mirrors of the cavity is mounted on a piezo-translator which allows to tune the resonance frequency of the cavity. An electronic servo loop controls the piezo and keeps the cavity in resonance with the incident light.[17] The fundamental light, 200mW from a dye laser at 486nm, is resonantly enhanced by a factor of about 50 and 4mW UV-light at 243nm is generated by the 6mm long crystal in an angle tuned geometry (55° phase match angle). The walkoff angle of 4.7° reduces the efficiency by a factor of 0.04 (B=16, $h_{mm}(16)=0.045$). Up to now there is no alternative crystal which is transparent at 243nm and allows phase matching with a temperature tuned noncritical 90° geometry. The experimental conversion factor $\gamma_{exp}=3.5*10^{-5}$ deviates by a factor 2.4 form the theoretically expected value of $\gamma_{the}=8.4*10^{-5}$ (d_{eff}=1.0pm/V, n=1.678). The difference may be caused by inhomogeneities in the crystal which distort the index of refraction such that perfect phase matching is not possible over the whole length of the crystal.

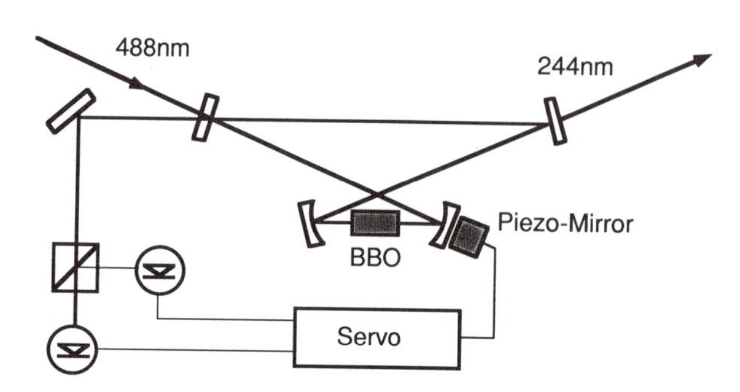

Figure 3 Singly resonant frequency doubling with an angle tuned ß-barium-borate crystal inside a bowtie resonator made of discrete elements.

At higher fundamental power the dielectric antireflection coatings of the crystal begin to develop additional absorption resulting in destruction if the power is further increased. Despite of that, the output power may be still enhanced by adjusting the cavity geometry to the walkoff. In a cavity with an elliptic beam cross section - which may be accomplished with cylindrical cavity optics - the overlap between the fundamental and the harmonic wave inside the crystal and hence the conversion efficiency may be improved while the intensity is slightly reduced. This idea seems attractive but has not been tested yet.

In the second example (Fig.4) the radiation of a semiconductor laser near 840nm is frequency doubled in a monolithic cavity made from potassium niobate. This material allows 90° phase matching for wavelengths around 850nm and 970nm exploiting two different nonlinear coefficients (d_{32} and d_{31}). In our example the fundamental wavelength was 840nm and the temperature -28.3°C. One side of the crystal is optically polished and acts as a plane mirror due to total internal reflection. The front and the rear facets with reflectivities of 93% and 99.8% are coated with dielectric layers . The doubler is placed inside a small vacuum housing and the temperature is controlled by thermoelectric elements. The temperature has to be stable within a few mK. From 90mW fundamental radiation it is possible to generate 22mW blue light at 420nm. In similar experiments conversion efficiencies up to 60% have been observed.[18]

In the case of a monolithic design it is hard to tune the frequency of the doubling cavity and thus the laser frequency has to be controlled. One may think of applying an electric field to tune the doubler cavity via the electrooptic effect, but the voltage which is

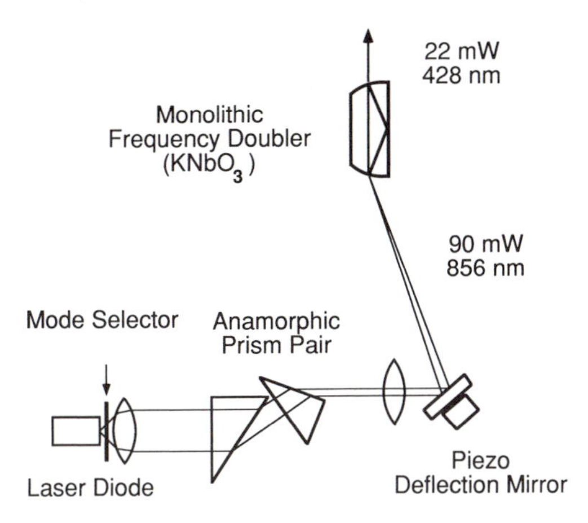

Figure 4. Singly resonant frequency doubling with a temperature tuned Potassium Niobate crystal shaped as a monolithic resonator.

needed to tune over the full frequency separation between two adjacent cavity resonances is too high to be practical. In addition the crystal develops irreversible defects already at a moderate field strength of 1000V/cm. Exploiting the temperature dependence of the index of refraction is less favorable because the temperature is already fixed by the phase matching condition.

Doubly Resonant Second Harmonic Generation

To increase the conversion efficiency in cases where 90° phasematching is not applicable or only little fundamental power is available one may resonantly enhance also the second harmonic light. Due to the backaction to the fundamental wave the power transfer depends linearly on the intensity of the harmonic light already generated. If the harmonic light is recycled by an optical resonator the conversion efficiency may in principle be increased by a factor equal to the power enhancement factor of the resonator (if saturation is neglected).

The theoretical expression for the harmonic output power P_{sh} is given by:

$$P_{sh} = \frac{1-R_{out}}{(1-\sqrt{1-L_{sh}})^2} \gamma P_c^2$$

with the reflectivity of the harmonic output coupler R_{out}, the round trip losses at the harmonic wavelength L_{sh} (including the output coupler), the conversion factor γ, and the fundamental power circulating inside the cavity P_c. To compete with the singly resonant technique this method requires high quality mirrors with little losses at the fundamental as well as at the harmonic wavelength.

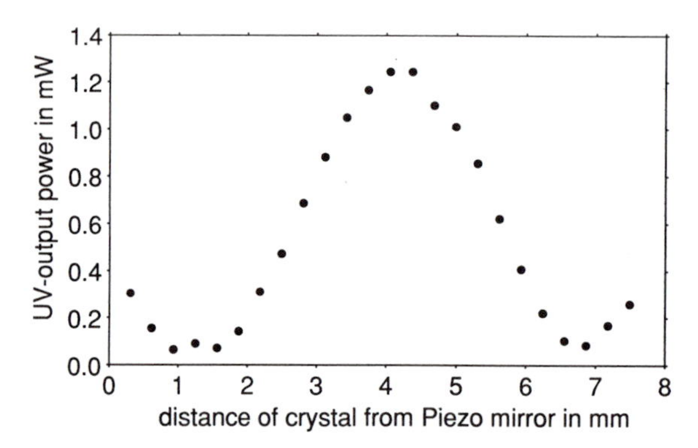

Figure 5. Doubly resonant second harmonic generation with ß-barium-borate. The standing wave geometry is compatible with walkoff.

For angle tuned crystals a standing wave resonator is the obvious cavity geometry (Fig.5). One of the two mirrors is mounted on a piezo translator and serves to control the resonance of the fundamental wave. To achieve resonance also for the harmonic wave one may exploit the dispersion of air. If the macroscopic mirror separation is chosen properly simultaneous resonance occurs at the same piezo tuning. Besides the phase matching condition and the two resonance conditions one also has to take care that the antinodes of the two standing waves overlap inside the crystal. Because of air dispersion this is not necessarily the case. As a consequence the nonlinear interaction is turned on and off periodically while moving the crystal along the cavity axes (Fig.6). In the given example the output of an argon ion laser at 488nm was doubled in an angle tuned 6mm long ß-barium-borate crystal. Despite of the walkoff the harmonic wave was generated in a

TEM_{00} cavity mode. It is interesting to note that for large walkoff the conversion depends only weakly on the size of the focus inside the cavity.[19] In the example the beam waist was only 0.25mm. This feature may become important in cases where optical damage and UV induced absorption allow only low intensities inside the crystal. The most critical parts in such a doubly resonant setup are the mirrors. Even though they were custom made, absorption and scattering were unacceptably high at both wavelengths (3% at 488nm and 15% at 243nm). Therefore the the results were comparable to those obtained with the singly resonant scheme, despite of the additional gain due to the harmonic resonance in the doubly resonant setup.

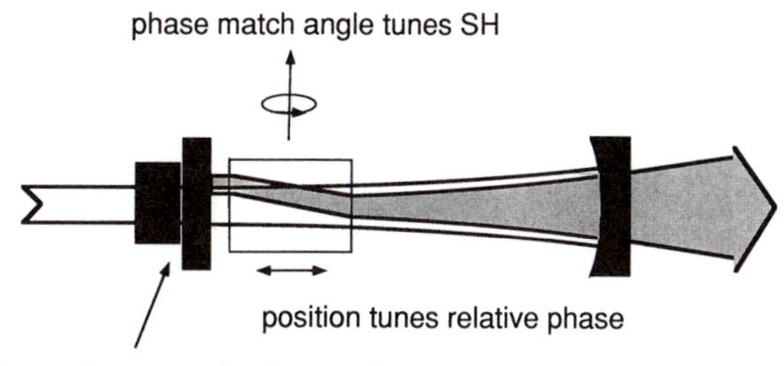

Figure 6. Doubly resonant second harmonic generation with ß-barium-borate: coupling constant variation due to air dispersion.

Second Harmonic Generation with Semiconductor Lasers

The high sensitivity of semiconductor lasers to optical feedback may be exploited to narrow their spectral linewidth and lock their output frequencies to the resonance of an external optical resonator. This is usually done by frequency selecting part of the laser output with a high finesse resonator and feeding it back into the laser.[20] To achieve stable locking only very little feedback is necessary (< 0.1%) . In fact laser diodes lock, more or less automatically, even to external ring cavities although, at first glance, there is no light reflected back into the laser. However, scattering at the mirror surfaces of the ring resonator couples the forward and the backward running mode and light builds up also in the reverse mode. The power in this counterpropagating mode can be quite large even for small values of the coupling between the two modes. To estimate this power, consider a ring cavity with an internal element, such as a weak reflector, that couples light from the forward running mode to the counterpropagating mode. Assume that R is the fraction of power coupled from one mode into the other and T is the fraction of power transmitted through the internal mode coupler. We define a quantity representing the losses of the coupling element by $A = 1 - R - T$. Furthermore, let R_1 denote the power reflectivity of the input coupler, and R_m the cavity reflectance parameter representing all power losses in the resonator other than the input coupler and the internal mode coupler. If we define a critical coupling parameter

$$R_0 = \frac{(1 - R_1 R_m T (1 - A))^2}{(1 + R_1 R_m T (1 - A))^2} (1 - A)$$

then for coupling values $R \leq R_0$ (weak coupling) the forward and reverse modes are resonant at the same frequency. In this case the ratio of power in the two modes at resonance can be written as

$$\frac{P_c}{P_f} = \frac{R}{(1 - \sqrt{R_1 R_m T})^2}$$

where P_c (P_f) is the power in the counterpropagating (forward running) mode. As R approaches R_0, the power ratio P_c / P_f approaches the value 1-A. If R exceeds R_0 (strong coupling) a splitting of the resonance frequencies occurs. In this regime at the resonance of the counterpropagating mode the power ratio P_c / P_f always has the value 1-A. R_0 can be rather small for typical values of $R_1 R_m$ (e.g. $A = 0$ and $R_1 R_m = 0.96$ yields $R_0 = 1.6 * 10^{-3}$).

This locking scheme still works with a crystal inside the cavity and allows the construction of a compact harmonic light source without electronic servo loop. This concept has been used for spectroscopy of the $5s_{1/2}$-$6p_{3/2}$ transition of Rubidium with a 90mW laser diode near 840nm and a temperature tuned Potassium Niobate crystal at -15°C.[21] The bowtie type cavity has been set up inside a vacuum chamber to isolate it from acoustic and thermal noise. The spectral purity of the source is given by the cavity stability and was sufficient to resolve Rubidium resonances with a line width of 1.7 MHz.

As a first step in the direction of a diode laser source for hydrogen spectroscopy we tested an analog setup with a laser diode at 972nm in combination with a 6mm long Potassium Niobate crystal.[22] Noncritical 90° phase matching is possible at -20°C. The experimental and theoretical conversion factors ($\gamma_{exp} = 3.5 * 10^{-3}$, $\gamma_{the} = 8.2 * 10^{-3}$) deviate by more than a factor of 3, probably due to thermal gradients. The cavity geometry was similar to that in Fig.3. The crystal was cooled by a stack of two peltier elements and placed inside a housing which was flooded with nitrogen from a gas bottle. A lid with two small openings

(Ø 1.5mm) for the laser beam protected the crystal from air humidity. Singly and doubly resonant versions were tested and it was possible to generate 2mW at 486nm from 20mW fundamental power. In these experiments a DBR laser (Distributed Bragg-Reflector)[8] was used. Different from the common Fabry-Perot type diodes the cavity of this laser consists of two microfabricated gratings. The wavelength selection due to the gratings restricts the laser to operate only on one specific longitudinal mode of the laser cavity and no mode hops have been observed even if the laser temperature varies over more than 40°C. Although less sensitive to optical feedback, optical locking is still possible and eased due to a wider acceptance range for the feedback strength. Recently a combination of a DBR laser with an integrated high power amplifier has been developed and will be commercially available next January. Up to 1.1W single mode radiation near 974nm are reported.[23]

A Proposed all Solid State UV-Source for (Anti-)Hydrogen Spectroscopy

The methods described above can be combined to construct a compact all solid state laser source for (anti)hydrogen spectroscopy (Fig.7). The light of a 1W DBR-laser at 972nm is coupled into a semi-monolithic ring resonator made from Potassium Niobate. One mirror is mounted on a Piezo-translator to enable cavity tuning without changing the temperature which is fixed by the phase matching condition. The laser is locked onto resonance of the doubler by optical feedback from the reverse mode or by electronic feedback.

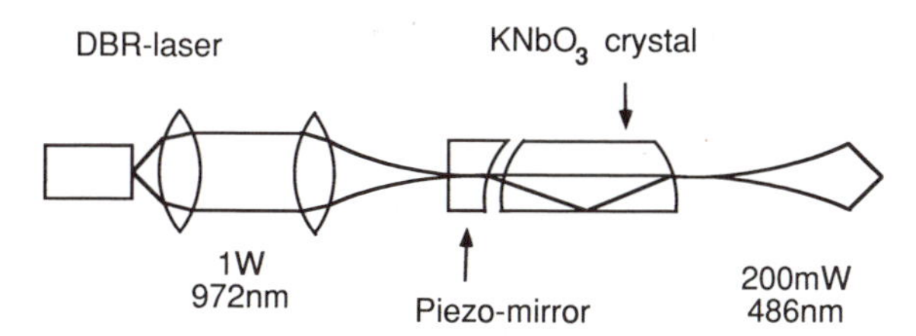

Figure 7. Semi monolithic frequency doubler for efficient doubling of 972nm with a temperature tuned Potassium Niobate crystal.

The output frequency is tuned by varying the length of the doubler cavity which has to be isolated carefully from vibrational noise since its stability determines the spectral purity of the output. To increase the passive stability the laser and doubler should be placed inside the same evacuated and temperature stabilized housing which can be kept at a size of a few centimeter. Laser linewidths of less than 100kHz should be possible. The locking behavior of a DBR-laser has not been extensively studied and even better results may be achieved. To compensate for long term drifts part of the output will be coupled into a very stable reference resonator comparable to the one already used in the present hydrogen experiment[6] and an electronic servo loop may keep the laser system in resonance with the reference cavity. If we extrapolate the experimental results of the test experiment described above to an input power of 1W the output power in the blue near 486nm may exceed 760mW. The theoretically expected maximum output is even higher (870mW), however, at this high power, effects like thermal lensing and imperfect phase matching due to temperature gradients must be taken into account. Even if the output is reduced by a factor 2 to 3 the remaining 200mW will be enough to generate several mW at 243nm with the doubler described above (Fig.3). This is comparable to what is obtained with conventional dye laser systems. If one goes to higher power ß-barium-borate suffers from UV induced losses. A purely oxygenic atmosphere reduces the effect and up to 30mW UV output has been observed with a passive ring doubler from 300mW fundamental input.[24] Inside an argon laser resonator even more than 1W at 244nm has been demonstrated.[25] Based on these experimental results the upper limit for an optimistic estimation of the maximum UV output would suggest 114mW at 243nm from 870mW at 486nm and 1W at 972nm.

REFERENCES

1. W. E. Lamb and R. C. Retherford, Phys. Rev. 79, 549 (1950).
2. B. Couillaud, Ph. Dabkiewicz, L. A. Bloomfield, and T. W. Hänsch, Opt. Lett. 7, 265 (1982).
3. R.G. Beausoleil, D.H. McIntyre, C.J. Foot, B. Couillaud, and T.W. Hänsch, Phys. Rev. A35 (1987) 4878.
4. M. Weitz, F. Schmidt-Kaler, T. W. Hänsch, Phys. Rev. Lett. 68 (1992) 1120.
5. C. Zimmermann, R. Kallenbach, T. W. Hänsch, Phys. Rev. Lett. 65 (1990) 571.
6. T. Andreae, W. König, R. Wynands, D. Leibfried, F. Schmidt-Kaler, C. Zimmermann, D. Meschede, and T.W. Hänsch, Phys. Rev. Lett. 68, 1923 (1992).
7. M. Haase, J. Qui, J. DePuydt, and H. Cheng, Appl. Phys. Lett. 59, 1272 (1991).
8. S. O'Brien, R. Parke, D. F. Welch, D. Mehuys and D. Scifres, Electron. Lett. 8, 1272 (1992).
9. T. W. Hänsch and C. Zimmermann, Spectroscopy of antihydrogen, in: "Antihydrogen", J.Eades,ed., Hyperfine Interactions,1992, to be published.
10. G.Gabrielse, X. Fei, K. Helmerson, S. L. Rolston, R. Tjoelker, T. A. Trainor, H. Kalinowsky, J. Haas, W. Kells, Phys. Rev. Lett. 57, 2504 (1986).
11. T. W. Hijmans, O. J. Luiten, I. D. Setija, J. T. M Walraven, J. Opt. Soc. Am. B6, 2235 (1989).
12. R. W. Boyd, 1992, "Nonlinear Optics", Academic Press, Boston.

13. F. Zernike and J. E. Midwinter, 1973, " Applied Nonlinear Optics",Wiley, New York.
14. G. D. Boyd and D.A.Kleinman, J.Appl. Phys. 39, 3597 (1968).
15. A. Ashkin, G. D. Boyd, and T. M. Dziedzic, IEEE J. Quantum Electron. QE-2,109 (1966).
16. H. Kogelnik und T. Li, Appl. Opt. 5, 1550 (1966).
17. T. W. Hänsch and B. Couillaud, Optics Comm. 35, 441 (1980).
18. W. J. Kozlovsky, W. Lenth, E. E. Latta, A. Moser, and G. L. Bona, Appl.Phys. Lett. 56, 2291 (1990).
19. C. Zimmermann, R. Kallenbach, T. W. Hänsch, and J. Sandberg Opt. Commun. 71, 229 (1989).
20. B. Dahmani, L. Hollberg, and R. Drullinger, Opt. Lett. 12, 876 (1987).
21. A. Hemmerich, D. H. McIntyre, C. Zimmermann, T. W. Hänsch, Opt. Lett. 15, 372 (1990) and US Patent No. 5,068,546, (26 November 1991).
22. C. Zimmermann and T. W. Hänsch, R. Byer, S. O'Brien and D. Welch, Appl. Phys. Lett. 61, 23 (1992).
23. D. F. Welch, private communication.
24. J. Sandberg, private communication.
25. R. Wallenstein, private communication.

DIODE LASERS AND METROLOGY

R. Fox[1], G. Turk[2], N. Mackie[1], T. Zibrova[3], S. Waltman[1], M.P. Sassi[4], J. Marquardt[5], A.S. Zibrov[3], C. Weimer[5], and L. Hollberg[1,5]

[1]National Institute of Standards and Technology
Boulder, CO, 80303, USA
[2]National Institute of Standards and Technology
Gaithersburg, MD. USA
[3]Lebedev Physical Institute, Moscow, Russia
[4]Instituto di Metrologia, Torino, Italy
[5]Department of Physics, University of Colorado
Boulder, CO 80303, USA

Introduction

At NIST in Boulder we have been pursuing an active research program developing diode-laser technology for scientific applications. Commercial diode lasers are readily available in a few wavelength bands in the red and near IR region of the spectrum. Our work has focused on the AlGaAs, InGaAlP, and InGaAsP lasers that operate at room temperature in the red and near IR region of the spectrum between 650 nm and 1.5 microns. These lasers have a number of recognized advantages, including: high efficiency, low cost, tunability, and moderate power levels (~1 to 100 mW). Increasing interest in applying diode lasers to science in general and spectroscopy in particular has stimulated a number of recent reviews on the subject.[1-4]

Unfortunately the distribution of wavelengths of commercial diode lasers only incompletely covers the potential spectral region. The available lasers are grouped into bands that meet specific commercial applications (such as laser printers, communication systems, CD players etc). In addition to the wavelength limitation the tuning range of any specific laser with temperature and/or injection current is discontinuous. Attempts to tune a diode laser to a specific wavelength are often frustrated by the tendency of these lasers to avoid the desired wavelength by jumping modes. The tuning characteristic of any given mode for a specific laser is somewhat reproducible but shows hysteresis and is easily perturbed by optical feedback. Usually once the laser reaches the desired wavelength it is very stable as long as the temperature, injection current and optical feedback are stable. However on a longer time scale there is some evidence of wavelength change as the laser ages.

Solid State Lasers: New Developments and Applications
Edited by M. Inguscio and R. Wallenstein, Plenum Press, New York, 1993

Wavelength Tuning and Anti-Reflection Coatings

Wavelength selective optical feedback may be used to almost eliminate diode laser tuning problems. The most popular technique is to use an optical grating to feedback to the output of the laser. This is an extended cavity configuration and will work to some degree with almost any laser, but it works best if the laser is single-mode (both spatial and longitudinal) and with a reduced reflectivity coating on the facet towards the grating[4-6]. Fortunately many of the higher power commercial diode lasers have a high reflectivity coating on the back facet and reduced reflectivity coating on the output facet. Obviously, feedback from the grating enhances the gain of this coupled cavity system at the feedback wavelength. The gain at other wavelengths is suppressed by the antireflection coating on the output facet. The lower the reflectivity the farther we can hope to tune the extended

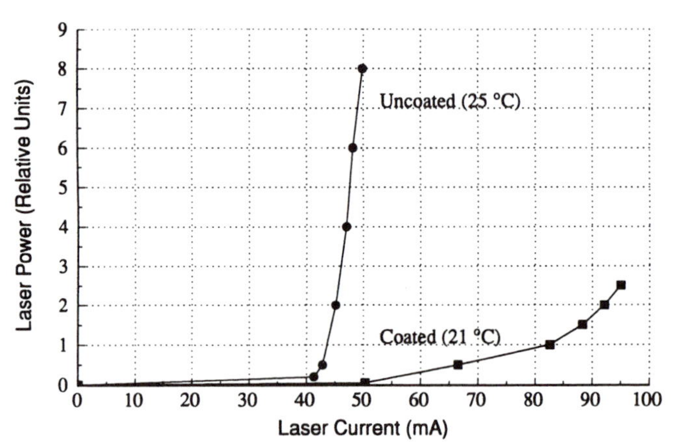

Fig. 1 Laser output power verses injection current for an AlGaAs diode laser as purchased, and with an anti-reflection coating on one facet of the laser.

cavity laser away from the gain peak of the solitary laser. We do not know that the optimum reflectivity is necessarily zero. Other factors such as linewidth, single mode operation or dynamical instabilities may turn out to be worse if the reflectivity is very low. However our experience suggests that the lower the reflectivity the better the extended cavity laser works. Reflectivities that we currently work with range from about 30 to 0.1 percent.

With these systems we can tune to the desired wavelength by tuning the grating and changing laser chip's temperature when necessary. Synchronously changing the laser cavity length while turning the grating will result in larger continuous tuning sweeps. However for finite facet reflectivities, the continuous tuning range near any given solitary chip mode depends on the type of semiconductor laser, the temperature, the grating dispersion, feedback power, injection current relative to threshold, tuning relative to the semiconductor gain curve, and the level of other parasitic optical feedback. For a typical AlGaAs system

Table I. Diode laser antireflection coating results. I_{th}/I_{tho} is the ratio of the laser's threshold current after coating to threshold before coating. The last column gives the calculated after-coating output facet reflectivity as a percentage.

Lasers and Coating Materials

Laser Type[10]	Half-Wave Correction	Coating Material	I_{th}/I_{tho}	% Front Facet Reflectivity
LTO26	YES	Y_2O_3	1.4	6.9
LTO26	YES	HfO_2	1.9	1.9
LTO26	YES	SiO	2.1	0.6
LTO26	NO	HfO_2	1.5	5.4
HLP1400	NO	Ta_2O_5	1.8	1.9
HLP1400	NO	Al_2O_3	1.4	9.2
ML2701	NO	SiO	1.6	3.8
ML4405	YES	HfO_2	1.4	1.0
TOLD9215	NO	HfO_2	1.7	2.0
TOLD9215	YES	HfO_2	2.1	0.6
TOLD9220	NO	HfO_2	1.7	2.0
TOLD9220	YES	HfO_2	2.0	0.1

we might have a continuous tuning range of ~50 GHz while the spacing between the solitary chip modes is about 150 GHz. The intermediate frequencies between the modes are accessible by changing the temperature of the laser chip. There is a great deal of published literature[7,8] on coating diode lasers and some of the results demonstrate[9] that if the output facet reflectivity is low enough the grating can control the laser wavelength without any mode jumps associated with the solitary chip modes.

In order to improve the operation of our extended cavity diode lasers we have been putting anti-reflection coatings on commercial diode lasers. Usually we do not have access to information on the actual semiconductor materials or geometry of these devices which makes it very difficult to design the appropriate coating. With each new laser type we are forced to make an educated guess and then empirically determine which coatings work best. While coating the laser we usually monitor the output power (from the back and/or front facet) as a function of the injection current and coating thickness. A typical power versus injection current plot for a laser both before and after coating is shown in figure 1.

An additional complication in coating diode lasers is that we are usually dealing with commercial lasers that already have some unknown optical/passivation coating on their facets. This coating makes it more difficult to achieve very low reflectance. Table I is a compilation of some of our anti-reflection coating results on commercial diode lasers. Our present best results are reflectance of ~ 0.1 %, and an increase in laser threshold of ~ 2.1 times.

Ultra-sensitive Detection

Now switching our discussion to applications, we can consider the prospects for using diode lasers for ultra-sensitive detection or other analytic applications.[3] The first thing we note is that the amplitude noise on diode lasers is very small which means we should be able to do direct absorption detection of absorbing species with very high sensitivity. Our first effort in this direction was to look at using diode lasers for laser enhanced ionization (LEI) in flames.[11] This work was done as a joint effort between NIST Boulder and NIST Gaithersburg. The very simple experimental system consisted of an

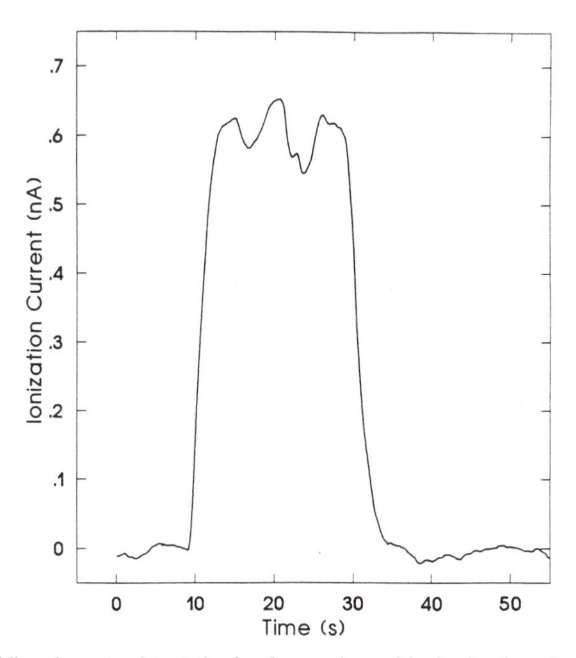

Fig. 2 10 ppb rubidium in water detected using laser-enhanced-ionization in a flame with a diode laser.

atmospheric pressure air-hydrogen (or acetylene flame) and water samples that contained the atoms of interest at low concentrations. We first measured cesium and rubidium detection limits because we could reach their resonance lines with convenient diode lasers. The water samples were aspirated into the flame where the diode laser light excited the atoms. The atoms were subsequently ionized by the flame and the electric field from a high voltage (~ -500 V) electrode. The laser beam was chopped and the current was detected synchronously with a lock-in amplifier. Experimental data for rubidium is shown in fig. 2 where the LEI signal is shown for a 10 ppb (parts/billion) concentration of rubidium in water. With just 2 mW of laser power this data gives a detection limit of 300 ppt (parts/trillion) atom concentration of rubidium in water.

These are just preliminary experimental results for a very simple, practical system that has not been optimized for ultimate detection sensitivity. This first demonstration does show that the technology is workable but it is limited by the present availability of diode laser wavelengths in the near-IR spectral region. However with blue light from frequency doubled diode lasers we could reach transitions in many more atomic species. This should greatly enhance the usefulness of diode laser LEI and other ultra-sensitive detection techniques.

Diode Laser fluctuations

One of the usual limitations to using lasers for high sensitivity detection of atomic species is the amplitude fluctuation of the detected laser light. This noise comes from fluctuations in the laser's output power, direction, and frequency. Fortunately the intrinsic fluctuation in the laser's output power is very low (roughly 10^{-6} for frequencies above 1 MHz). These fluctuations can be reduced by the use of electronic feedback.[12] Within the electronic servo bandwidth, the amplitude noise can be reduced to ~3 dB above the shot noise level with a 50% expenditure of power for the feedback loop. However even with smaller fractions of the laser power used for the servo, significant reduction of the amplitude noise is possible.

But unfortunately with diode lasers we have the additional problem that when the frequency of the laser is scanned there is usually a relatively large systematic change in the output power. This is in addition to the usual etalon and other multi-path effects that often interfere with low-level absorption signals. The affect that this residual amplitude modulation has upon detection limits can be greatly diminished by suitable demodulation techniques.

Fast detectors - frequency difference measurements

Using optical locking[13-15] and/or electronic feedback techniques it is possible achieve very narrow linewidths with diode lasers. To effectively use the resolution that is available from narrow linewidth lasers it is necessary to be able to first measure, and second control the laser's center frequency with a precision that is comparable with the linewidth. For example, an optically-locked diode laser with a linewidth of ~3 kHz has a potential resolution of ~1 part in 10^{11}. It takes a very high accuracy frequency reference to control the center frequency with that level of precision. This can be achieved with the best quality optically-contacted Fabry-Perot reference cavities or with standards-quality atomic/molecular resonances. Even with one of these good references to lock to we often want to measure or even scan the laser's frequency relative to the reference. This can be done with RF frequency-offset-locking techniques as pioneered by Hall and collaborators.[16]

We have been exploring the use of high speed Schottky diodes with diode lasers for measuring large optical frequency intervals and for controlling the laser frequency with high precision. The use of very small area Schottky diodes to detect very high frequency laser beat-notes in the visible has been demonstrated by Daniel et al.[17,18] In our present near-infrared experiments we optically-lock two 830 nm diode lasers to the same confocal reference cavity. The beat note between these two lasers is then detected with the fast Schottky diode. By increasing the difference frequency between the two lasers we can measure the useful signal-to-noise ratio and bandwidth of the Schottky diodes. In our initial experiments we measure a signal-to-noise ratio of ~60 dB with a resolution bandwidth of 10 kHz for beat-notes in the "base-band" region from DC to about 25 GHz. The detector's DC optical responsivity at 830 nm is about 0.2 mA/mW. Based on the Schottky's capacitance and series resistance the detectors intrinsic millimeter-wave bandwidth should be about 2 THz. In order to look for the very high speed response we used the Schottky as a harmonic mixer to generate the harmonics of a 47 GHz klystron. The signal-to-noise of the laser beat note signal as detected relative to harmonics of the klystron is plotted in fig 3. This data is plotted as a function of the difference frequency between the two lasers and the data points correspond to the various klystron harmonics.

The signal-to-noise ratio at low frequencies in this data is presently limited by a frequency noise pedestal that remains on the two optically locked diode lasers. We believe that the apparent roll-off towards higher frequencies is a combination of the Schottky diode's harmonic generating efficiency and some optical roll-off, but this has not been confirmed yet. With our preliminary results we know we can see beat notes between diode lasers with good signal-to-noise out to at least 400 GHz and with appropriate local oscillators we are hopeful that this can go much higher.

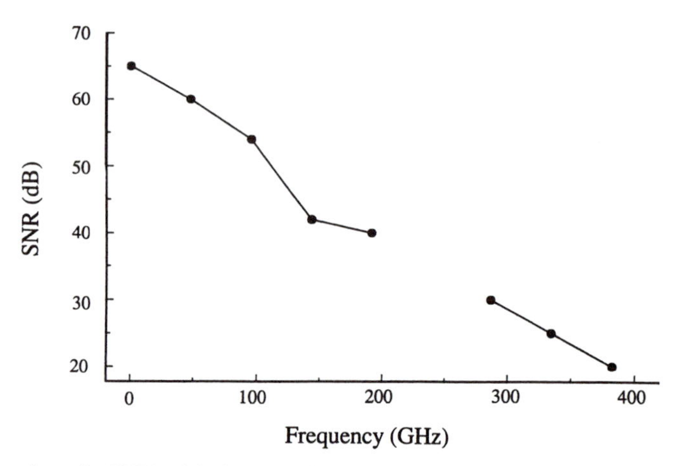

Fig. 3 Signal-to-noise ratio (SNR) of the beat-note between two optically-locked diode lasers and the klystron harmonics, as a function of the laser-laser difference frequency. The klystron was operated at about 47 GHz. The data point corresponding to the 5th klystron harmonic is not shown.

Visible Wavelength / Frequency References

The relatively new red diode lasers (with wavelength bands between 630 and 690 nm) used in conjunction with narrow transitions in the alkalis and alkaline-earths atoms provide us with the opportunity for significantly improve visible wavelength/frequency references. Very nice results have already been achieved with Ba (791 nm)[19], and Sr (689 nm)[20], while we have been pursuing Ca (657 nm)[12]. Calcium is attractive because of its very narrow (400 Hz natural width) transition, because it can be readily laser cooled, and because it is well established as a reference wavelength.[21-24] There is a growing need for an improved visible wavelength reference as evidenced by recent precision measurements of atomic hydrogen transitions.[25] Because of the relative simplicity of the diode lasers there is a realistic hope for a portable length/frequency transfer standard based on calcium with

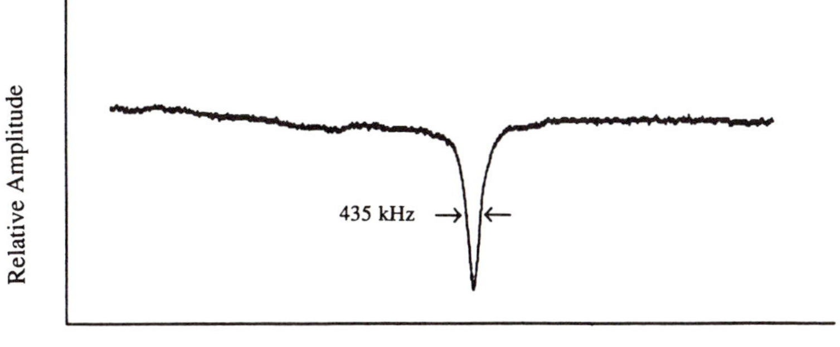

Fig. 4. Calcium Saturated absorption spectrum of the 657 nm line taken with a diode laser. The laser was locked to a cavity and the cavity scanned over the line. A precise frequency scale for the horizontal axis was provided by modulation sidebands of the laser that are not shown in this scan.

very high accuracy. Figure 4 shows a recent saturated absorption signal that we have obtained froma hot calcium cell and 500 μW of diode laser power. Similar calcium linewidths have been obtained with this same diode laser and the PTB calcium beam (in collaboration with V. Velichansky, F. Riehle, and J. Helmcke). The narrowest saturated absorption line-widths that we have observed with our diode laser and a hot calcium cell are less than 150 kHz. These widths are presently limited by transit broadening and wave-front curvature effects due to the inhomogeneities in the laser's spatial mode.

ACKNOWLEDGEMENTS

We would like to acknowledge the many contributions to this effort made by V. Velichansky, J. Bergquist, K. Evenson, and J. L. Hall. This work is supported in part by AFOSR and NASA.

References

1. J. Camparo, The diode laser in atomic physics, *Contemp. Phys.* 26, 443, (1985).
2. M. Ohtsu and T. Tako, Coherence in semiconductor lasers, *in:* "Progress in Optics XXV," E. Wolf ed., Elsevier, p. 193, (1988).
3. J. Lawrenz, K. Niemax, A semiconductor diode laser spectrometer for laser spectrochemistry, *Spectrochimica Acta.* 44B, 155, (1989).
4. C. Wieman and L. Hollberg, Using diode lasers for atomic physics, *Rev. Sci. Inst.* 62, 1, (1991).
5. E.M. Belenov, V.L. Velichansky, A.S. Zibrov, V.V. Nikitin, V.A. Sautenkov, V.A. Uskov, Methods for narrowing the emmision line of an injection laser, *Sov. J. Quant. Elect.* 13, 792, (1983).
6. A. Akul'shin, V. Bazhenov, V. Velichansky, M. Zverkov, A. Zibrov, V. Nikitin, O. Okhotnikov, V. Sautenkov, N. Senkov, E. Yurkin, *Sov. J. Quant. Elect.* 16, 912, (1986).
7. I.P. Kaminow, G. Eisenstein, and L.W. Stulz, Measurement of the Modal Reflectivity of an Antireflection Coating on a Superluminescent Diode, *IEEE J. Quant. Elect.* 19, No. 4, 493, (1983).
8. H. Ukita, K. Mise, and Y. Katagiri, Simple Measurement of the Reflectivity of Antireflection-Coated Laser Diode Facets, *Jap. J. Appl. Phys.* 27, L1128, (1988).

9. P. Zorabedian, W.R. Trutna Jr., and L.S. Cutler, Bistability in Grating-Tuned External Cavity Semiconductor Lasers, *IEEE J. Quant. Elect.* 23, 1855 (1987).
10. Mention of specific commercial laser products does not constitute an endorsement but is made to clarify the particulars of our experiments. Other devices may be better suited to this type of application.
11. G.C. Turk, J.C. Travis, J.R. DeVoe and T.C. O'Haver, Laser Enhanced Ionization Spectrometry in Analytical Flames, *Anal. Chem.* 51, No. 12, 1890 (1979).
12. L. Hollberg, R. Fox, N. Mackie, A.S. Zibrov, V.L. Velichansky, R. Ellingsen, and H.G. Robinson, Diode Lasers and Spectroscopic Applications, *in:* "Tenth International conference on Laser Spectroscopy," M. Ducloy, E. Giacobino and G. Camy, p. 347, World Scientific, (1992).
13. B. Dahmani, L. Hollberg, R. Drullinger, Frequency stabilization of semiconductor lasers by resonant optical feedback, *Opt. Lett.* 12, 876, (1987).
14. Ph. Laurent, A. Clairon and Ch. Breant, Frequency Noise Analysis of Optically Self-Locked Diode Lasers, *IEEE J. Quant. Elect.* 25, No. 6, p. 1131, (1989).
15. H. Li and H.R. Telle, Efficient Frequency Noise Reduction of GaAlAs Semiconductor Lasers by Optical Feedback from an External High-Finesse Resonator, *IEEE J. Quant. Elect.* 25, No. 3, 257, (1989).
16. J. Hough, D. Hils,M. D. Rayman, Ma L.-S.,L. Hollberg, and J. L. Hall, Dye-Laser Frequency Stabilization Using Optical Resonators, *Appl. Phys. B 33*, 179, (1984).
17. H.-U. Daniel, B. Maurer and M. Steiner, A broadband Schottky Point contact mixer for visible laser light and microwave harmonics, *Appl. Phys. B* 30, 189, (1983).
18. J.C. Bergquist, H.-U. Daniel, A wideband frequency-offset-locked dye laser spectrometer using a Schottky barrier mixer, *Opts. Comm.* 48, 327, 1984.
19. A.M. Akulshin, A.A. Celikov and V.L. Velichansky, Nonlinear Doppler-free spectroscopy of the 6^1S_0-6^3P_1 intercombination transition in barium, *Optics Comm.* 93, 54 (1992).
20. G.M. Tino, M. Barsanti, M. de Angelis, L. Gianfrani and M. Ingusicio, Spectroscopy of the 689nm intercombination line of strontium using and extended-cavity InGaP/InGaAlP diode laser, *Appl. Phys. B* 55, 397, (1992).
21. J.C. Bergquist, R.L. Barger and D.J. Glaze, *in:* "Laser Spectroscopy IV," H. Walther and K.W. Rothe eds., Springer-Verlag, p. 120, (1979).
22. J. Helmcke, A. Morinaga, J. Ishikawa and F. Riehle, Optical Frequency Standards, *IEEE Trans. Inst. Meas.* 38, 524, (1989).
23. N. Beverini, F. Giammanco, E. Maccioni, F. Strumia, and G. Vissani, Measurement of the calcium $^1P_1$-$^1D_2$ transition rate in a laser-cooled atomic beam, *J. Opt. Soc. Am. B* 6, No. 11, p. 2188, (1989).
24. T. Kurosu, F. Shimizu, Laser Cooling and Trapping of Calcium and Strontium, *Jap. J. Appl. Phys.* 29, L2127, (1990).
25. T. Andreae, W. König, R. Wynands, D. Leibfried, F. Schmidt-Kaler, C. Zimmermann, D. Meschede, and T.W. Hansch, Absolute Frequency Measurement of the Hydrogen 1S-2S Transition and a New Value for the Rydberg Constant, *Phys. Rev. Lett.* 69, 1923, (1992).

SEMICONDUCTOR DIODE LASERS IN ATOMIC SPECTROSCOPY

G. M. Tino,[1] M. de Angelis,[1] F. Marin,[2] and Massimo Inguscio[2]

[1] Dipartimento di Scienze Fisiche dell' Università di Napoli
Mostra d'Oltremare Pad.20, I-80125 Napoli, Italy

[2] European Laboratory for Nonlinear Spectroscopy (LENS)
Largo E. Fermi 2, I-50125 Firenze, Italy

INTRODUCTION

The use of lasers in experiments of atomic and molecular spectroscopy opened new possibilities to the investigation of these fundamental components of matter. The well known properties of intensity, spectral purity, and directionality of the laser radiation, gave rise to completely new techniques. Both sensitivity and spectral resolution improved by several orders of magnitude with respect to conventional spectroscopy. Of course, due to the central role played by the laser in such experiments, the scientific achievements strictly followed the technical development of the laser sources. Until recently, dye lasers dominated the scenery. They were the only lasers which could be tuned to any particular wavelength in a wide spectral range spanning from the ultraviolet to the near-infrared. However, dye lasers are very expensive and require a large pump laser (an Ar^+ or Kr^+ ion laser for cw dye lasers). This restricted the use of such lasers and made it virtually impossible to think of experiments in which more than one or at most two lasers were needed. In the experimental set-ups there was very often an evident unbalance between the size of the sample cell, usually a simple glass bulb or an electrical discharge, and the laser used to investigate it.

Solid State Lasers: New Developments and Applications
Edited by M. Inguscio and R. Wallenstein, Plenum Press, New York, 1993

Another laser with a wide wavelength tuning range is the semiconductor diode laser (SDL). Although it was invented back in 1961, long before the dye laser, it did not contribute very much to the early achievements in laser spectroscopy[1]. At the beginning, diode lasers had to be operated at liquid nitrogen temperatures, emitted over several cavity modes, and were generally unreliable. In recent years, however, semiconductor diode lasers have been rapidly improving in power, spectral purity, and wavelength coverage, mainly for commercial applications. They can now emit cw at room temperature producing tens of milliwatts on a single mode. Considering their cheapness, small size, and ease of operation, they will become, and we shall show that in many cases they already are, a real alternative to dye lasers for spectroscopic applications. Indeed, they show some specific properties which make them suitable sources of radiation for very high resolution and/or extremely high sensitivity experiments.

In this paper, we show the possibilities offered by semiconductor diode lasers in atomic spectroscopy by describing some experiments that we performed recently using semiconductor diode lasers operating in the visible (650-690 nm) and near-infrared (750-850 nm) regions. In particular, we illustrate the results we obtained in an accurate investigation of near-infrared transitions of atomic oxygen and of the visible intercombination line of strontium. We discuss the spectral resolution, the accuracy of frequency measurements, and the detection sensitivity achievable with diode lasers.

CHARACTERISTICS AND OPERATION OF DIODE LASERS

The basic structure and operation of semiconductor diode lasers are in principle very simple and easy to understand. However, real devices can have very complex structures whose development required important technological advances. By referring to Hanke's contribution to this volume for an overall discussion of recent advances in this field, we will limit ourselves to a sort of brief description of a user's understanding of diode lasers operation, focusing on those characteristics which are more important for spectroscopy.

The diode laser basically works as a p-n junction biased in the forward direction. The recombination of electrons and holes taking place at the junction leads to the emission of radiation whose frequency depends on the energy gap between the conduction and the valence bands. Different from LEDs, in SDLs the reflectivity of the cleaved facets of the diode is high enough to have stimulated emission dominate.

In practice, for cw operation of the laser it is also necessary to confine both the carriers and the light in a well defined region in order to have a high enough gain. This is accomplished by what is called a double-heterostructure. In a double-heterostructure laser the active region is sandwiched between two layers of semiconductors with a larger bandgap. This prevents the carriers to flow out of the active region. Furthermore, because a larger bandgap corresponds to a smaller index of refraction, this structure also acts as a waveguide for the radiation which is confined in the same region. If the chip is grown with a structure such that this confinement is achieved only in the longitudinal direction (the direction in which the current is injected) we have gain guided lasers (for the transverse direction the carriers confinement is given only by the limited region in which the current is injected). If the active region is limited in all directions by a higher bandgap material the laser is called index-guided.

Recently, also quantum-well and multiple-quantum-well lasers (see contribution by R. Cingolani to this volume) have become commercially available, with lower threshold currents and higher output power compared with normal double-heterostructure SDLs.

The above description of the principle of operation of laser diodes, although a little naif, allows us to understand some of the characteristics which are relevant for their use in spectroscopy experiments.

Emission wavelength: this is one of the most important properties for a source of radiation to be used in spectroscopy. Such a source should be tunable to any particular wavelength of interest in a wide interval. In particular, since most of the important atomic transitions are in the visible and near-infrared region of the spectrum, in the following we will focus on SDLs emitting in this region.

As mentioned above, the emission wavelength of semiconductor diode lasers depends on the bandgap of the semiconductor forming the active region. In the visible and near-infrared region, depending on the particular semiconductor, we can have for example, emission of light in the 1100-1650 nm interval ($In_{1-x}Ga_xAs_yP_{1-y}$ semiconductor), in the 700-890 nm interval ($Ga_{1-x}Al_xAs$), or in the 630-690 nm interval (AlGaInP). In practice, the set of wavelengths available from the laser diodes presently available on the market is much smaller than this and is far from being continuous. The reason for that is twofold. First, although semiconductors can be produced with energy gaps corresponding to the emission of radiation from the visible to the infrared, they can work as laser material only if they satisfy some requirements. In particular, only direct semiconductors can give rise to laser emission (in direct semiconductors the minimum of the conduction band and the maximum of the valence band correspond to the same value of the wave vector) because direct radiative transitions are possible. Also, the material must have good optical and electrical characteristics in order to minimize losses. For this reason, for example, room temperature operation was only recently demonstrated for diode lasers emitting in the green-yellow region. The second main factor which limits the wavelengths presently available from commercial laser diodes is not related with fundamental laws of physics; it depends on the fact that companies produce only laser diodes emitting at wavelengths which are important for commercial applications. For example, laser diodes are produced emitting in the 630-690 nm interval (for laser pointers, bar code readers, optical data storage devices), around 780 and 850 nm (CD players, laser printers, fiber optic communications), and near 1300 and

1500 nm (fiber optic communication systems). High power diode lasers are also produced at particular wavelengths for pumping solid state lasers and amplifiers.

From the above discussion it is clear that if a particular region of the spectrum must be investigated, first of all a laser must be found which is made of the proper semiconductor material. Then, because the laser's wavelength depends also on the diode's temperature, a given laser can typically be tuned by about 15-20 nm by changing its temperature. However, a change in the temperature of the diode leads to a change of the frequency of the cavity modes, because of the change of the optical length of the cavity, and to a shift of the gain curve. The temperature dependencies of these two effects are different and, as a consequence, the wavelength changes in a discontinuous and rather unpredictable way.

The wavelength tunability and control is therefore one of the major problems in using SDLs for spectroscopy. In the following, we will show that this problems can be reduced by using optical feedback techniques, which allow to enlarge the tuning interval and to avoid the wavelenght "jumps".

In the experiments described in the following, we used AlGaAs/GaAs diode lasers emitting in the 750-890 nm region and AlGaInP/GaInP diode lasers emitting in the 650-690 nm region.

Spectral linewidth: This is an important characteristic of laser sources, especially for high resolution spectroscopy applications. As for other characteristics of diode lasers, it is rapidly improving following progresses in diodes fabrication. In particular, index-guided lasers can now be found in several wavelength ranges; in proper operating conditions, they essentially oscillate in a single transverse and longitudinal mode. However, the emission linewidth is still large compared, for example, to the intrinsic linewidth of important atomic transitions. Several factors contribute to the SDLs emission linewidth[2], but the most fundamental is the one given by the Schawlow-Townes formula (modified by the introduction of a factor α which takes into account the dependence of the refractive index on the carrier density)[3]:

$$\Delta\nu_{FWHM} = \frac{h\nu}{8\pi P_0}\left(\frac{c}{nL}\right)^2\left[aL+\ln\left(\frac{1}{R}\right)\right]\ln\left(\frac{1}{R}\right)n_{sp}\left(1+\alpha^2\right) \quad \propto \quad Q_c^{2} \tag{1}$$

where P_0 is the output power, L is the cavity length, R is the facets reflectivity, a gives the distributed losses in the cavity, c/n the group velocity, and n_{sp} is the spontaneous emission factor which is of the order of unity. In the case of a typical AlGaAs diode laser (P_o=10mW, L=300µm, R=0.3, α~4-5), this formula leads to a linewidth of several MHz. Also, smaller power-independent factors contribute to the emission linewidth.

From the above formula it is apparent that the emission linewidth can be reduced by increasing the cavity length L or, more generally, by increasing the cavity quality factor Q_c. This is the idea leading to the use of external optical cavities for the line narrowing of SDLs, as discussed in the following.

Amplitude noise: as in the case of the frequency noise, the intrinsic stability of the structure of SDLs reduces the sources of technical noise which affect, for example, dye lasers. On the other hand, the properties of the active medium give rise to a peculiar AM and FM noise spectrum. In the low frequency range, which is an important range for typical absorption spectroscopy experiments, the amplitude noise is typically only 10-20 dB above the shot noise level. This means that the power fluctuations can be as low as one part in 10^7 for a 1 Hz detection bandwidth, which makes the diode lasers extremely good sources for high sensitivity spectroscopy. However, larger fluctuations can be found at frequencies around 2-3 GHz due to the lasers relaxation oscillations.

The importance of the small amplitude noise for high sensitivity spectroscopy is mentioned in the following and discussed in more detail in the contribution by K.Ernst et al. and in the one by L.Hollberg et al.. In the latter, methods to further reduce the amplitude noise are also discussed.

FREQUENCY STABILIZATION OF DIODE LASERS

Although commercial diode lasers usually have a rather broad emission bandwidth which makes them not suitable for very high resolution spectroscopy, they can be easily stabilized by means of electronic or by optical feedback techniques[4].

In the first case, electronic feedback is used for the stabilization of the laser frequency to a frequency reference by changing the injection current to the diode. This method can allow a considerable reduction of the emission linewidth but, in fact, it is not very popular for several reasons. First, this line-narrowing technique requires very fast electronics because the FM noise extends to high frequencies. Second, a change in the injection current produces changes both in the emitted frequency and in the emitted power, so that an additional amplitude noise is introduced. Finally, this stabilization method does not give any control on the emission wavelength and does not allow to avoid the mode "hopping" problem mentioned above.

A drastic reduction of the emission linewidth of the laser can be more easily accomplished using optical feedback, and this is in fact the most common method. As mentioned above, the basic idea in this case is to reduce the intrinsic linewidth of the laser by using a cavity with a Q much higher than the one of the solitary laser diode. In addition, if a wavelength selective optical element is used in such a cavity, this system may also allow a control of the emission wavelength. Two main schemes have been devised to achieve stabilization and control of the emission frequency of a diode laser: one is coupling through optical feedback to a high-Q Fabry-Perot interferometer[5], the other is the operation of the diode laser in an external cavity. They correspond to two different regimes, weak-feedback and strong-feedback regimes, respectively, in which the compound laser diode - external cavity system results in a stable single mode operation.

In our work, we mainly used the latter scheme. Strong feedback from an external diffraction grating allowed us to obtain a reduction by about two orders of magnitude of the emission linewidth and to achieve a wavelength tuning range of several nanometers at fixed temperature. Compared with the other methods, one of the drawbacks of using external-cavity systems for the frequency stabilization of SDLs is that they usually require the diode facets to be antireflection coated in order to increase the amount of light coupled into the diode, as discussed in L. Hollberg's contribution to this volume. However, most of the average- and high-power diode lasers available on the market are already provided with a high reflectance coating on the back facet and with a reduced reflectance coating on the output facet. In fact, in this work we used such diode lasers in pseudo-external cavities without any modification to the commercial diodes.

Around 780 nm we mainly used Sharp LTO24 MD lasers; in the free-running operation their output power is about 20 mW in a single mode with a linewidth of 20-30 MHz. The lasers used in the 850 nm region were Sharp LTO15MD, Mitsubishi ML5101A emitting 20÷30 mW, and STC LT50A-03U multiple-quantum-wells lasers. The output power rating for the STC lasers is 50 mW with an injection current of 120 mA. They emitted in a single mode with less than 10 MHz linewidth. In the visible, we used Nec and Toshiba lasers. The work reported here was done with the Toshiba TOLD9140 which emits ~ 20 mW at 690 nm.

The design of the extended-cavity lasers is shown in Fig.1. The first order diffracted beam from a grating, mounted in the Littrow configuration, was fed-back into the laser diode.

The diode was attached to a small copper block in thermal contact with a Peltier element. By proper insulation and with a constant current flowing through the Peltier junction, a passive stability on the order of 10 mK was achieved. Using active stabilization, the temperature was controlled to better than 1 mK. The diode was fed by a low-noise current source (Laser Optronic) or by a battery. The 9 cm-long external cavity consisted of a

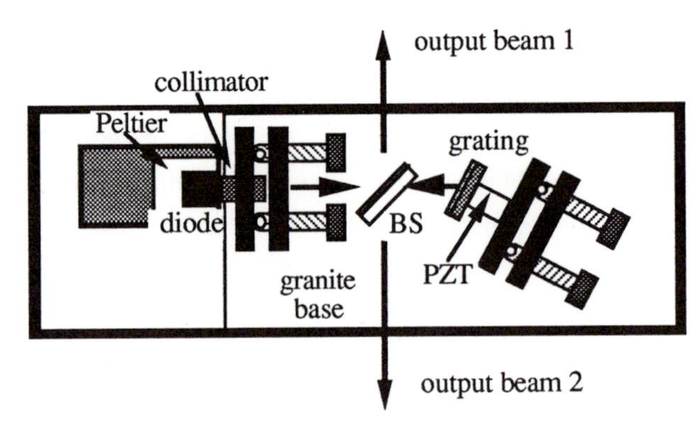

Figure 1 Design of the pseudo-external-cavity laser using a commercial diode laser; frequency stabilization is achieved by optical feedback from a diffraction grating. BS: beam-splitter, PZT: piezoelectric transducer.

collimating lens with numerical aperture = 0.5 (Melles Griot), a 1200 lines/mm ruled diffraction grating blazed for 750 nm (Edmund Scientific), and a 20-30% reflectance beam splitter as output coupler. The proper choice of the reflectance of the beam splitter, that is of the feedback level, may depend on the particular laser diode. Also, a higher reflectivity beam-splitter can be used to couple more light out of the cavity; this is obtained, however, at the cost of a reduced tuning range. It is also possible to omit the beam splitter and to use only the zero order from the grating as the output. In this case, however, it is not easy to control the amount of power in the output beam, which depends on the efficiency of the grating.

The lens and the grating were held by commercial mounts (Lees Corporation) with high precision tilters for the alignment. Granite or steel baseplates were used.

The procedure for the alignment of the cavity was the following: first, the laser diode was mounted so that the output beam hit near the center of the grating and was polarized in the direction perpendicular to the grating grooves in order to increase the grating reflectivity (a higher resolution is obtained, instead, if the beam is polarized parallel to the lines; because of the elliptical shape of the beam, a larger number of lines is illuminated in this case). Then, the lens position was adjusted until the beam appeared the same diameter over a path of several meters. The grating was rotated and tilted to direct the first order diffracted beam back into the laser. By operating the diode at a current slightly above the threshold current, the good alignment of the cavity was indicated by an increase of the output power. Sometimes, we also used a Michelson λ-meter during the alignment process. When the laser is frequency narrowed, the longer coherence length gives a higher contrast of the interference fringes; also, the wavelength can be changed by rotating the grating. This monitoring system was used, for example, for the alignment of cavities including STC lasers, which have no internal photodiode.

The extended-cavity lasers are typically tunable, by simply rotating the grating, over a range of ~ 10 nm in a coarse manner. In fact, because of the finite reflectivity of the chip's facets, the longitudinal modes of the solitary laser diode remain essentially unchanged. The feedback from the grating allows to select one of these modes and to drastically suppress other side-modes. Fine tuning to the frequency of interest can then be achieved by slightly changing the temperature and the injection current of the diode. Continuous frequency scans up to ~ 10 GHz were accomplished by synchronously sweeping the length of the cavity, by means of a piezoelectric transducer on the grating, and the injection current. Light from the intracavity beam-splitter and the weak zero-order reflection from the grating were used for the experiments. The output power depended, of course, on the type of laser diode and on the reflectance of the beam splitter. The 20% reflectance beam splitter coupled out two beams; the maximum output power was, for example, 4 and 3 mW for the Sharp diodes and about 8 and 6 mW for the STC diodes. The effect of the increased Q of the laser cavity was qualitatively evident from the recordings of the transmission of a Fabry-Perot interferometer as the laser frequency was scanned (Fig.2). Broad peaks (20÷50 MHz depending on the particular laser and operating conditions) were recorded by blocking the return beam from the

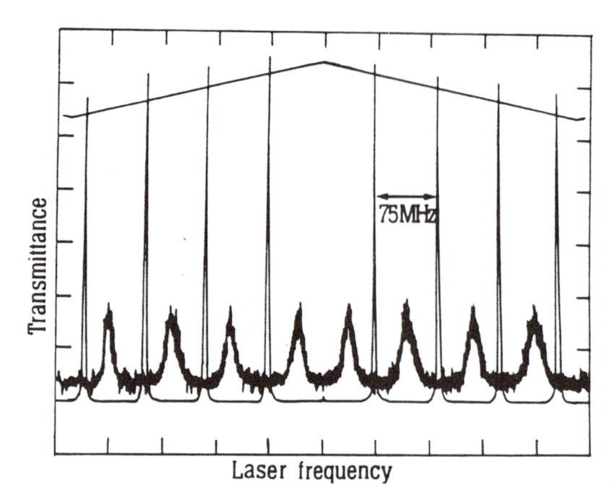

Figure 2. Transmission spectrum of a 75 MHz free-spectral-range Fabry-Perot interferometer as the frequency of the diode laser is scanned without optical feedback (broad peaks) and in presence of optical feedback (narrow peaks).

grating, while much narrower peaks (~ 1 MHz) were obtained in presence of the optical feedback. A more accurate analysis of the spectral properties of the extended cavity lasers was then performed by mixing two similar lasers in a fast photodiode. By observing the heterodyne spectrum for two frequency stabilized near-infrared lasers, we found that in the millisecond time scale the lasers exhibited a linewidth of less than 50 kHz; on a longer time scale, acoustic noise increased the linewidth to about 400 kHz, which is not bad, considering that the cavities were only covered with a lucite box. Such a linewidth did not affect the resolution achieved in the experiments performed in this work; therefore no more efforts were done to suppress the acoustic jitter.

HIGH RESOLUTION SPECTROSCOPY WITH SEMICONDUCTOR DIODE LASERS

Spectroscopy of near-infrared transitions of atomic oxygen

In atomic oxygen, we investigated the 3s $^5S_2$-3p $^5P_{1,2,3}$ transitions at 777 nm, using Sharp LTO24 lasers, and the 3s $^3S_1$-3p $^3P_{1,2,0}$ transitions at 845 nm, using STC LT50A lasers. The Doppler broadening of the lines was eliminated by means of high-resolution spectroscopy schemes; Fig. 3 shows the scheme of the apparatus which allowed us to implement different saturation and polarization spectroscopy techniques[6].

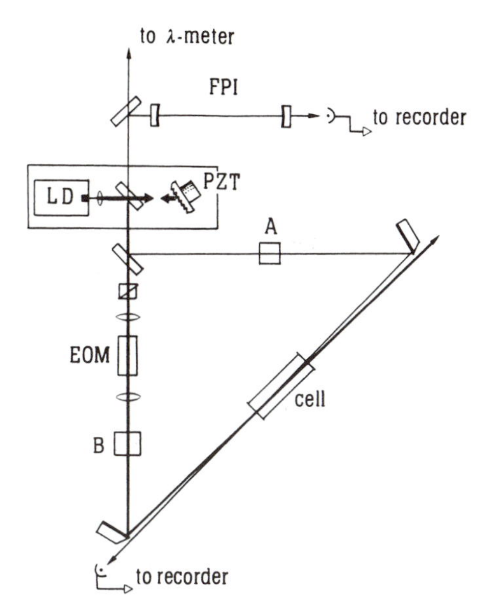

Figure 3. Scheme of the experimental apparatus used to record Doppler-free spectra by saturation or polarization spectroscopy methods. A and B indicate different optical devices depending on the detection technique. A high-finesse Fabry-Perot interferometer (FPI) is used for the frequency calibration of the spectra. EOM: electro-optic modulator.

In particular, by polarization spectroscopy techniques, we obtained a drastic reduction of the collisional broadening of the lines which allowed us to estimate their radiative linewidth.

Using isotopically enriched samples, we could directly observe the isotope shift and the hyperfine structure of the lines. Fig. 4 shows indeed the spectra relative to the 3 $^5S_2$-3 $^5P_1$ transition, recorded by saturation spectroscopy in an isotopically enriched sample.

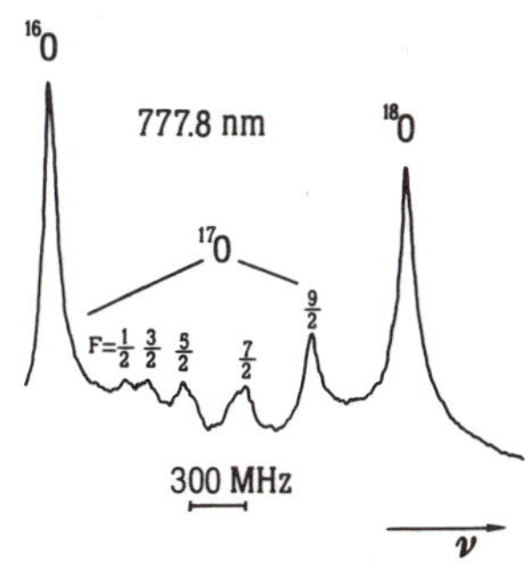

Figure 4. Doppler-free spectra of the 3 $^5S_2$-3 $^5P_1$ transition of atomic oxygen recorded by saturation spectroscopy using an isotopically enriched sample. Five peaks are observed for ^{17}O, corresponding to the hyperfine structure of the 3 $^5S_2$ level (F=9/2,...,1/2).

The transmission fringes of a 75-MHz free spectral range Fabry-Perot interferometer (which are not shown in the figure) were recorded simultaneously and provided the calibration of the laser frequency scan. In addition to the ^{16}O-^{18}O isotope shift, the hyperfine structure of ^{17}O can be observed, with five main components corresponding to the structure of the $^5S_2$ level (I=5/2).

Our main purpose was indeed to look for a nuclear volume effect in the isotope shift of these transitions, which start from the lowest excited s-levels[7]. Due to the small size of the expected effect we had to perform very accurate measurements of the frequency separations amongst the lines of the three stable isotopes ^{16}O, ^{17}O, and ^{18}O. Possible errors due to unproper calibration of the reference interferometer or nonlinearity in the laser frequency scans were eliminated using a heterodyne spectroscopy scheme[8]. Two independent lasers were locked to the sub-Doppler resonances of different isotopes and mixed in a fast photodiode. The scheme of the experimental apparatus is shown in Fig. 5. It consisted of two similar set-ups for saturation spectroscopy, a fast photodiode (300 ps risetime), and a spectrum analyzer (HP 8592A). The lasers were frequency-narrowed by means of the grating extended-cavity, and temperature stabilized to better than 1 mK, as described above. Atomic oxygen was produced in two pyrex cells by means of radio frequency discharges.

Each laser beam was split into two parts of different intensities and sent in opposite directions into the sample cell. The changes in the probe beam intensity were detected with good S/N ratio using a photodiode. Derivative signals were obtained by applying a 780 Hz modulation to the laser frequency and using phase sensitive detection. Both first and third derivative signals could be easily detected. In the case of the first derivative, the sub-Doppler signals were superimposed on a wide background slope. This affected the precision of the locking of the laser frequency to the line centers. Indeed the servo loop works around zero voltage and thus a systematic shift can occur. The background is instead strongly reduced using 3f detection.

The servo loop (bandwidth ~10 Hz) used for frequency-locking each of the laser, consisted of a lock-in, an integrator, and a high voltage amplifier; the output of the integrator was fed back to the high-voltage amplifier, which controlled the grating position by means of the piezoelectric transducer. The intrinsic frequency-stability of such a laser system is determined by the increased Q of the optical cavity, which reduces the fast frequency fluctuations; only a relatively slow electronics is then required to correct long term drifts. The beat note between two lasers was observed using the spectrum analyzer.

The isotope shifts were directly measured from the frequency of the beat note between two lasers locked on the relevant isotope resonances and combined in the fast photodiode. The broadening of the beat note due to the FM modulation of the lasers was minimized by in-phase modulation. The center frequency was determined considering the middle point between the frequencies corresponding to the half maximum on the two sides of the curve. Since absolute frequency measurements are not precise enough with our spectrum analyzer, we used a synthesizer (HP 8341B) for calibration. Measurements were repeated several times for different discharge conditions. The beat frequency reproducibility was within 2 MHz in the case of 1f locking and within a few hundred kilohertz for 3f locking. The data obtained by the heterodyne technique resulted more accurate by one order of magnitude with respect to the ones we had obtained in previous experiments, where the Fabry-Perot interferometer was used for frequency calibration.

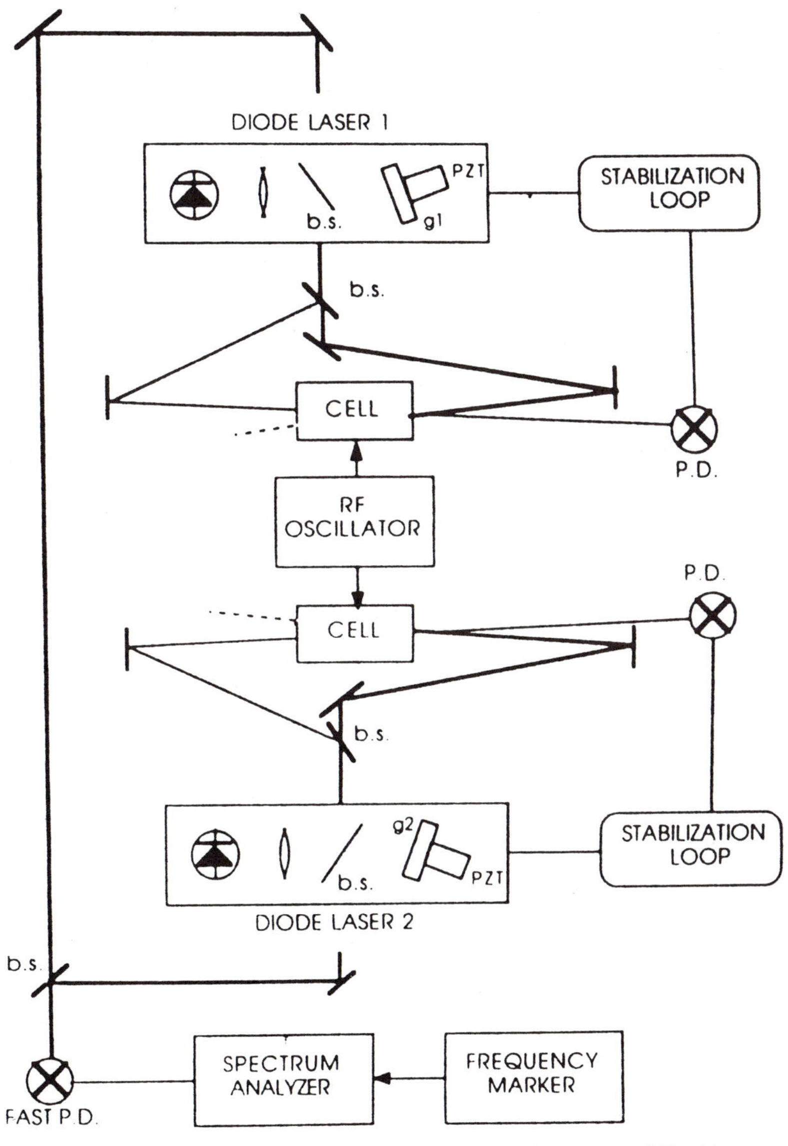

Figure 5. Experimental set-up used for heterodyne spectroscopy. Two frequency-stabilized lasers are locked on two relevant transitions and mixed in a fast photodiode. Spectral structures can be accurately measured by measuring the frequency of the beat note. P.D.: photodiode; b.s.: beam-splitter.

The above results demonstrated the possibility of very accurate measurements of spectral structures by beating of extended-cavity diode lasers locked to sub-Doppler resonances. The simplicity of the method is based on the versatility and low cost of this kind of lasers, which makes the simultaneous operation and frequency-locking of different sources a rather simple matter. The use of faster detectors for heterodyne measurements and the demonstration of frequency stabilization of visible diode lasers makes it possible to extend this technique to new spectral regions and to wider frequency separations.

Spectroscopy of the visible intercombination line of strontium

It is only a few years now that visible diode lasers have been commercially available. The first visible diode lasers operating at ~ 670 nm at room temperature were gain-guided and emitted on several modes. Also the available output power did not exceed 3 mW. With the development of index-guided lasers, single mode operation was achieved but still with a linewidth of several tens of megahertz. The extension of the optical feedback schemes to this class of lasers was not straightforward and often required the deposition of an antireflection coating on the output facet of the diode.

We demonstrated that diode lasers emitting at 690 nm at room temperature (in particular the Toshiba TOLD9140 SDL) can be used in an extended cavity configuration without any additional coating treatment with a reduction of the emission linewidth to less than one megahertz. Using this laser with a grating external cavity, we investigated the narrow (8 kHz) intercombination line of strontium at 689.448 nm.[9] Amongst the alkaline-earth elements, Ca and Ba intercombination lines have also been investigated with diode lasers[10,11]; indeed, because of their narrowness, these transitions are important for laser-frequency stabilization in the optical range.

Strontium was produced by sputtering in a hollow cathode discharge sustained by argon. With a discharge current of 100 mA and an argon pressure of 1 Torr, we observed a 1% absorption signal corresponding to a density of about 10^{13} atoms/cm^3.

The Doppler-free resonance was observed by means of saturation and polarization spectroscopy. Thanks to the low amplitude noise of the diode laser, the small saturation dip (~ 5 %) was observed in real time on an oscilloscope (Fig. 6). The minimum linewidth we observed by polarization spectroscopy was 6 MHz FWHM.

In addition to the intrinsic interest of this line for metrogical reasons, it can also provide a significative test for the actual linewidth of the laser. In fact, after subtracting the contribution due to pressure broadening and finite angle between pump and probe beam, we found that the contribution of the laser linewidth to the observed width of the line was less

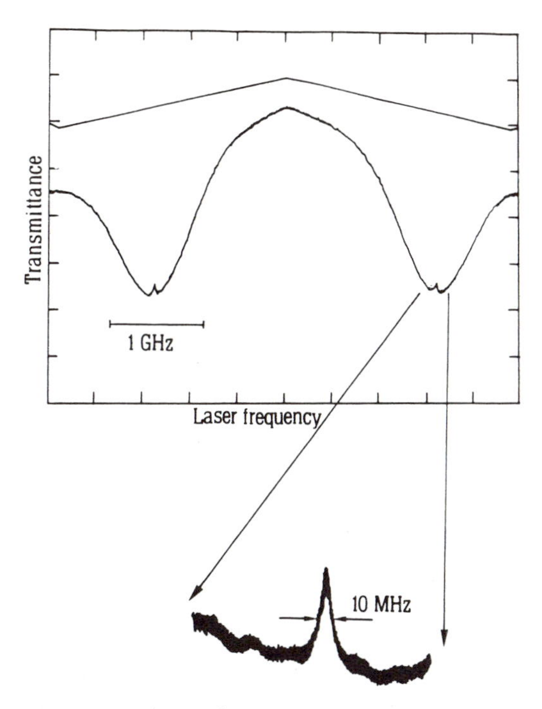

Figure 6. Absorption profile of the 5s $^1S_0$-5p $^3P_1$ intercombination transition of strontium at 689.4 nm. The laser beam was retroreflected through the cell in order to produce a saturation dip, which can be observed on the top of the Doppler profile.

than 1 MHz. A more precise determination requires heterodyning of two similar lasers or, even better, the observation of the intercombination line in an atomic beam.

Work is actually in progress in these directions. Preliminary results obtained in the atomic beam can be found in Ref.12 .

In the same work, we also observed the P63 R70 (8,4) iodine transitions and resolved their hyperfine structure by means of polarization spectroscopy. Since the frequency of these lines is very close (~ 4 GHz) to that of the Sr line, they can provide an interesting frequency marker at this wavelength.

HIGH SENSITIVITY SPECTROSCOPY WITH DIODE LASERS

As already mentioned above, one of the most important characteristics of diode lasers is their low amplitude noise. Because of the intrinsic stability of their structure, the diode lasers exhibit an amplitude noise which is considerably lower than that of a dye laser. In the low frequency range, the noise is typically 10÷20 dB above the shot noise level. This makes them ideal radiation sources to perform absorption spectroscopy. In fact, in the experiments reported above, although resolution was the major issue, a high detection sensitivity was also required. In the case of oxygen, because of the small concentration of atoms produced in the relevant excited states. In the case of strontium, because of the weakness of the investigated transition.

High sensitivity spectroscopy techniques are important not only for fundamental physics but also for some important applications such as environmental monitoring. As already mentioned, this aspect is discussed in more detail elsewhere in this volume. It turns out that a noticeable improvement in the detection sensitivity can be achieved with respect to simple absorption spectroscopy if heterodyne detection schemes are used. In these techniques, another characteristic of diode lasers can be exploited, that is the possibility of fast frequency modulation. Indeed, by modulation of the SDL injection current an emission spectrum can be produced with modulation sidebands up to frequencies of a few GHz. Although the presence of feedback from an external optical cavity generally decreases the index of frequency modulation by the injection current, we recently observed that by choosing proper values of the modulation frequency and of the external cavity length, modulation sidebands up to 4.5 GHz can be produced in a grating stabilized laser. A systematic investigation is in progress in order to investigate the possibility of combining the advantages offered by the presence of the external cavity in terms of frequency stabilization and control, with the high sensitivity achievable by heterodyne detection schemes.

CONCLUSIONS

From the results reported in this paper it is apparent that the diode lasers can be a real alternative to dye lasers for several atomic physics experiments. They can potentially cover the visible and near-infrared region where most of the important atomic transitions are found. At present, however, a drawback is the limited spectral range covered by the diodes which are commercially available. This problem is partially reduced by the optical feedback techniques, as the one described above, which allow to slightly enlarge the spectral range covered by a diode laser and, more important, to eliminate the gaps in the tuning. In particular, the demonstration of frequency control of the visible diode lasers by means of external cavities makes also these lasers available for spectroscopy. Also, blue and near-ultraviolet radiation can be produced by frequency-doubling diode lasers (see contribution by K.Zimmermann).

The spectral emission properties can also be improved by optical feedback techniques and the emission linewidth can be drastically reduced down to levels which allow also very narrow transitions to be investigated. This is obtained at the cost of a slightly larger size and complexity of the system. However, it is worth mentioning that with the progress of integrated optics, it becomes possible to integrate the grating with the diode both for near-infrared and for visible lasers.

The limited output power of diode lasers, on the other hand, does not represent a serious problem for several reasons. First, the power emitted by SDLs, especially quantum-wells SDLs, starts to be comparable with the output power of dye lasers. Furthermore, as demonstrated also in this work, high sensitivity spectroscopic techniques can be developed which only require low laser power. Under this respect, the possibility of fast frequency modulation and the low amplitude noise are particularly important characteristics of the diode lasers.

Finally, two more advantages of diode lasers must be emphasized; one is the high efficiency and low cost of these lasers which do not need large power supplies nor cooling water. This allows experiments to be performed which could not even be conceived before. An example is given by the experiments of cooling and trapping of atoms by laser radiation, which can now be performed also in very difficult conditions such as in micro-gravity laboratories[13]. The other advantage of using diode lasers for spectroscopy is the ease of operation, a characteristic which anybody who used a ring dye laser will certainly appreciate.

REFERENCES

1. J. C. Camparo, The diode laser in atomic physics, Contemp. Phys. 26, 443 (1985).
2. M. Ohtsu and T. Tako, Coherence in semiconductor lasers, in: "Progress in Optics XXV", E. Wolf ed., Elsevier, Amsterdam (1988).
3. D.Welford and A.Mooradian, Output power and temperature dependence of the linewidth of single-frequency cw (GaAl)As diode lasers, Appl. Phys. Lett. 40, 865 (1982).
4. C. E. Wieman and L. Hollberg, Using diode lasers for atomic physics, Rev. Sci. Instrum. 62, 1 (1991).
5. B. Dahmani, L. Hollberg, and R. Drullinger, Frequency stabilization of semiconductor lasers by resonant optical feedback, Opt. Lett. 12, 876 (1987).
6. M. de Angelis, M. Inguscio, L. Julien, F. Marin, A. Sasso, and G. M. Tino, Saturation spectroscopy and velocity-selective optical pumping of oxygen using an (AlGa)As diode laser, Phys. Rev. A 44, 5811 (1991).
7. F. Marin, C Fort, M. Prevedelli, M. Inguscio, G.M. Tino, and J. Bauche, Hyperfine structure and isotope shift of the 3s-3p transitions of atomic oxygen, Zeit. Phys. D, in press.
8. F. Marin, P. De Natale, M. Inguscio, M. Prevedelli, L. R. Zink, and G. M. Tino, Heterodyning of AlGaAs lasers: direct frequency measurement of the isotope shift in the oxygen atom, Opt. Lett. 17, 148 (1992).
9. G. M. Tino, M. Barsanti, M. de Angelis, L. Gianfrani, and M. Inguscio, Spectroscopy of the 689 nm intercombination line of strontium using an extended-cavity InGaP/InGaAlP diode laser, Appl. Phys. B 55, 397 (1992).
10. L. Hollberg, V. Velichansky, R. Ellingsen, A. Zibrov, and G.M. Tino, Observation of a Doppler free resonance on the intercombination transition of calcium with a tunable diode laser, in "Proc. 14th Int. Conf. Coher. and Nonlin. Optics", p. 137 (1991); L. Hollberg, R. Fox, N. Mackie, A.S. Zibrov, V.L. Velichansky, R. Ellingsen, and H.G. Robinson, Diode lasers and spectroscopic applications, in "Proc. 10th Int. Conf. Laser Spectroscopy", M. Ducloy, E. Giacobino, G. Camy eds., p.347, World Scientific (1992).
11. A.M. Akulshin, A.A. Celikov, and V.L. Velichansky, Nonlinear Doppler-free spectroscopy of the 6^1S_0-6^3P_1 intercombination transition in barium, Opt. Commun. 93, 54 (1992).
12. F.S.Pavone, G.Giusfredi, A.Capanni, M. Inguscio, M.de Angelis, and G.M. Tino, Narrow-linewidth visible diode laser at 690 nm. Spectroscopy of the SrI Intercombination line, in "Proc. of the Int. Conf. on Frequency Stabilized Lasers and their Applications", Boston, November 1992, to be published.
13. C. Salomon et al., to be published.

OVERTONE MOLECULAR SPECTROSCOPY WITH DIODE LASERS

Krzysztof Ernst[1] and Francesco Pavone[2]

[1]Institute of Experimental Physics, Warsaw University
Hoza 69, 00681 Warsaw, Poland
[2]European Laboratory for Non-linear Spectroscopy
(LENS) Lgo E.Fermi 2, 50125 Florence, Italy

INTRODUCTION

All fundamental vibrational transitions of molecules are in the infrared region. At the same time only few molecules have electronic absorption bands in the visible. Such situation is not favourable for performing molecular spectroscopy measurements in the visible range which is very convenient for several reasons (tunable laser sources, sensitive detectors). On the other hand, due to transparency of molecular gases for visible radiation, the Sun light can easily penetrate through the atmosphere what is of the primary importance for all kind of life on the Earth.

The vibrational motion of a molecule besides of the fundamental frequencies, corresponding to different normal vibrational modes ν_i, contains also overtone frequencies $2\nu_i, 3\nu_i \ldots$ Actually overtone frequencies are not exactly two or three times the frequency of the fundamental. They are slightly less. For convenience, however, it is conventionally written $2\nu_i, 3\nu_i$. In addition, in polyatomic molecules combination vibrations as $\nu_i + \nu_k, \nu_i + 2\nu_k$ are also present.

Absorption lines corresponding to overtone and combination transitions are weak enough from the point of view of the atmosphere transparency, but at the same time they may be sufficiently strong for spectroscopic applications. However, such measurements require suitable laser sources and sensitive detection.

It has to be emphasized that the intensity of successive overtone and combination transitions decreases very rapidly with increasing number of vibrational modes and/or quanta involved. For instance, in the case of HCl molecule the intensity of the fourth overtone is more than four orders of magnitude weaker than that of the fundamental transition.

Recent progresses in semiconductor diode lasers[1]and in particular their output power, reliability, low cost, room temperature operation, and large spectral coverage have been of great importance for a continously increasing use both in pure and applied spectroscopy. In particular, the extension of operation from the infrared to the visible, have really made accessible the wide and important field of absorption measure-

Solid State Lasers: New Developments and Applications
Edited by M. Inguscio and R. Wallenstein, Plenum Press, New York, 1993

ments concerning the overtone molecular transitions. Various applications are possible with this sources in the field of the high resolution spectroscopy and high sensitivity detection. One of them is real time, non contact polution measurement.

As it was already mentioned, moving from the infrared fundamental vibrational transitions to the overtones in the visible, absorption coefficients decrease several orders of magnitude. On the other hand, in comparison with laser sources operating in the infrared including lead salt diodes, semiconductor diode lasers in the visible (and in the near IR) offer the advantage of much simpler operation and much better amplitude stability. Moreover, remote sensing of atmospheric species could be more useful in the visible than in the infrared because of the reduced opacity of the atmosphere.

TECHNIQUES DEVELOPED TO DETECT VERY WEAK ABSORPTIONS

In order to reach a satisfactory sensitivity in absorption spectroscopy several detection methods have been proposed and successfully applied. The simplest possible technique is the direct absorption measurement which consists on sweeping the frequency and detecting the signal against the constant background. The sensitivity of such a method is obviously very low. We can improve it by modulating the amplitude of the light source, but we are still limited by background contribution.

In order to understand the motivation for developing specialized techniques for direct absorption spectroscopy (i.e., the direct measurement of the optical attenuation of a light beam through an absorbing sample) let us consider some general requirements concerning the signal-to-noise ratio (SNR).

The signal is usually a detector photocurrent and may be written as $S = kI$, where k represents the net attenuation of the laser intensity I incident upon the absorption sample. Noise contributions may be separated into three terms: N_e- originating from the detection electronics, and then independent of I, N_o—detector shot noise proportional to $\sqrt{I}$ and N_sI - contribution of amplitude – fluctuation background proportional to I. The SNR may then be written[2]

$$SNR = \frac{kI}{[N_e^2+(\beta\sqrt{I})^2+(N_sI)^2]^{1/2}}$$

As the intensity I increases, the SNR increases proportionally to I until the I-dependent terms in the denominator exceed N_e. The SNR will then saturate at some value of I that depends on the relative values of β and N_s. As we can see the laser light amplitude modulation can not solve our problem, as this imparts a time dependence to I, which is common to both numerator and the denominator in the equation for SNR. However, by using frequency modulation, one may effectively modulate the signal kI without modulating the noise terms. In view of high sensitivity measurements this method has been developed in three different ways.

I. Wavelength Modulation (WM) Spectroscopy[3,4,5]

The single frequency laser is modulated at a relatively low frequency ($10^3 Hz$), small compared to the width of the spectroscopic feature of interest, so that the absorption is probed simultaneously by a number of sidebands. WM is a sensitive form of derivative spectroscopy. Unlike direct absorption methods where signal is detected as a change against a constant background, in WM spectroscopy the signal arises from the difference in the absorptions of different sidebands. Therefore, the sensitivity and spectral resolution are greatly enhanced, provided that the laser shows no or very little variation of intensity with wavelength and responsivity of the detector is independent of λ. Consider the current C from the detector, C =IGT, where I is the laser light intensity, T is the transmissivity of the sample, and G is responsivity of the detector. We can then write for the detector signal

$$\frac{1}{C}\frac{dC}{d\lambda}\Delta\lambda = \frac{1}{I}\frac{dI}{d\lambda}\Delta\lambda + \frac{1}{G}\frac{dG}{d\lambda}\Delta\lambda + \frac{1}{T}\frac{dT}{d\lambda}\Delta\lambda$$

where $\Delta\lambda$ is the wavelength modulation depth. Only the last term in the above expression should give a contribution to the signal. Unfortunately dI/$d\lambda$ is not negligible in the case of diode lasers and gives an unwanted contribution to the noise level.

II. Frequency Modulation (FM) Spectroscopy[6,7]

FM spectroscopy is a kind of extension of WM spectroscopy to much higher frequencies ($\approx 10^8 Hz$). Frequency modulation produces sidebands which are widely spaced in frequency so that the spectral feature of interest can be probed by only one sideband at a time. Viewed in frequency space, the spectral distribution of the modulated laser field consists of the strong carrier at ω_c and two sidebands of the same amplitude but 180^0 out of phase, displaced by the angular modulation frequency ω_m from the carrier. When there is no absorption present, the beat signal at ω_m between the carrier and the upper sideband exactly cancels with the beat signal between the carrier and the lower sideband. If, however, the laser frequency is tuned over an absorption, so that one of the sidebands is absorbed, the imperfect cancellation of two beats produces a FM signal.

If the modulation frequency does not exceed the absorption linewidth, the technique is usually called High Wavelength Modulation (HWM).

III. Two-Tone Frequency Modulation (TTFM)[8,9]

If we wish to investigate broad spectral features, such as absorption lines broadened by atmospheric pressure to $2 - 3$ GHz, we must have correspondingly high modulation frequencies. The TTFM technique was applied in order to reduce the detection bandwidth requirement since detectors with bandwidth in GHz range are not easily available.

In the TTFM, the laser is modulated simultaneously at two distinct but closely spaced angular frequencies $\omega_1 = \omega_m + \Omega/2$ and $\omega_2 = \omega_m - \Omega/2$ ($\Omega/\omega_m < 10^{-3}$). The TTFM absorption signal arises from the difference between the absorption of the carrier and the sum of the absorptions of the sidebands. It is thus approximately analogous to a second derivative signal whereas single-tone FM is analogous to a first derivative signal. In both cases, however, there is no background signal if there is no absorption. The most attractive feature of TTFM resides in the fact that the signal is detected at Ω instead of ω_m.

Unfortunately pure frequency modulation is rarely achieved. There is always some residual amplitude modulation (RAM) present, especially in diode lasers where any change in the injection current changes not only the frequency of the laser but also its output power. It is evident that RAM is undesirable feature in all kinds of frequency modulation spectroscopy. It gives rise to a background signal with accompanying noise, even when there is no absorption present, and thus limits the detection sensitivity.

It is worthwhile to mention here that a wide variety of laser techniques allowing high sensitivity and high resolution have been proposed and successively applied. Let us give few examples as laser-induced fluorescence (LIF)[10], optoacoustic spectroscopy(OA)[11], optogalvanic spectroscopy (OG)[11], resonance ionisation spectroscopy (RIS)[12], and laser-intracavity absorption[13]. With the exception of direct absorption methods described above, all these techniques monitor some indirect effects of the optical absorption.

Any qualitative comparison of each of these methods would be extremely difficult since their applicability depends essentially on specific aims and experimental conditions of the desired measurements. However, two important advantages of the extracavity direct absorption should be emphasized: simple callibration procedure and remote-sensing possibilities. All of them are very important from the point of view of environmental studies.

OVERTONE SPECTROSCOPY

Several overtone transitions of various molecules have been already studied using various light sources and different detection techniques. Table 1 contains selected information concerning molecules with overtone or combination absorption lines in two spectral ranges ($630 - 690$ and $750 - 900 nm$) covered by diode lasers. As one can see only in a few recent works diode lasers have been applied. Other data were obtained with dye lasers or even with methods of pre-laser spectroscopy. Some overtone transitions in the selected ranges refer to reported numerical values and can be considered as potential candidates for future experimental research.

Table 1. Overtone and combination transitions in spectral ranges covered by diode lasers.

Molecule	absorp. line (nm)	transition	laser and det.tech.	sensitivity other inf.	reference
H_2O	818	$2\nu_1 + \nu_2 + \nu_3$	Diode	min.det.abs.	14
			WM	$3.1x10^{-4}$	
			FM	$2.8x10^{-4}$	
			TTFM	$2.4x10^{-6}$	
	789	$\nu_2 + 3\nu_3$	Diode FM	$3x10^{-4}$	15
	818-831	$2\nu_1 + \nu_2 + \nu_3$	Diode FM		15
HI	820	$v = 6 \Leftarrow 0$	Diode		16

Table 1. (cont'd)

C_2H_2	789	$\nu_1 + 3\nu_3$	Diode	min.det.conc.	17
			WM	0.2ppm per km	
	640	$5\nu_3$	Dye	intra-cav.	18
	670	$\nu_1 + \nu_2 + \nu_3$	OA		
	848	$2\nu_1 + \nu_2 + \nu_3$			19
CH_4	886	$3\nu_1 + \nu_3$	Diode	min.det.abs.	20
			LWM	$4.5x10^{-7}$	
			HWM	$9.7x10^{-8}$	
			TTMF	$6.4x10^{-8}$	
	860	$2\nu_1 + 2\nu_3$	Diode	$3x10^{-5}$	21
	682	$2\nu_1 + 3\nu_3$		White cell	22
	782	$2\nu_1 + \nu_2 + 2\nu_3$			23
C_2H_6	741	$5 \Leftarrow 0$	OA		24
		CH str.			
NH_3	790	$4\nu_1$	Diode	min.det.abs.	21
		$2\nu_1 + 2\nu_3$	FM	$3x10^{-4}$	
	792	$4\nu_1$	Diode		25
	647	$5\nu_1$	Dye	intra-cav	26
		$4\nu_1 + \nu_3$			
	647	$4\nu_1 + \nu_3$	photoplate		27
C_6H_6	865-	$4 \Leftarrow 0$	opto-thermal		28
	877	CH str.			
H_2O_2	793	$5\nu_{OH} + \nu_{OO}$	Dye		29
HCl	640	$v = 6 \Leftarrow 0$	Dye	intra-cav	30
	750	$v = 5 \Leftarrow 0$	OA		
CO_2	869	$5\nu_3$			31
	789	$2\nu_2 + 5\nu_3$			
	783	$\nu_1 + 5\nu_3$			
C_3H_4	787	$4\nu_1$	OA		24
	737	$5 \Leftarrow 0$			
HCN	790	$4\nu_3$			32
CO	804	$v = 6 \Leftarrow 0$			33
NO	800	$v = 7 \Leftarrow 0$			33
HF	877	$v = 3 \Leftarrow 0$			33
	675	$v = 4 \Leftarrow 0$			33

Results obtained for H_2O and CH_4 were used for comparing the sensitivity of different frequency modulation techniques. In the first experimental work[14] dedicated to H_2O the authors found that the difference in sensitivity between FM and TTFM was about two orders of magnitude. In other work[20] the minimum detectable absorption was measured using methane - the gas of obvious interest because of environmental applications. A component of a combination band at $886nm$ has been chosen as a test transition, mainly because of its strength. Two recordings are shown in Figure 1:a) pure absorption signal at pressure of 100 Torr corresponding

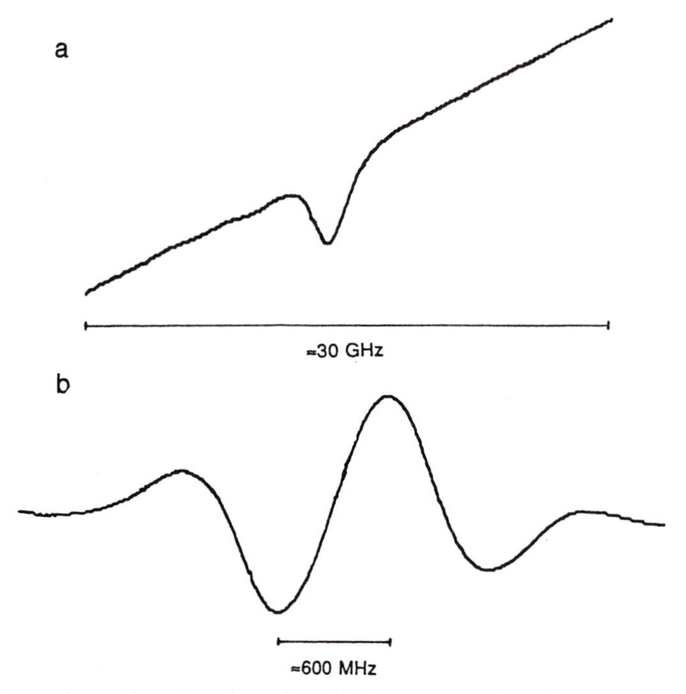

Figure 1.*a*) Pure absorption signal on the third overtone of methane at 100 Torr in $1.5m$ pathway. *b*) Derivative lineshape of a two-tone recording at 500 mTorr.

to an absorption of $3.6x10^{-2}$($1.5m$ pathway, $100Hz$ detection bandwidth, S/N ratio of about 20); *b*) derivative lineshape of a two-tone signal at pressure of 500 mTorr corresponding to an absorption of $1.8x10^{-4}$($1.5m$ pathway, $1Hz$ detection bandwidth) with a measured S/N ratio of a few thousands.

Minimum detectable absorption for the HWM and TTFM techniques turned out to be of the same order of magnitude, while for the value referred to LW modulation it is about one order of magnitude less, as indicated in Table 1. In the last case the detection limit value is determined by the laser amplitude $1/f$ excess noise, while in the first two cases this noise contribution is negligible with respect to other ones as detector induce shot noise, thermal and RAM noises. The detection limits well agree with the calculated "quantum limited" values based on measured laser power, modulation index, noise of the electronic components, and other parameters of the apparatus.

In most of the measurements with diode lasers the frequency modulation was produced by modulating the injection current. In the case of C_2H_2 the diode laser was used in an extended cavity configuration. Low frequency modulation could be then produced by both changing the injection laser current and/or varying the external cavity length (by means of a PZT). With the extended cavity configuration we have measured a reduction of the linewidth (to less than 1 MHz) and better short term stability (100 ms). Another advantage of such a configuration is that wavelength modulation by varying the cavity length leads to the lower level RAM noise than direct injection current modulation.

The experimental apparatus used for C_2H_2 overtone absorption measurements[17] is schematically shown in Figure 2. The extended cavity configuration is outlined in Figure 3. The first order beam diffracted from a grating was fed-back into the laser diode. The tilting of the grating allows to select one of the longitudinal modes within tuning range of about $20nm$.

Molecular transitions could be observed either in pure absorption or in a phase sensitive detection scheme. In the latter case, the lock-in derivative signals were

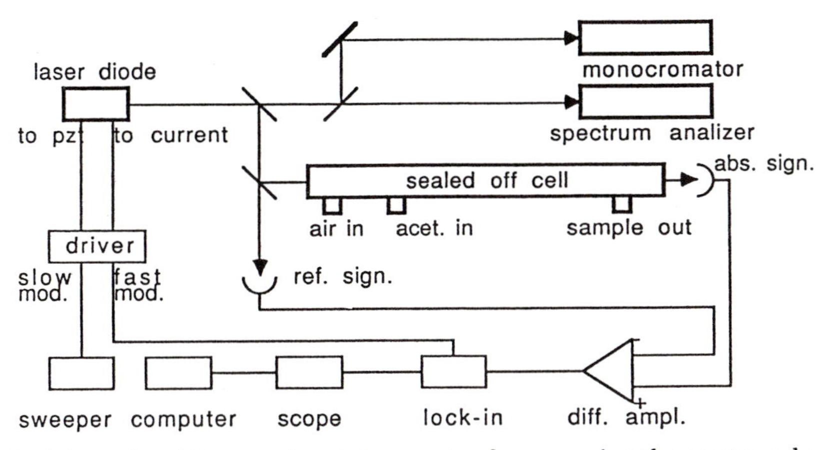

Figure 2. Schematic of the experimental apparatus for measuring the overtone absorption in acetylene.

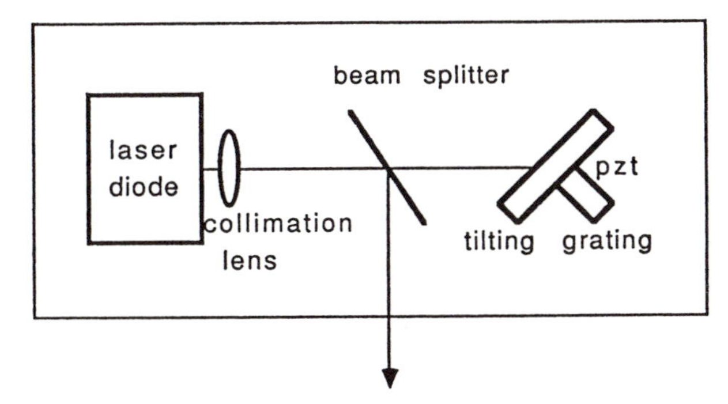

Figure 3. Schematic of the extended cavity diode laser configuration using a grating for wavelength control.

recorded by introducing a kHz modulation on the cavity length. The derivative signal for the $P(11)$ component of the $12646.966 cm^{-1}$ transition is shown in Figure 4 at two different gas pressures. We measured a S/N value of a few thousands and an absorption of about 6% at 30 Torr (trace a), and $S/N = 6$ at 36 mTorr (b). The latter measurement gives a minimum relative absorption of 10^{-6} which is comparable with that obtained by means of TTMF[14]. This result is also consistent with the recent report[34] where WM and TTFM techniques have been compared for lead salt diodes.

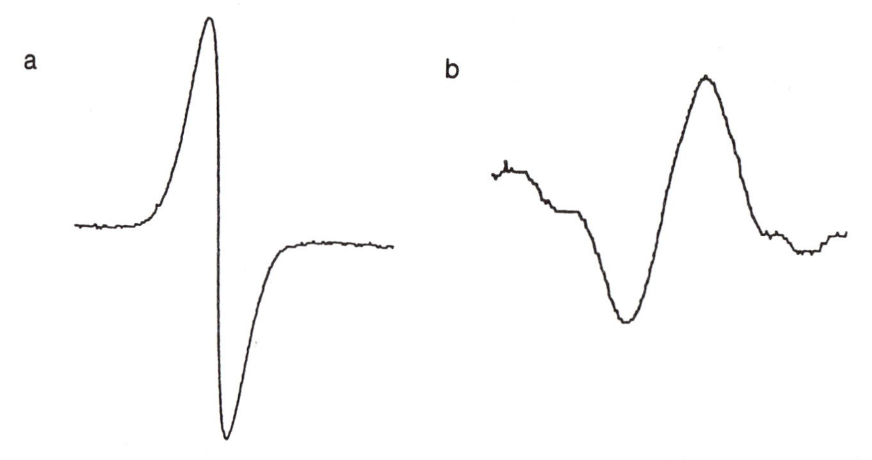

Figure 4. Derivative signal for the $P(11)$ component at 10 Torr (a) and 36 mTorr (b).

Regarding possible environmental trace gases detection we have also extracted the minimum acetylene pressure in presence of air at atmospheric pressure still keeping a signal with a S/N ratio of a few units. We are able to detect 0.1 ppm per km of acetylene in air.

The narrow linewidth emission obtained with the extended cavity configuration has also permitted to perform the lineshape analysis measuring self and air pressure broadening parameter for two different components of the combination band. The Pade method [35] has been used to extract the collisional Lorentzian contribution from the Voigt profile and thus the pressure broadening parameter. Air broadening measurements have been performed on the P(11) and R(5) components. Our results are reported in Table 2 together with earlier measurements of other components of C_2H_2 as well as recent measurements of pressure broadening for H_2O, CH_4 and NH_3 using diode lasers. As one can see the air broadening coefficients for overtone transitions are very close to that for the fundamental transition.

For many practical applications it is requested to monitor gas traces in open-path configuration in the atmosphere. For this reason the knowledge of air broadening parameters can be useful. However, some problems arise for such experiments at atmospheric pressures. Widths of absorption lines due to pressure broadening are of the order of GHz what leads to the substantial decrease of the detection sensitivity. Moreover, the overlapping of different lines may also occur and it has to be taken into consideration.

Table 2. Pressure broadening parameters.

Molecule	Transition	γ (MHz/Torr) air	γ (MHz/Torr) self	Refer.
C_2H_2	$789(\nu_1 + 3\nu_3)$			20
	P(11)	8.0	11.0	
	R(5)	9.0		
	$862(\nu_2 + 3\nu_3)$			36
	R(7)		12.3	
	$849(2\nu_1 + \nu_2 + \nu_3)$			36
	R(7)		9.9	
	ν_5			
	R(5)	7.0		37
	P(11)	6.2		
H_2O	$822(2\nu_1 + \nu_2 + \nu_3)$	5.9		15
CH_4	$860(2\nu_1 + 2\nu_3)$	3.2	3.7	21
NH_3	$790(2\nu_1 + 2\nu_3)$		21.7	21
	$792(4\nu_1)$		27.2	21

As for the sensitivity of diode laser systems destinated to monitoring the atmosphere one should think about its improvement. The stabilization of the diode lasers operating at wavelength below $1\mu m$ could also be important for use in conjunction with efficient pulsed solid state laser (such as Titan Saphire) operating in the same region. This can be interesting in the frame of LIDAR[38,39] measurements, where amplification or injection seeding control could make possible to perform detection at larger distances with an improved sensitivity and an easier spectral control.

References

1. C.Wieman and L.Hollberg, Rev.Sci.Instr., 62:1(1991).
2. M.Gehrtz, G.Bjorklund and E.Whitaker, JOSA, B2:1510(1985).
3. J.Telle and C.Tang, Appl.Phys.Lett., 24:85(1974).
4. E.Moses and C.Tang, Opt.Lett., 1:115(1977).
5. P.Pokrowsky, W.Zapka, F.Chu and G.Bjorklund, Opt.Commun., 44:175(1983).
6. G.Bjorklund, Opt.Lett., 5:15(1980).
7. G.Bjorklund, M.Levenson, W.Lenth and C.Ortiz, Appl.Phys., B32:145(1983).
8. G.Janik, C.Carlisle and T.Gallagher, JOSA, B3:1070(1986).
9. D.Cooper and R.Warren, JOSA, B4:470(1987).
10. W.Demtroder, Laser Spectroscopy, Springer Verlag, 1981, p.416.
11. K.Ernst and M.Inguscio, Rivista del Nuovo Cimento, 11: no.2 (1988).
12. G.Hurst, M.Payne, S.Kramer and J.Young, Rev.Mod.Phys., 51:767 1979).
13. L.Pakhomycheva, E.Sviridenkov, A.Suchkov, L.Titova and S.Churilov, JETP Lett., 12:43(1970).
14. L.Wang, H.Riris, C.Carlisle and T.Gallagher, Appl.Opt., 27:2071(1988).

15. A.Lucchesini, L.Dell'Amico, I.Longo, C.Gabbanini, S.Gozzini and L.Moi, Il Nuovo Cimento, 13D:677(1991).
16. F.Matsushima, S.Kakihata and K.Tagaki, J.Chem.Phys., 94:2408 1991).
17. F.Pavone, F.Marin, M.Inguscio, K.Ernst and G.di Lonardo, Appl.Opt., in print.
18. G.Scherer, K.Lehmann and W.Klemperer, J.Chem.Phys., 78:2817 (1983).
19. G.Funke and E.Lindholm, Z.Physik, 106:518(1937).
20. F.Pavone and M.Inguscio, submitted to Appl.Phys.B.
21. A.Lucchesini, I.Longo, C.Gabbanini, S.Gozzini and L.Moi, Proc. of SPIE Conf. "High Performance Optical Spectrometry", Warsaw, June 1992.
22. L.Giver, J.Quant.Spectrosc.Rad.Transfer, 19:311(1978).
23. H.Vedder and R.Mecke, Z.Physik, 86:137(1933).
24. M.Crofton, C.Stevens, D.Klenerman, J.Gutow and R.Zare, J.Chem.Phys., 89:7100(1988).
25. K.Nakagawa and T.Shimizu, Jap.Journ.Appl.Phys., 26:L1697(1987).
26. E.Antonov, E.Berik and V.Koloshnikov, J.Quant.Spectrosc.Rad. Transfer, 22:45(1979).
27. Siu-Hung Chao, Phys.Rev., 150:27(1936).
28. M.Scotoni, C.Leonardi and D.Bassi, preprint.
29. X.Luo, P.Fleming, T.Seckel and T.Rizzo, J.Chem.Phys., 93:9194 (1990).
30. K.Reddy, J.Mol.Spectr., 82:127(1980).
31. A.Adel and D.Dennison, Phys.Rev., 43:716(1933).
32. R.Badger and J.Binder, Phys.Rev., 37:800(1931).
33. D.Proch, J.Wanner, Report IPP IV/17 of Max Planck Institut fur Plasmaphysik, Garching, 1971.
34. D.Bomse, A.Stanton and J.Silver, Appl.Opt., 31:718(1992).
35. P.Minguzzi and A.di Lieto, J.Mol.Spectr., 109:388(1985).
36. Y.Ohsugi and N.Ohashi, J.Mol.Spectr., 131:215(1988).
37. D.Lambot and G.Blanquet, J.Mol.Spectr., 136:86(1989).
38. S.Svanberg, NATO-ASI Applied Laser Spectroscopy, San Miniato Italy, September 1989, and references therein.
39. P.Rairoux, These, EPFL, Lausanne, 1991.

OPTICAL FREQUENCY METROLOGY WITH SOLID STATE LASERS

Thomas Andreae, Wolfgang König, Robert Wynands, Theodor W. Hänsch

Max-Planck-Institut für Quantenoptik
D - 8046 Garching b. M.

Due to their high intrinsic stability, solid state lasers can be powerful tools for the precise measurement of optical frequencies. The rapid increase in reacent years of the spectral bandwidth covered by semiconductor lasers offers new perspectives foroptical frequency measurements. In the visible up to now only an interferometric comparison with the I_2 stabilized HeNe laser, a secondary optical frequency standard, has been possible, and the accuracy was limited by the errors introduced by dispersion, wavefront, and calibration problems occurring in an interferometric measurement. By contrast, a direct frequency comparison, the counting of the zero crossings of the beat signal between two optical frequencies, can be performed with radio frequency accuracy, that is to better than 1 Hz. An absolute measurement of optical frequencies therefore only is limited by the frequency standard.

This contribution first gives a brief overview of optical frequency standards; the second and third sections deal with methods of transfering the standard´s accuracy to other optical frequencies. The last part of this paper gives an example for the need of frequency metrology in basic research, the measurement of the 1S-2S transition frequency in atomic hydrogen, and its importance for the system of fundamental constants.

Solid State Lasers: New Developments and Applications
Edited by M. Inguscio and R. Wallenstein, Plenum Press, New York, 1993

Up to now optical frequency standards have been based on the coincidence of saturable absorber resonances in a molecule with the gain profile of an amplifying laser medium [1,2]. Such combinations are the CO_2 laser with an OsO_4 absorber (λ = 10.32 μm, ν = 29.054 THz), the HeNe laser with a CH_4 absorber (λ = 3.39 μm, ν = 88.376 THz), the HeNe laser with an I_2 absorber (λ = 633 nm, ν = 473.612 THz, and λ = 612 nm, ν = 489.880 THz), a dye laser with a I_2 absorber (λ = 576 nm, ν = 520.207 THz), and the Ar^+ ion laser with a I_2 absorber (λ = 515 nm, ν = 582.491 THz). Except the CO_2/OsO_4 system, all of these are subject to the recommendations for the realization of the meter given by the CIPM† in 1983 [3]. Among these, the CH_4 stabilized HeNe laser may today be considered the most accurate. Systems resolving the hyperfine structure of the $F_2^{(2)}$ component of the ν_3 band in the P(7) methane line achieve a reproducibility of the absolute laser frequency of the order $\Delta\nu/\nu = 10^{-14}$. A stability in terms of the Allan parameter [4,5] for τ = 1 - 100 sec of $\sigma(\tau) = 4 \cdot 10^{-15}$ has been demonstrated. For further details and references, see the review by Bagayev and Chebotayev [6]. With recent advances in the selection of cold molecules, thereby reducing the line shift and broadening due to the second order Doppler effect, a frequency reproducibility of even better than 10^{-14} can be expected [7]. This would be better than the uncertainty of the Cs atomic clock, the current primary frequency standard, having an uncertainty of some parts in 10^{14} at a frequency of 9192631770 Hz [8].

One challenging task in establishing an optical frequency standard is the measurement of the absolute frequency by comparison with the primary microwave frequency standard. The frequency of the HeNe/CH_4 laser, for example, is about four orders of magnitude higher than the frequency of the Cs clock, and the HeNe/I_2 frequency at λ = 633 nm is even more than five times that of the HeNe/CH_4. (Not only the absolute frequencies, but also laser difference frequencies in the visible are difficult to measure. The change in wavelength of 0.1 nm at λ = 633 nm corresponds to a change in frequency of $\Delta\nu$ = 75 GHz.) For such a comparison the desired frequency can be synthesized step by step with a so-called frequency (synthesizing) chain. In each step the input frequency is multiplied by a factor n, and a suitable coherent source near the n-th harmonic is compared with the exact n-th harmonic. If it is possible to synchronize the phase of the two oscillators, the full accuracy of the input frequency is transferred to the oscillator at the n-th harmonic. These oscillators are therefore called transfer oscillators. The frequency of the transfer oscillator is then used as the input frequency for the next

† CIPM - Comité International des Poids et Mesures

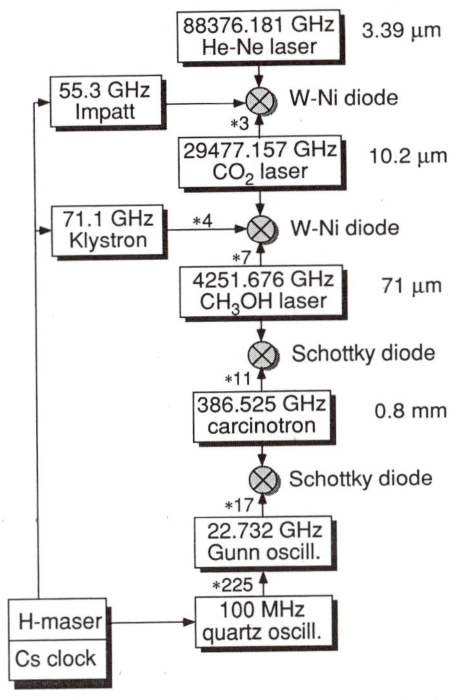

Figure 1. The PTB frequency chain. [12]

step in the chain. The problem is to find efficient harmonic mixing devices and coherent sources with matching frequencies, so they are suitable as transfer oscillators.

In the microwave domain and the far infrared up to some THz, Schottky point contact diodes and Josephson junctions are used as efficient harmonic mixers [9]. They allow the generation of up to the 100th harmonic of a microwave, and the direct frequency measurement of a far infrared laser has been demonstrated [10]. Frequencies up to 200 THz can be synthesized with the help of metal-isolator-metal (MIM) point contact diodes [11]. As an example the frequency chain used to measure the absolute frequency of the HeNe/CH_4 laser at the PTB is shown in fig. 1 [12]. Other frequency chains connecting various optical frequencies with the Cs clock are reviewed in the book by Vanier and Audoin [1].

Beside these optical frequency standards based on saturated absorption, new types of standards based on narrow transitions in laser cooled atoms or ions appear promising. One of these uses a single ion in a Paul trap [14]. Within the scope of this summer school we can focus on the work of Madej and Sankey on a single Sr^+ ion [15]. Sr^+ has an energy level system typical for ions useful as frequency standards (see fig. 2). A fast $S_{1/2}$-$P_{1/2}$ resonance enables laser cooling of the ion in the trap, whilst the electric-quadrupole-allowed $S_{1/2}$-$D_{5/2}$ clock transition has a natural linewidth of only 0.4 Hz. The situation is complicated by the eventual loss of the ion to the metastable $D_{3/2}$ state; a third laser at λ = 1092 nm therefore has to drive the $P_{1/2}$-$D_{3/2}$ transition. This system has the advantage that all relevant transitions are accessible to solid state lasers. The cooling transition at 421.7 nm can be excited with a frequency doubled AlGaAs diode laser at 843 nm; the clock transition at 674 nm is within the bandwidth of the new semiconductor lasers operating around 670 nm and the 1092 nm can be generated with a Nd^{3+}-doped fiber laser.

A realization of such a frequency standard, using the disruption of the resonance fluorescence while the ion is in the D-state, has to account for the on/off nature of the detection process ("quantum jumps"). Therefore time sequencing of various operational functions will be necessary to record the line profile of the

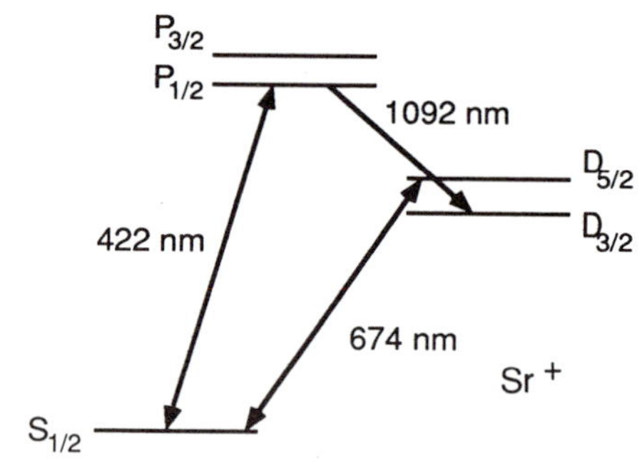

Figure 2. Sr^+ eregy levels.

clock transition, and stabilization to the line center will require a certain integration time. For optical frequency standards based on other ions, the reader is refered to Blatt et. al. [16].

The second new type of frequency standard uses laser-cooled atoms in an atomic beam or fountain. Recent advances have been made with the Mg- [17] and Ca-atoms [18]. In both of these systems a $^1S_0$-$^3P_1$ intercombination line with a natural linewidth of the order of 100 Hz is used as the clock transition. Laser cooling is performed via a $^1S_0$-$^1P_1$ fast dipole transition. In the Ca experiment the reduction in transit time broadening achieved by laser-cooling allowed the recoil splitting to be resolved in saturated absorption. This experiment has the advantage that both transitions are accessible to diode lasers. For the cooling transition (λ = 423 nm) the laser must be frequency-doubled, whereas the excitation of the clock transition (λ = 657 nm) may be possible with a diode specialy selected from diodes produced for λ = 670 nm.

Besides the 1S-2S transition in hydrogen, the Ca transition is the only one of the possible new frequency standards where some progress has been made towards an absolute frequency measurement. To realize this, the PTB frequency chain (fig. 1) is used as far as the CO_2 laser at ν = 29.5 THz. Against this, a second CO_2 laser (ν = 28.5 THz) is compared, operating at nearly the 16th subharmonic of the Ca frequency. The 16th harmonic can be generated using a color center laser at the 4th harmonic and a diode laser at the 8th harmonic as transfer oscillators [19].

TRANSFER OF ACCURACY

Since the accepted optical frequency standards are very isolated points on the frequency axis, the comparison of an arbitrary optical frequency to a known standard

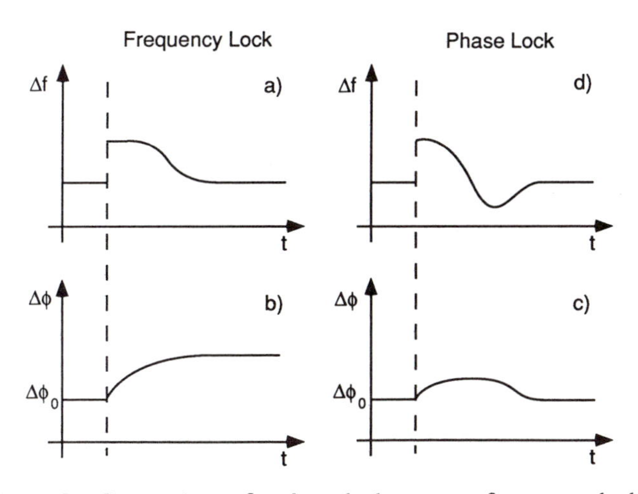

Figure 3. Comparison of a phase-lock versus a frequency-lock.

is in most cases cumbersome. One essential technique, already mentioned above, is the phase-locking of two oscillators.

The example of a sudden frequency jump of one oscillator explains the principle of phase-locking in comparison to a frequency synchronization (see fig. 3). A frequency lock pulls back the frequency to its original value (fig. 3a). After a frequency jump, this process accumulates a phase difference which can easily reach several cycles. A frequency measurement at this point, the counting of the zero crossings of a periodical signal, would miss these "cycle slip", and give a counting error (fig. 3b). The phase-lock instead pulls back the phase to its original value (fig. 3c). Before the frequency settles at its original value, it is swinging below it, to account for the transient higher frequencies. In this case no cycles are lost (fig. 3d). The theoretical background of phase-locking is given in the book by F. M. Gardner [20]; herein the concepts are developed for the radio frequency domain, but they are easily transferred to optical phase-locking.

Optical phase-locking is based on the beat signal of the two laser frequencies. The current in a photodiode is proportional to the total intensitiy:

$$\begin{aligned} I &= a\left[E_1 \cos(\omega_1 t) + E_2 \cos(\omega_2 t + \varphi_2(t))\right]^2 \\ &= \frac{a}{2}\left(E_1^2 + E_2^2\right) + aE_1E_2 \cos\left((\omega_1 - \omega_2)t - \varphi_2(t)\right). \end{aligned}$$

Terms with frequencies $2\omega_1$, $2\omega_2$, and $\omega_1+\omega_2$ are too high to be detected by the photo diode and have been omitted. The first term is a DC-current, the second the desired beat signal. One possibility to realize an optical phase-lock is to maintain $\omega_1=\omega_2$. In this case the time dependent signal depends only on the phase difference $\varphi_2(t)$. This $\cos(\varphi_2(t))$ has two zero crossings, and can serve as an error signal for a servo loop. In such a servo loop the total signal of the photo diode is fed into a differential amplifier to subtract the DC-current [21]. The remaining signal, the phase error, is amplified and fed back to the phase/frequency tuning element of one of the lasers. The disadvantage of this homodyne phase-lock is the sensitivity of the servo loop to high electronic 1/f flicker noise at low frequencies and voltage drifts.

A more reliable realization is a heterodyne phase-lock. In this case the AC coupled photo diode signal is compared with a highly stable reference frequency (fig. 4). This comparison can be performed with a RF mixer/phase detector, whose IF output is propotianal to the product of the two input signals, having two components with the sum and the difference frequency, respectively. Now the frequency condition for phase-locking is ω_1-ω_2=ω_{LO}, and the error signal is of the form $\cos(\varphi_2(t)+\varphi_{LO})$. The two oscillators do not have exactly the same frequency,

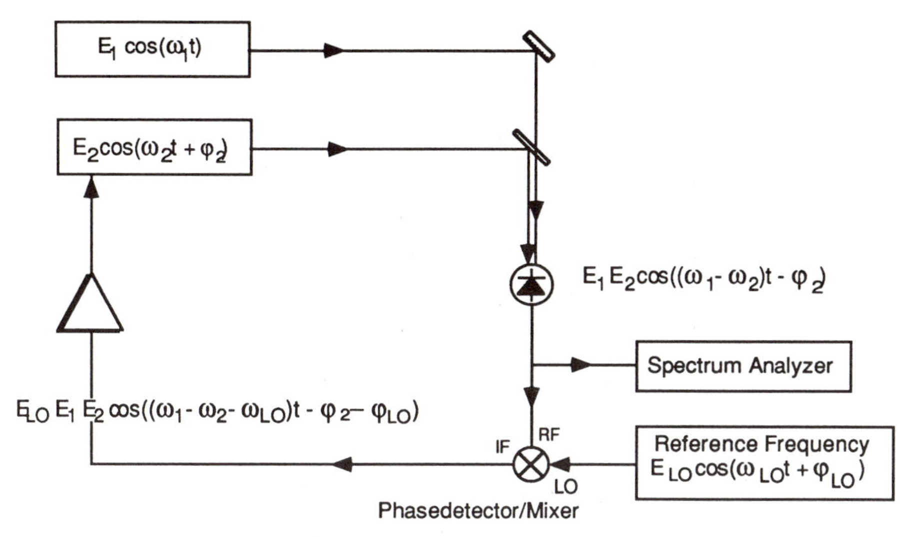

Figure 4. Experimental setup for a heterodyne phase-lock.

but since the reference frequency can be made highly stable and is well known, this is not of any importance for the transfer of stability and accuracy.

How does one know if two lasers are phase-locked? The beat spectrum of two independent lasers with uncorrelated fluctuations is the convolution of the two individual laser linewidths†. Each of them is dominated by various noise sources. In the case of two phase-locked lasers, having a constant phase relation, the beat spectrum collapses to a δ-peak, only limited by the bandwidth of the spectrum analyzer. Fig. 5 shows a beat spectrum between two phase locked diode lasers with different loop gains. Beside the carrier one can see the gain dependent "servo bumps", originating from positive feedback of disturbances with Fourier frequencies near the unity gain frequency of the servo loop. In fig. 5b the loop gain is more or less optimum, whereas in fig. 5a the gain is too small and the noise level between the carrier and the servo bumps is increased. With higher servo gain, fig. 5c, the amplitude of the servo bumps is increased, leading to an increased residual phase error $\Delta\varphi_{2rms}$. The relative power in the carrier is given by the relation [22]

$$\eta = \exp\left(-(\Delta\varphi_{rms})^2\right).$$

A more detailed description of optical phase-lock loops is given by Hall et. al. [23] and Ramos and Seeds [24].

† The resolution bandwidth and scan time of the spectrum analyzer have to be taken into account in addition.

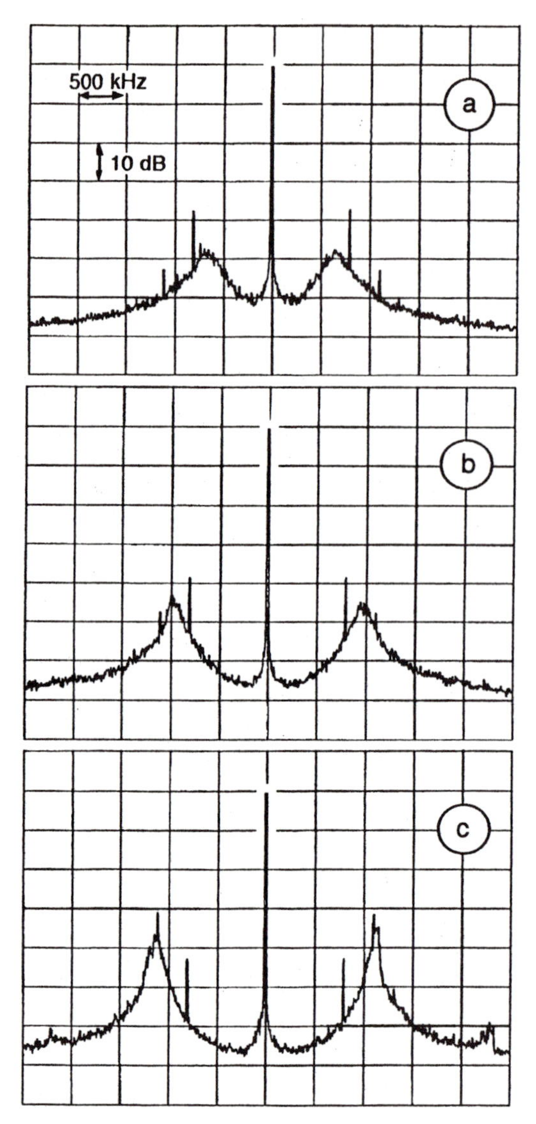

Figure 5. Heterodyne beat signals between two phase-locked diode lasers, each of them prestabilized by optical feedback from a confocal Fabry-Perot cavity (see fig. 9). RBW 3 kHz, scantime 17 sec.

MEASURERING DIFFERENCE FREQUENCIES

One can use the heterodyne phase-locking technique to account for large difference frequencies between a known frequency and the one to measure. In such a measurement the maximum possible difference frequency is limited by the bandwidth of the detector and the mixer. In the visible, sophisticated silicon diodes have a bandwidth of up to 60 GHz, corresponding to $\Delta\lambda = 0.15$ nm at 850 nm

A detector/mixer combination with a bandwidth of several hundred GHz is a GaAs/W Schottky point-contact diode. In the Schottky-contact, formed by a metallic whisker pressed against a semiconductor surface, the different workfunctions of the two materials, and the stable space charge in the semiconductor, lead to a potential barrier at the junction [25]. For a difference frequency measurement the two laser frequencies are colinearly focused onto the Schottky-contact. A photo effect in the GaAs substrate generates a current oscillating with the beat frequency, which is rectified at the highly nonlinear current-voltage (I-V) characteristic of the junction. This beat frequency can be measured directly only up to some GHz because of the high inductance of the tungsten whisker at higher frequencies.

To detect higher frequencies, microwave radiation of known frequency is also coupled into the junction. For this purpose the whisker serves as an antenna. Due to the nonlinearity of the I-V-characteristic, the microwave is mixed with the beat frequency, and the remaining difference frequency can be measured directly. In addition higher harmonics of the microwave are generated at the junction, and higher mixing products can also be used to measure laser beat frequencies. Berquist and Daniel [26] have demonstrated a frequency-offset-locking of two dye lasers by about 234 GHz, using the sixth harmonic of the output of a Q-band klystron and a Schottky-barrier diode. With a higher fundamental frequency of the microwave oscillator, the bandwidth can be further improved.

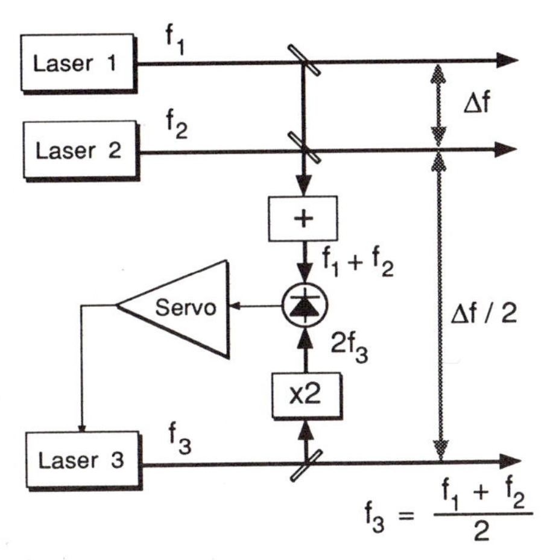

Figure 6. Setup of an optical frequency divider.

If the frequency difference is too large to be measured directly, one can use the technique of optical difference frequency division [27]: a scheme is shown in fig. 6. The lasers 1 and 2 have the difference frequency Δf. A division of Δf by two can be realized using a third laser and two nonlinear optical processes. Lasers 1 and 2 are frequency summed, whereas laser 3 is frequency doubled. Phase-locking the second harmonic to the sum frequency yields a difference frequency of $\Delta f/2$ between laser 3 and laser 1 and 2 respectively. Repeating this procedure n times results in a remaining difference frequency of $\Delta f/2^n$, which can be counted directly.

In contrast to the previous mentioned method, there is no limitation in difference frequency for such divider stages, and a bisection of 530 THz has been demonstrated by Wynands et. al. [28]. The lasers 1 and 2 were a HeNe laser, $\nu = 88.4$ THz ($\lambda = 3.39$ μm), and a dye laser with $\nu = 618.4$ THz ($\lambda = 484$ nm), which is the seventh harmonic of the HeNe laser, and the frequency summation was performed in a $LiNbO_3$ crystal. Laser 3 was a AlGaAs diode laser with $\nu = 353.5$ THz ($\lambda = 848$ nm), the fourth harmonic of the HeNe laser. The doubling was done in a $KNbO_3$ crystal.

If one needs to chain several stages, the only chance of succeeding (and of affording it) is to use simple and relatively inexpensive lasers like diode lasers. In the range of $\lambda = 850$ nm AlGaAs diode lasers are powerful devices, and in $KNbO_3$ second harmonic generation and sum frequency mixing is possible in the same crystal at the same time. A bisection of a 4.5 THz interval has been demonstrated in this way [28].

THE ABSOLUTE FREQUENCY MEASUREMENT OF THE 1S-2S TRANSITION IN ATOMIC HYDROGEN

The spectral lines of the hydrogen atom, the simplest among the stable atoms, are very well understood [29]. The 1S-2S transition in atomic hydrogen, with a natural line width of only 1.3 Hz, is an exciting challenge from the fundamental point of view [30]. A comparison of the 1S-2S transition frequency with that of the 2S-4S transition can be used in a determination of the 1S-Lamb-shift, one of the most stringent tests of quantum electrodynamics (QED) in a bound system [31]. A measurement of the absolute frequency of the 1S-2S transition, together with the 1S-Lamb-shift, leads to a determination of the Rydberg constant, an important cornerstone in the system of fundamental constants, and the principal scaling factor of any spectroscopic transition.

A frequency chain to measure the 1S-2S transition frequency is shown in fig. 7. The concept of this chain is based on the frequency relation of nearly 1:28 between the HeNe/CH_4 optical frequency standard and the 1S-2S transition frequency. Since the HeNe/CH_4 frequency is about 302 GHz higher than the 28th sub harmonic of the hydrogen transition, in the scheme of fig. 7 this leads to a difference frequency of Δf = 1.059 THz at the fourth harmonic of the standard.

The right hand side of fig. 7 shows the 1S-2S spectrometer with the dye laser at λ = 486 nm. This spectrometer is described in more detail by C. Zimmermann [32] as well as by M. Weitz [33], and in Ref. [34]. To excite the two photon transition at λ = 243 nm the dye laser frequency is doubled. Therefore the frequency relation

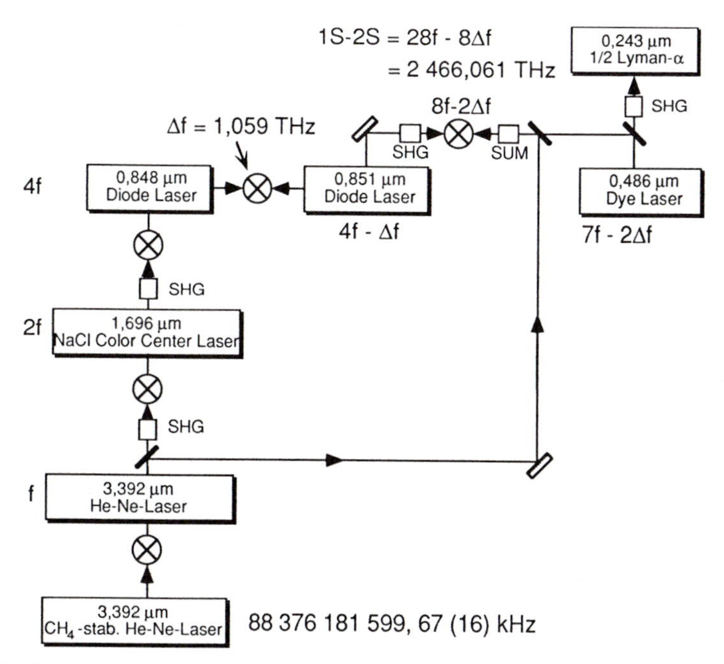

Figure 7. The frequency chain to measure the 1S-2S transition frequency with respect to the HeNe/CH_4 optical frequency standard.

between the standard and the dye laser is nearly 1:7. To compare the dye laser frequency with the standard, the eighth harmonic of the HeNe/CH_4 laser is generated in three steps of second harmonic generation (SHG, left hand side of fig. 7). On the other side of the figure the frequencies of the dye - and HeNe/CH_4 laser are summed. A difference frequency of about 1.06 THz at $\lambda \approx 850$ nm can be generated using the divider stages mentioned above, so that the remaining difference at the eighth harmonic can be counted directly.

The transportable HeNe/CH_4 standard was built in the group of Prof. V. P. Chebotayev and Prof. S. N. Bagayev in the Institute of Laser Physics, Novosibirsk,

Russia. The intracavity absorption cell is filled with CH_4 at a pressure of about 3 mTorr, resulting in a linewidth of the inverted Lamb dip of 100 kHz (FWHM). To lock the laser to the molecular transition, its frequency is modulated over the resonance, and its first derivative gives the error signal for the servo system. During ten days in November 1991 the standard was calibrated at the Physikalisch-Technische Bundesanstalt (PTB) in Braunschweig by comparison with the Cs atomic clock using the PTB frequency chain (see fig. 1). The measured frequency was 88 376 181 599.67 ± 0.15 kHz ($\Delta\nu/\nu = 1.8 \cdot 10^{-12}$). The stability in terms of the Allan parameter $\sigma(\tau)$ for averaging times of 10 sec < τ < 100 sec is $\sigma(\tau) < 6 \cdot 10^{-14}$.

Since the output power of the HeNe/CH_4 standard of only 0.4 mW is not sufficient to drive two nonlinear processes, a second HeNe laser with an output power of 20 mW is phase-locked to the CH_4-stabilized laser. For the phase-lock a digital phase-frequency detector is used, similar to the one described by Hall et. al. [23]. Half of the HeNe power is used to generate the second harmonic (SHG) in a $AgGaSe_2$ crystal, and a power of about 10 nW at 1.7 μm is achieved. Combined with 1 mW from the color center laser (CCL) on a glass plate beam splitter a beat signal with 40 dB of signal-to-noise ratio is obtained and used to reliably phase-lock the CCL.

The setup of the CCl is shown in fig. 8. A NaCl:OH^- crystal at 77K is pumped with 4 W from a Nd:YLF laser at 1054 nm and 200 mW at 514 nm from an Ar^+ ion laser. To obtain single-mode operation the laser cavity is of a bowtie ring type. With two etalons and a Lyot filter as frequency-selective elements and an optical diode for unidirectional operation, a power of more than 500 mW at 1.696 μm is achieved. A slow and a fast piezo transducer as well as an intracavity electrooptic modulator are used as servo elements for the different Fourier components of the error signal in the phase-locked loop. Here the error signal is obtained with a

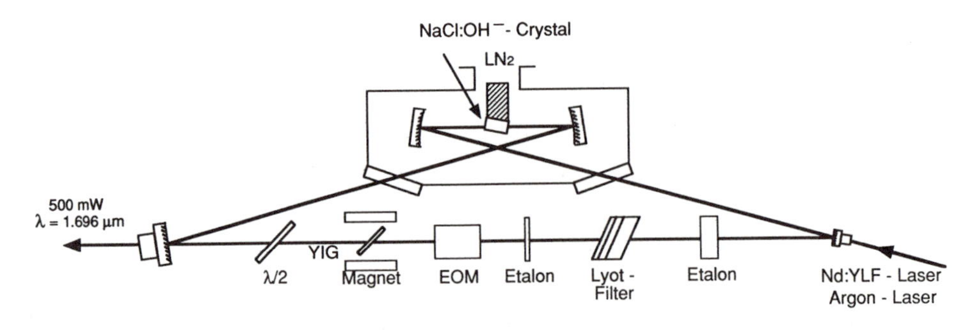

Figure 8. The setup of the color center laser.

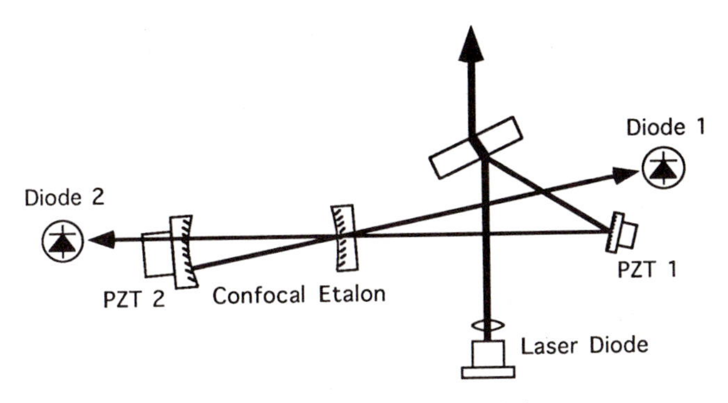

Figure 9. The setup of the diode laser.

second digital phase/frequency detector. The second harmonic of the CCL radiation is generated in a $LiIO_3$ crystal, cut for a phasematching angle of 18°. A power of 0.4 μW at 850 nm was generated.

The AlGaAs diode laser has an output power of 50 mW, and the frequncy is stabilized by optical feedback from a confocal Fabry-Perot etalon [35](see fig. 9). A 60 dB Faraday isolator [36] protects the system from back-reflections, mostly stemming from the doubling crystal. The beat signal with the second harmonic of the CCL is mixed with a local oscillator frequency, provided by a synthesizer, in an RF phase detector. The low-frequency component of the error signal is fed to the feedback cavity PZT, whilst the high-frequency components are used to control the injection current of the laser diode. The residual phase error in this phase-locked loop is about 0.1 rad rms, and during measurements this beat signal was observed continously on a spectrum analyzer. The second harmonic of the diode laser radiation is generated in the a-c principal plane of a $KNbO_3$ crystal at a phase-matching temperature of -5.4 °C. A power of more than 6 μW at 424 nm is easily achieved. At this point one should mention that this 424 nm radiation at the eighth harmonic of the HeNe/CH_4 is the highest frequency to be known with an accuracy in the order of 10^{-12}.

Finally, the sum frequency of the blue dye laser radiation (80 mW) and the second half of the HeNe radiation is generated in a $LiNbO_3$ crystal. Ninety degree phasematching is obtained at a temperature of 267 °C, resulting in a power of 200 nW near 424 nm. Combined with the second harmonic of the diode laser, this is enough to achieve a beat signal with a signal-to-noise ratio of 30 dB. With the help of a tracking filter, this residual difference frequency can be counted directly.

A preliminary measurement used the modes of the highly stable Fabry-Perot reference resonator of the 1S-2S spectrometer to account for the difference frequency of $2\Delta f = 2.1$ THz at $\lambda = 486$ nm [37]. The mode spacing of this resonator was measured beforehand at several frequencies between the seventh harmonic of the HeNe/CH_4 standard and one quarter of the 1S-2S transition frequency with a precision of 0.4 Hz [38]. At a frequency of 616.51 THz (λ = 486 nm), the measured mode spacing was 332383433.1(4) Hz.

The uncertainty of the final result, f_{1S-2S} = 2466061413.182(45) MHz, $\Delta\nu/\nu$ = $1.8 \cdot 10^{-11}$, was dominated by the interpolation procedure of the resonator drift (42.5 kHz), and the residual uncertainty due to the mode spacing of the reference resonator (10.2 kHz). Nonetheless, this is an 18-fold improvement over the best [39] of the previous measurements [39-41] (see fig. 10a).

The relationship between the frequency and the Rydberg constant is given by the Dirac theory, corrected for the finite nuclear mass and Lamb shifts, and a new

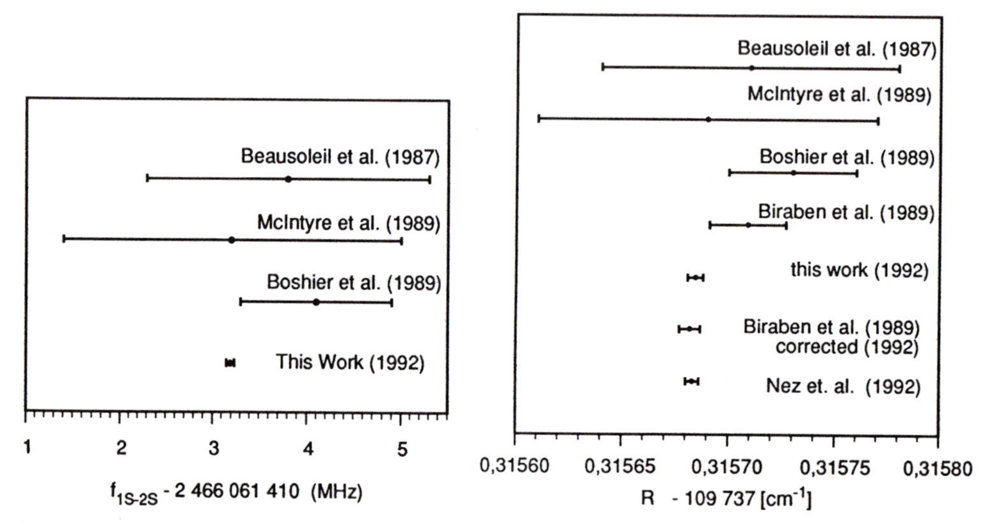

Figure 10. a) Comparisom of the recent measurement of the 1S-2S transition frequency with the previous ones [37,39-41]; b) The thereof deduced value for the Rydberg constant, compared to the value obtaind by other groups [37, 39-41, 43,45].

value of $R_\infty(H)$ = 109737.3156841(42) cm^{-1} is deduced from this measurement. Using the value for the hydrogen-deuterium isotope shift of measured recently [42] ($\Delta\nu_{1S-2S}$ = 670994.337(22) MHz), a second value for the Rydberg constant, $R_\infty(D)$ = 109737.3156862(84) cm^{-1}, can be obtained. The weighted mean of these two lead to the final result of

$$R_\infty = 109737.3156845(37)\ cm^{-1}.$$

The uncertainty of $3.4 \cdot 10^{-11}$ is largely due to that in the 1S-Lamb-shift. Corrections due to the Lamb-shift, the electron-proton mass ratio m_e/m_p, and the proton charge radius are now becoming more important than the experimental accuracy of the 1S-2S energy separation. The new value for R_∞ is more accurate by a factor of five than the previous most precise value [43], from which it differs by two standard deviations. A comparison of the recent published values for R_∞ is shown in fig. 10b. With a new determination of the frequency of the iodine stabilized HeNe standard at λ = 633 nm, which was found to be red shifted by 133 kHz with respect to the value determined in 1982 [44], the Paris group has confirmed this deviation from the former accepted value [45]. The Paris result of 1989 [43] was corrected with the new HeNe/I_2 frequency, and is also shown in fig. 10b.

An accuracy of the Rydberg constant to a level of some parts in 10^{12} seems possible in the next years. A phase coherent measurement of the 1S-2S transition, already in preparation in Garching, will transfer the full accuracy of the HeNe/CH_4 standard to this fundamental transition, resulting in an improvement in accuracy of an order of magnitude. D. Leibfried in Garching is working on a new experiment to determine the 1S-Lamb-shift with a tenfold reduced uncertainty. Finally R. S. VanDyck, who has made the best measurement of m_e/m_p to date [46], expects with his improved experimental setup to see an improvement in accuracy of m_e/m_p by a factor of ten in the near future [47].

Since all other hydrogen transitions may be predicted with the improved accuracy of the Rydberg constant, provided the theory of quantum electrodynamics is correct, the hydrogen atom offers a new set of reference frequencies throughout the ultraviolet, visible, infrared, and microwave region. With an even further improved value for the Rydberg constant, the time standard itself might in the future be related to the fundamental constants, as is already the case for the voltage standard due to the Josephson-effect [48], and the resistance standard due to the quantum-Hall-effect [49]. The second would thus be defined with the respect to a suitable transition in hydrogen, accessible to a very precise theoretical understanding, in contrast to the current arbitrary choice of the cesium hyperfine transition.

ACKNOWLEDGMENT

We gratefully acknowledge much help and advice from V. P. Chebotayev, G. Kramer, H. R. Telle, and J. L. Hall. We also thank M. Weitz for his assistance with the determination of the Rydberg constant, D. Meschede for many fruitful discussions and help, as well as T. Freegarde for carfully reading the manuscript. The 1S-2S frequency measurement has been supported in part by the Deutsche Forschungs - gemeinschaft.

REFERENCES

1. J. Vanier and C. Audoin, “The Quantum Physics of Atomic Frequency Standards”, Adam Hilger, Bristol, Philadelphia, 1989.
2. V. S. Letokhov, V. P. Chebotayev, “Nonlinear Laser Spectroscopy”, Springer, Berlin, Heidelberg, New York, 1977.
3. "Documents Concerning the New Definition of the Metre", Metrologia **19**, 163 (1984).
4. D. W. Allan, Proc. IEEE **54**, 221 (1966).
5. J. A. Barnes , A. R. Chi, L. S. Cutler, D. J. Healey, D. B. Leeson, T. E. McGunigal, J. A. Mullen, Jr., W. L. Smith, R. L. Sydnor, R. F. C. Vessot, G. M. R. Winkler, IEEE Trans. Instr. Meas. IM-**20**, 105 (1971).
6. S. N. Bagayev, V. P. Chebotayev, Sov. Phys. Usp **29** , 82 (1986).
7. S. N. Bagayev, E. V. Baklanov, V. P. Chebotayev, A. S. Dychkov, Appl. Phys. B **48**, 31 (1989).
8. C. Audoin, Metrologia **29**, 113 (1992).
9. D. J. E. Knigt, P. T. Woods, J. Phys. E: Scientific Instruments **9**,898 (1976).
10. T. G. Blaney, D. J. E. Knigt, E. K. Murray LLoyd, Opt. Comm. **25**, 176 (1978).
11. K. M. Evenson, D. A. Jennings, F. R. Petersen, J. S. Wells, in “Laser Spectroscopy III”, J. L. Hall, J. L. Carlsten, ed., Springer Ser. Opt. Sci. **7**, Springer, Berlin, Heidelberg, New York 1977, p. 56.
12. C. O. Weiß, G. Kramer, B. Lipphardt, E. Garcia, IEEE J. Quantum Electron. **QE-24** , 1970 (1988); und G. Kramer, B. Lipphardt, C. O. Weiß, IEEE Frequency Control Symposium, Boulder 1992, to be published.
13. G. Kramer, C. O. Weiß, B. Lipphardt, in “Frequency Standards and Metrology”, A. De Marchi , ed., Springer Verlag, Berlin, Heidelberg, New York, 1989.
14. The spectroscopy of trapped ions is reviewed by R. C. Thompson, to be published in Adv. Atomic, Mol. and Opt. Phys.
15. A. A. Madej, S. D. Sankey, Opt. Lett. **15** , 634 (1990).
16. R. Blatt, P. Gill, R. C. Thompson, J. Mod. Opt. **39**, 193 (1992).
17. U. Sterr, K. Sengstock, J. H. Müller, D. Bettermann, W. Ertmer, Appl. Phys B **54**, 341 (1992).
18. A. Witte, Th. Kister, F. Riehle, J. Helmke, J. Opt. Soc. Am. B **9**, 1030 (1992).

19. H. Schnatz, S. Ohshima, H. R. Telle, F. Riehle, Conference on Precision Electromagnetic Measurements, Paris 1992.
20. F. M. Gardner, "Phaselock Techniques", John Wiley & Sons, New York, 1979.
21. G. Wenke, S. Saito, Jap. J. Appl. Phys. **24**, L908 (1985).
22. H. E. Hagemeier, S. R. Robinson, Appl. Opt. **18**, 270 (1979).
23. J. L. Hall, M. Long-Shen, G. Kramer, IEEE J. Quantum Electron. **QE-23**, 427 (1987).
24. R. T. Ramos, A. J. Seeds, Electron. Lett. **26**, 389 (19909.
25. S. M. Sze, "Semiconductor Devices - Physics and Technology", John Wiley & Sons, New York, 1985.
26. J. C. Berquist, H.-U. Daniel, Opt. Comm. **48**, 327 (1984).
27. H. R. Telle, D. Meschede, T. W. Hänsch, Opt. Lett. **15**, 532 (1990).
28. R. Wynands, T. Mukai, T. W. Hänsch, "Coherent bisection of optical intervals as large as 530 THz", zu Veröffentlichung akzeptiert bei Opt. Lett.
29. G. W. Series, ed., "The Spectrum of Atomic Hydrogen: Advances", World Scientific Publishing Co Pte Ltd, Singapore, 1988.
30. G. F. Bassani, M. Inguscio, T. W. Hänsch, ed.,"The Hydrogen Atom", Springer Verlag, Berlin, Heidelberg, New York, 1989.
31. M. Weitz, F. Schmidt-Kaler, T. W. Hänsch, Phys. Rev. Lett. **68**, 1120 (1992).
32. C. Zimmermann, this volume.
33. M.Weitz, this volume.
34. C. Zimmermann, R. Kallenbach, T. W. Hänsch, Phys. Rev. Lett. **65**, 571 (1990); F. Schmidt-Kaler, D. Leibfried, C. Zimmermann, T.W. Hänsch, to be published.
35. É. M. Belenov, V. L. Velichanskii, A. S. Zibrov, V. V. Nikitin, V. A. Sautenkov, A. V. Uskov, Sov. J. Quant. Electron. **13**, 792 (1983); B. Dahmani, L. Hollberg, R. Drullinger, Opt. Lett. 12, 876 (1987); see also L. Hollberg, this volume.
36. R. Wynands, F. Dietrich, D. Meschede, H. R. Telle, Rev. Sci. Instrum. Dec. 1992.
37. T. Andreae, W. König, R. Wynands, D. Leibfried, F. Schmidt-Kaler, C. Zimmermann, D. Meschede, T. W. Hänsch, Phys. Rev. Lett. **69**, 1923 (1992).
38. R. Kallenbach, B. Scheumann, C. Zimmermann, D. Meschede, T. W. Hänsch, Appl. Phys. Lett. 54, 1622 (1989); D. Leibfried, F. Schmidt-Kaler, M. Weitz, T. W. Hänsch, excepted for publication Appl.Phys. B.

39. M. G. Boshier, P. G. E. Baird, C. J. Foot, E. A. Hinds, M. D. Plimmer, D. N. Stacey, J. B. Swan, D. A. Tate, D. M. Warrington, G. K. Woodgate, Phys. Rev. A **40**, 6169 (1989).
40. R. G. Beausoleil, D. H. McIntyre, C. J. Foot, E. A. Hildum, B. Couillaud, T. W. Hänsch, Phys. Rev. A **35**, 4878 (1987).
41. D. H. McIntyre, R. G. Beausoleil, C. J. Foot, E. A. Hildum, B. Couillaud, T. W. Hänsch, Phys. Rev. A **39**, 4591 (1989).
42. F. Schmidt-Kaler, T. Andreae, W. König, D. Leibfried, L. Ricci, M. Weitz, R. Wynands, C. Zimmermann, T. W. Hänsch, ICAP XIII, München 1992, and F. Schmidt-Kaler, D. Leibfried, M. Weitz, T. W. Hänsch, to be published
43. F. Biraben, J. C. Garreau, L. Julien, M. Allegrini, Phys. Rev. Lett. **62**, 621 (1989).
44. D. A. Jennings, C. R. Pollock, F. R. Petersen, R. E. Drullinger, K. M. Evenson, J. S. Wells, J. L. Hall, H. P. Layer, Opt. Lett. **8**, 136 (1983).
45. F. Netz, M. D. Plimmer, S. Bourzeix. L. Julien, F. Biraben, R. Felder, ICAP XIII, München 1992; F. Netz, M. D. Plimmer, S. Bourzeix. L. Julien, F. Biraben, R. Felder, O. Acef, J. J. Zondy, P. Laurent, A. Clairon, M. Abed, Y. Millerioux, P. Juncar, Phys. Rev. Lett. **69**, 2326 (1992).
46. R. S. VanDyck, F. L. Moore, D. L. Farnum, P. B. Schwinberg, Bull. Am. Phys. Soc. **31**, 244 (1986).
47. R. S. VanDyck, priv. communication, ICAP XIII, München 1992.
48. R. Pöpel, Metrologia **29**, 135 (1992).
49. A. Hartland, Metrologia **29**, 175 (1992).

FREQUENCY STABILIZED TI:SAPPHIRE LASER FOR HIGH RESOLUTION SPECTROSCOPY OF ATOMIC HYDROGEN

M. Weitz, F. Schmidt-Kaler, and T. W. Hänsch

Max-Planck-Institut für Quantenoptik, 8046 Garching, Germany

ABSTRACT

Advances in the development of tunable solid state laser sources in the near infrared have been essential for the recent dramatic progress in high resolution spectroscopy of atomic hydrogen. We discuss here the frequency stabilization, by means of a radio-frequency sideband technique, of a 972 nm Ti:sapphire laser to an external, highly stable optical resonator, and its application to excitation of the hydrogen 2S-4S,4D transitions. By comparing the optical frequencies of the 1S-2S and 2S-4S,4D two-photon transition frequencies, the 1S ground state Lamb shift has been accurately determined.

1. INTRODUCTION

Atomic hydrogen is, because of its simplicity, well suited for stringent tests of some of the fundamental laws of physics.[1] Indeed, it was the discovery of the $2S_{1/2}$-$2P_{1/2}$ Lamb shift[2] in 1947 which provoked the development of quantum electrodynamics (QED), perhaps the most successful theory in modern physics despite of its unsettling concepts.[3] On the way to a more stringent test of QED on a bound system, the 100 MHz natural linewidth of the nearby 2P state is now limiting the accuracy of radio-frequency measurements of the 2S

Solid State Lasers: New Developments and Applications
Edited by M. Inguscio and R. Wallenstein, Plenum Press, New York, 1993

Lamb shift.[4] Recent dramatic advances in high-resolution laser spectroscopy allow *optical* Lamb shift measurements with radio-frequency precision using transitions of much narrower natural linewidth.

For optical Lamb shift measurements, the extremely sharp 1S-2S two-photon transition is particularly intriguing (see fig. 1). It has a natural linewidth of only 1.3 Hz and Doppler-free two-photon spectroscopy of a cold atomic beam,[5-7] has reached a resolution of better than 1 part in 10^{11}. Further, since QED effects scale with $1/n^3$, the 1S Lamb shift is the largest in atomic hydrogen. By comparing a quarter of the 1S-2S frequency with the frequencies of the 2S-4S and 2S-4D transitions, the main contributions described by the simple Rydberg formula may be eliminated, and the 1S Lamb shift can be deduced from the relatively small residual frequency difference (about 5 GHz) which can be directly counted using radio-frequency techniques.[8-10] The natural linewidths of the 2S-4S and 2S-4D two-photon transitions, 0.69 and 4.4 MHz, are well below that of the 2P state. At the expense of the simplicity of the frequency comparison with the 1S-2S transition, and with higher sensitivity to stray electric fields, transitions with even smaller natural linewidths to the Rydberg levels nS and nD ($n>4$) can also be excited.[11]

Together with experimental values of the Lamb shifts, an absolute measurement of the 1S-2S transition frequency can yield an improved value for the Rydberg constant.[12] Such a measurement, together with the determination of the Rydberg constant, is described by T. Andreae, W. König, R. Wynands, and T. W. Hänsch elsewhere in this volume.

A major part of the experimental effort in experiments of this kind deals with the generation of the necessary intense and extremely monochromatic light. The 1S-2S transition is excited by Doppler-free two-photon spectroscopy with two ultraviolet photons at 243 nm. Since at this wavelength there is no suitable cw laser source, frequency doubling of visible light is necessary, as has been discussed by C. Zimmermann, A. Hemmerich, and T. W. Hänsch in this volume (see also Ref. 13). In the current experiment a dye laser operating at 486 nm is frequency doubled in a nonlinear BBO crystal.

Generation of sufficient intense light near 972 nm for excitation of the Doppler-free two-photon transitions 2S-4S and 2S-4D was until recently not readily possible. Laser dyes, when available, become

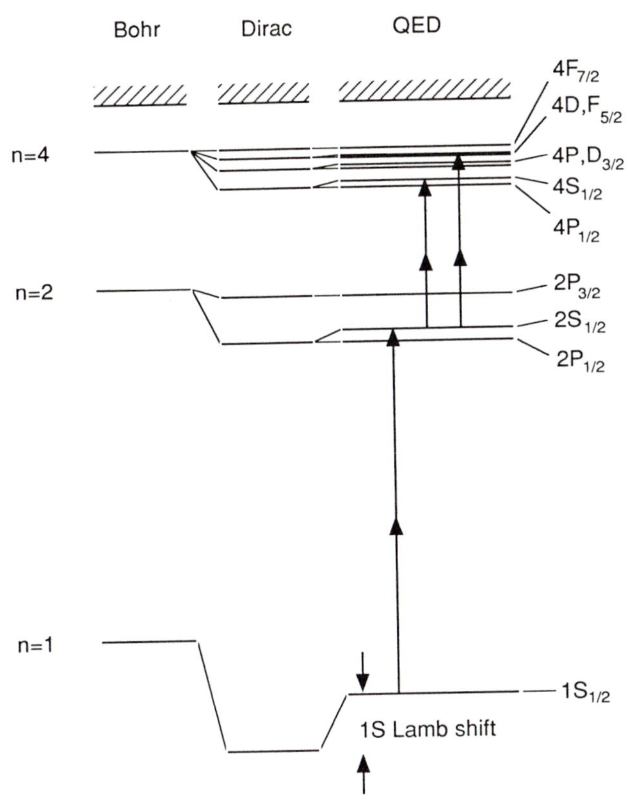

Fig. 1. Simplified level diagram of the hydrogen atom. The fine structure and quantum electrodynamic corrections to Bohr's atomic model are shown enlarged and not to scale; the hyperfine structure has been neglected. The two-photon transitions 1S-2S, 2S-4S and 2S-$4D_{5/2}$ are marked.

rather inefficient at wavelengths slightly above the visible spectral range. This situation changed dramatically with the introduction of tunable solid state lasers containing transition metal ions like the system Ti:sapphire.[14]

In the future, a Ti:sapphire laser with two frequency doubler stages could also excite the hydrogen 1S-2S transition. Such a laser would offer a better intrinsic frequency stability and easier maintenance than a dye laser. For the first stage frequency doubling of a Ti:sapphire laser at 972 nm should be capable of producing a few hundreds of milliwatts of blue light at 486 nm. Up to 650 mW of blue light have already been generated by frequency doubling of a Ti:sapphire laser at a shorter wavelength.[15] The second stage from 486 nm to 243 nm has already been demonstrated, where with 300 mW blue light from a dye laser some milliwatts of UV light at 243 nm could be generated. A similar scheme for excitation of the 1S-3S transition is also under construction in Paris. Here the difficulty arises that the produced 205 nm is close to the absorption edge of the nonlinear BBO crystal, leading to a very low damage threshold.[16]

In section 2 we report on the frequency stabilization of a Ti:sapphire laser, which turns out to be considerably simpler than the stabilization of a dye laser with its rapid fluctuating dye jet. In section 3 we report on the precise optical measurement of the hydrogen ground state Lamb shift with the help of this laser source.

2. FREQUENCY STABILIZED TI:SAPPHIRE LASER

In the Ti:sapphire crystal the mechanically and thermally very stable sapphire host crystal (Al_2O_3) has been weakly doped with the transition metal ion Ti^{3+}. The energy levels of the Ti^{3+}-ion are strongly broadened by the crystal lattice field, so that tunable laser operation is possible in the wavelength range between 680 and 1100 nm. Since the 4-level system absorbs between 450 and 600 nm, argon ion lasers are obvious pump sources.

We use a commercial prototype Schwartz Electro-Optics Ti:sapphire ring laser, which has been stabilized very well to an external high finesse cavity to ensure the necessary small linewidth and frequency stability for high resolution spectroscopy. Fig. 2 shows the resonator setup consisting of the laser crystal, two curved and two flat mirrors, an optical diode (a terbium gallium garnet rotator and quartz back rotator) for unidirectional operation and two etalons and a Lyot filter to select the desired longitudinal cavity mode. In contrast to the commercial version, the uncoated single etalon

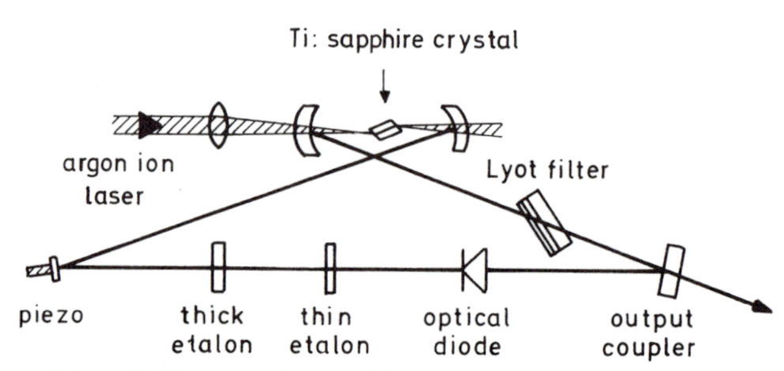

Fig. 2. Setup of the Ti:sapphire laser resonator.

(1 mm) was replaced by two uncoated 0.5 and 3 mm thick etalons, which improve the stability at higher pump powers above some 7 W. This laser delivers typically 1 W single frequency output near 972 nm at 10 W all-line pump power from the argon ion laser.

The observed free running linewidth of the laser is a few MHz. The Schawlow-Townes linewidth is well below one Hz, but technical noise is broadening the line. Active frequency stabilization can compensate for these frequency fluctuations; for a review see Hamilton.[17] The noise is mainly due to acoustics and amplitude fluctuations of the argon ion laser, which are converted into frequency fluctuations via the temperature dependent refractive index of the laser crystal. The frequency spectrum of these fluctuations drops off well below 100 kHz. Compared to dye lasers, solid state lasers, apart from their easier maintenance, offer the advantage that there is considerably less high frequency noise. This is due to the fact that there is no rapidly fluctuating dye jet with phase excursions of some radians in the few 100 kHz range. In dye lasers, fast electro-optic phase modulators with unity gain frequencies above a MHz must be used to compensate for this high frequency noise.[18] In contrast, for frequency stabilization of a Ti:sapphire laser much slower servo loops will yield good results.

In our experiment, the frequency of the Ti:sapphire laser is stabilized to a mode of a very stable reference cavity (fig. 3). This cavity consists of a 45 cm long Zerodur spacer with mirrors of high reflectivity (finesse 2000) optically contacted to the ends. The spacer is

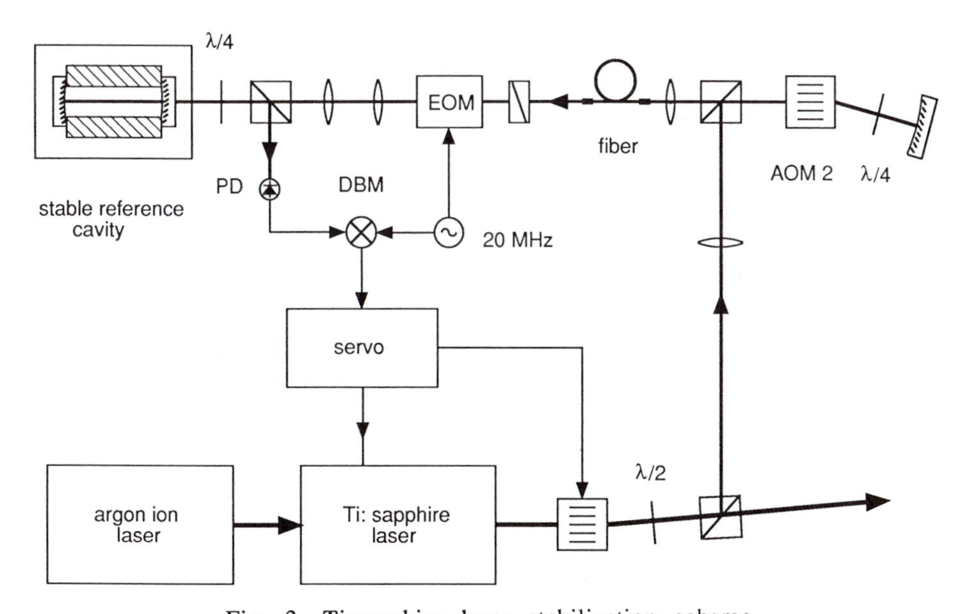

Fig. 3. Ti:sapphire laser stabilization scheme.

suspended by soft springs in a temperature-controlled vacuum chamber in order to minimize the transport of acoustic and vibration noise. Since the frequency of the resonator is fixed, the laser is tuned by varying the drive frequency of an acousto-optic modulator (AOM 2) between laser and reference cavity. The modulator is used in double pass in order to compensate for the drive frequency-dependent deviation angle.

To detect deviations of the laser frequency from the reference resonator, an error signal is generated via a radio-frequency sideband technique.[19] This error signal is amplified and applied to frequency control elements, which compensate the frequency fluctuations of the laser. The frequency error $\Delta\nu_0$ of the laser relatively to the reference cavity at the Fourier-frequency f is reduced in this closed servo loop:

$$\Delta\nu_C(f) = \frac{\Delta\nu_0}{1+V(f)},$$

where V(f) denotes the frequency-dependent total gain of the servo system. Thus, the aim is to obtain the highest possible gain of the servo system at all Fourier frequencies where there is a significant noise level. Since the largest frequency error occurs for small frequencies, the servo gain decreases for higher frequencies. An important condition for the stability of a servo system is that, near the point with gain 1 ("unity gain frequency"), the frequency dependent gain decreases with less than 12 db per octave.[20] Following the Kramers-Kronig relation, a slope of 12 db per octave would correspond to a phase shift of 180^0 and thus imply an oscillation of the servo system. Fig. 4 shows the frequency-dependent gain of our servo system.

Stabilization of a Ti:sapphire laser has previously been achieved simply by feeding the amplified error signal to a piezo transducer on which a laser mirror was mounted.[21] In this way unity gain frequencies of about 10 kHz could be realized, which were limited ultimately by the piezo´s mechanical resonances. For small laser powers, when the pump laser can be operated in single frequency mode, this kind of stabilization reaches rms linewidths slightly above one kHz. At higher pump laser powers however the linewidth dramatically increases. The beat signal between the frequency-doubled Ti:sapphire laser, although then stabilized to the stable reference cavity, and an even more highly stabilized dye laser, grows to a width of about 200 kHz. The pump laser amplitude noise, which increases with pump power, adds to the

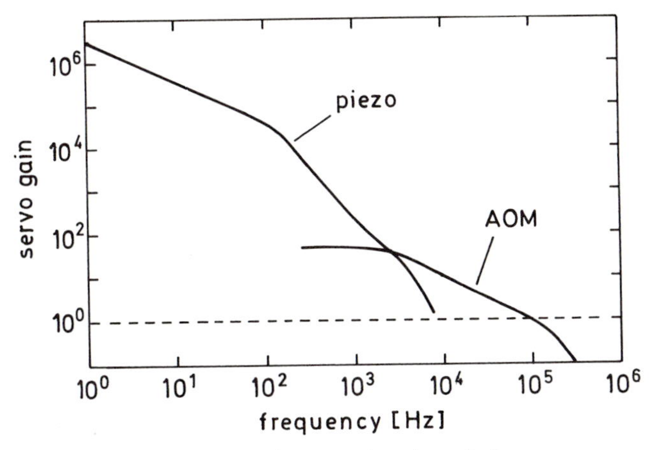

Fig. 4. Frequency-dependent total gain of the servo system.

frequency noise of the Ti:sapphire laser. A Fourier analysis of the error signal shows that the major contribution to the remaining noise occurs between 10 and 50 kHz, i.e. beyond the piezo´s servo bandwidth.

To compensate for high frequency noise at larger pump powers the high frequency component of the error signal is fed to a voltage controlled oscillator which drives an external acousto-optic frequency shifter (AOM 1) (see also Refs. 22-24). No precautions were necessary to compensate for the varying deflection angle of the modulator at different VCO frequencies. The typical angle variation is calculated to be only about 10 μrad, which is more than two orders of magnitude below the beam divergence.

The servo bandwidth that can be obtained with such a stabilization scheme is limited by the transit time of the acoustical wave through the modulator of the order of 1 μs. Whereas the high frequency fluctuations of a dye laser (some 100 kHz) cannot be compensated with such a simple control system,[25] the servo bandwidth of about 150 kHz is sufficient to suppress the phase fluctuations of a Ti:sapphire laser to well under 1 radian for Fourier frequencies above a few kHz.

The spectral noise density of the stabilized Ti:sapphire laser as determined from the error signal is shown in fig. 5. The rms value of the spectral noise density is 700 Hz at 1 W output power, which is the laser rms linewidth relative to the cavity. Absolute frequency stability can only be determined from comparison with an independent frequency reference. The actual laser linewidth was determined via the beat signal between the frequency doubled Ti:sapphire laser and the even more stable blue dye laser to be about 10 kHz.

These experiments show that for solid state lasers very good frequency stabilization is possible without the use of a fast electro-optic modulator. With a similar stabilization system, phase locking of a Ti:sapphire laser has already been achieved by J. Hall and collaborators.[26]

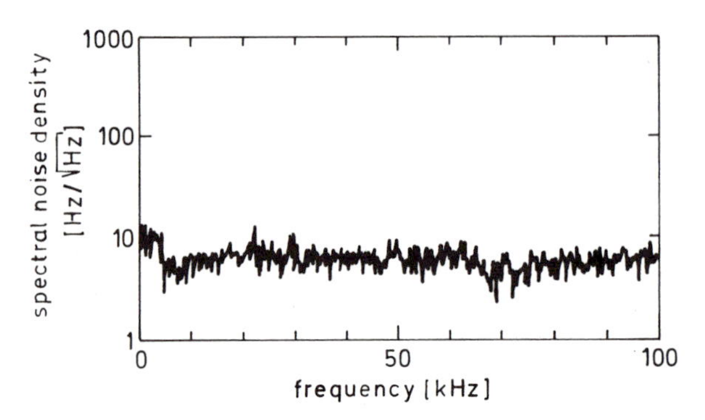

Fig 5. Spectral noise density of the stabilized Ti:sapphire laser relative to the reference cavity.

3. LAMB SHIFT MEASUREMENTS OF ATOMIC HYDROGEN

The 1S Lamb shift has been measured via a direct comparison of the two-photon transitions 2S-4S/4D and 1S-2S (see fig. 1).[5,9,10] The experimental setup for the measurement of the hydrogen 1S Lamb shift is shown in fig. 6.

For the 1S-2S spectrometer[5-7] we employ a ring dye laser operating at 486 nm, which is stabilized using a radio-frequency sideband technique to a reference cavity with gyroscope-quality mirrors. These mirrors are optically contacted to the same Zerodur spacer as used for the Ti:sapphire laser reference cavity. A barium β-borate (BBO) crystal inside an external buildup cavity produces about 2 mW of UV-light near 243 nm which is coupled into a further linear buildup cavity inside the vacuum chamber of an atomic beam apparatus. A cold nozzle emits a beam of hydrogen atoms in the 1S ground state collinear to the standing wave UV-light field. Atoms are excited by Doppler-free two-photon transitions into the metastable 2S state. At the end of the atomic beam the metastable atoms enter the

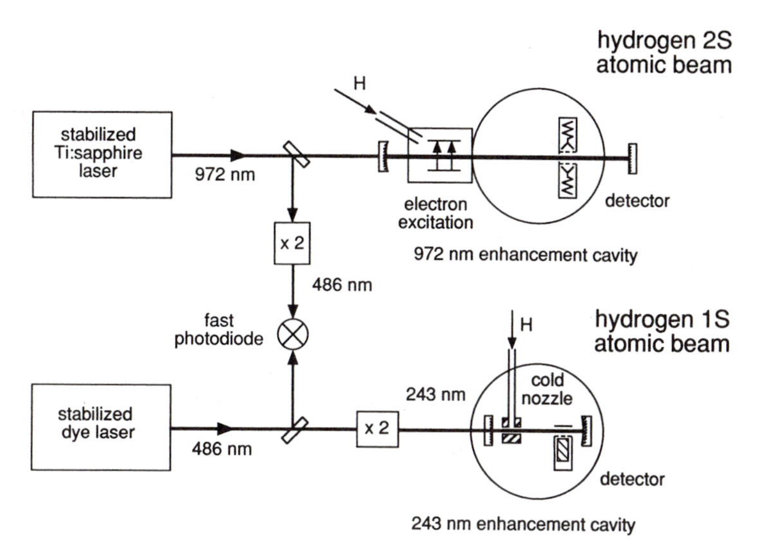

Fig. 6. Experimental setup for measurement of the hydrogen 1S Lamb shift.

detector where a small electric field mixes the 2S and 2P state and thus quenches the excited state. A photomultiplier detects the resulting Lyman-α fluorescence. The observed linewidth of about 10 kHz at 243 nm corresponds to a resolution of 1 part in 10^{11}, limited by transit time broadening. The residual second order Doppler-effect is below 1 kHz. The frequency of the blue dye laser is locked to the maximum of the 1S-2S signal with the help of a personal computer.

The stabilized Ti:sapphire laser near 972 nm is coupled into a standing wave cavity of the hydrogen 2S atomic beam apparatus, giving a circulating power of some 50 W on a mean waist of 0.5 mm. The beam of metastable 2S atoms is produced using a method similar to that of Lamb[2] and also Biraben[11] by electron impact excitation of hydrogen atoms. The 2S atoms are directed along the laser axis and detected at the end of the atomic beam via a quench field and two potassium iodine coated channeltrons sensitive to the Lyman-α fluorescence. The count rate is reduced when the laser is on resonance, since the excited 4S and 4D atoms cascade quickly via an intermediate P-state with 95% probability into the 1S ground state. Typical spectra are shown in fig. 7 with fits to theoretical line shapes. The maximum signal decrease is about 5% for 2S-4S and about 20% for 2S-4D. For atoms which travel directly along the laser axis, the latter transition especially is highly saturated. A typical spectrum is

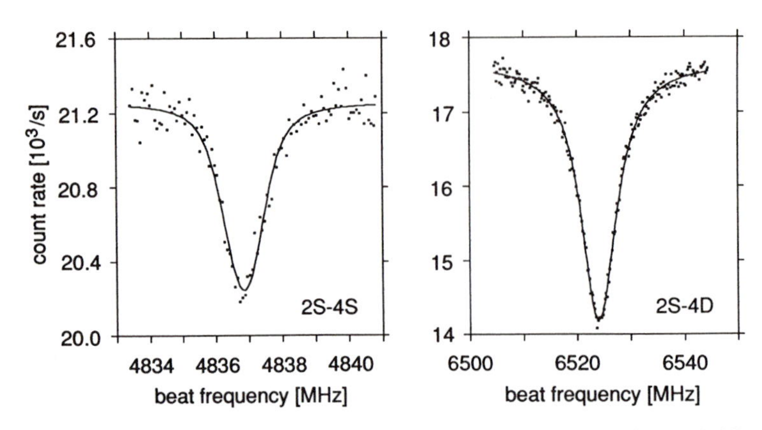

Fig. 7. Hydrogen two-photon spectra $2S_{1/2}(F=1)$-$4S_{1/2}(F=1)$ (left) and $2S_{1/2}(F=1)$-$4D_{5/2}$ (right). The beat frequency has been measured with the dye laser locked to the $1S_{1/2}(F=1)$-$2S_{1/2}(F=1)$ transition.

obtained by averaging 40 scans during 20 minutes of measuring time.

For comparison of the 1S-2S and 2S-4S/4D transition frequencies, part of the infrared light is send into a $KNbO_3$ crystal and the resulting blue second harmonic is then mixed with the radiation of the dye laser on a fast photodiode. The beat signal Δ is monitored with a counter and a spectrum analyzer.

The 2S-4S and 2S-4D line shapes are slightly asymmetric and their ac Stark shift is no longer growing strictly linearly at high laser power, since atoms which travel directly along the laser axis are significantly depleted. Therefore, our spectra have been fitted to line profiles calculated by Garreau et al.[27] by numerically summing over the contributions from all possible atom trajectories to correct for the ac Stark effect. A small residual dependence of the fitted line center on laser power is accounted for by recording spectra at different light powers and extrapolating to zero power.

The 1S Lamb shift now can be derived from the measured beat frequency Δ (see also Table 1):

$$\Delta = (E_{4S,4D}-E_{2S}) - \frac{1}{4}(E_{2S}-E_{1S})$$

$$= \Delta_0 + \Delta_{HFS} + \frac{1}{4}L_{1S} - \frac{5}{4}L_{2S} + L_{4S,4D}\ ,$$

where L denotes the Lamb shift of the indicated state. Relativistic reduced mass corrections are contained in Δ_0 and the hyperfine structure corrections in Δ_{HFS}. For the difference $L_{2S_{1/2}} - L_{2P_{1/2}}$ we take the value 1057.845(9) MHz obtained by radio-frequency measurements.[4] For the small Lamb shifts of the 2P, 4S and 4D states we rely on calculated values. At the present level of accuracy, since the 1S Lamb shift is the largest in hydrogen, one can still rely on these values. When future measurement precision exceeds the accuracy of radio-frequency measurements, it will be reasonable to compare the value of $1/4\ L_{1S} - 5/4\ L_{2S} + L_{4S}$ with theory directly.

Our final result for the hydrogen 1S Lamb shift, L_{1S} = 8172.84(9) MHz, agrees with the theoretical value of 8172.94(6) MHz, which is predicted if the proton charge radius is taken to be 0.805(11) fm.[28] Using a more recent contradicting measurement,[29] the theoretical value increases by 150 kHz, lessening the agreement between experiment and theory. The result for the deuterium 1S Lamb shift is L_{1S} = 8184.03(16) MHz and agrees with the theoretical prediction of 8184.10(9) MHz. With the use of the recently very precisely measured hydrogen-deuterium isotope shift of the 1S-2S transition,[5,30] we calculate for the hydrogen 1S Lamb shift a weighted mean value of L_{1S} = 8172.86(8) MHz. Fig. 8 shows a comparison with

Table 1. Determination of the 1S Lamb shift (typical data).

	$2S_{1/2}$-$4S_{1/2}$ [MHz]	$2S_{1/2}$-$4D_{5/2}$ [MHz]
Extrapolated beat frequency (2S-4S/4D)-1/4(1S-2S)	4836.136(28)	6523.623(28)
Corrections:		
(1) lineshape	0.000(12)	0.000(11)
(2) dc Stark effect	0.000(2)	0.000(3)
(3) reference cavity drift	0.000(4)	0.000(4)
(4) second-order Doppler shift:		
1S-2S (room temp.)	-0.016(1)	-0.016(1)
2S-4S/4D	0.036(2)	0.036(2)
Hyperfine structure Δ_{HFS}	-38.837(3)	-33.511(0)
Dirac and relativistic reduced mass contributions Δ_0	-3928.707(0)	-5752.645(1)
$1/4\ L_{1S} - 5/4\ L_{2S} + L_{4S,4D}$	868.612(31)	737.487(31)
$5/4\ (L_{2S_{1/2}} - L_{2P_{1/2}})$[a]	1322.306(11)	1322.306(11)
$5/4\ L_{2P_{1/2}} - L_{4S,4D}$	-147.722(3)	-16.581(3)
$1/4\ L_{1S}$ Lamb shift	2043.196(33)	2043.212(33)

[a]Experimental $2S_{1/2}$-$2P_{1/2}$ Lamb shift.[4]

the theoretical value and other recent measurements.

With an uncertainty of $1.0 \cdot 10^{-5}$, this optical measurement exceeds the accuracy of earlier measurements by an order of magnitude and rivals and complements radio-frequency measurements of the 2S Lamb shift as one of the best tests of quantum electrodynamics on a bound system. An indirect comparison of the Garching 1S-2S absolute frequency with the Paris 2S-8S/8D transition frequency via frequency chains and intermediate frequency standards of L_{1S} = 8172.804(83) MHz agrees well with our result.[11]

Substantial further improvements in precision should be possible by using the optically excited slow metastable beam of the 1S-2S spectrometer for excitation into the 4S state. The 2S-4S transition can also be detected by its blue 4S-2P fluorescence, which requires a light detection system on the entire length of the metastable beam, but avoids the strong background of unexcited metastable atoms.[31] We also are constructing optical divider stages (T. Andreae, W. König, R. Wynands, and T. W. Hänsch, this volume) to compare the 1S-2S transition frequency with the narrower 2S-8S or even 2S-16S,.. Rydberg transitions.[32] The 2S-8S transition frequency is about 5/16 of the 1S-2S transition frequency; the two frequencies can be compared using two optical divider stages.

More precise measurements of the proton charge radius would permit a test of QED with yet higher precision; measurements so far

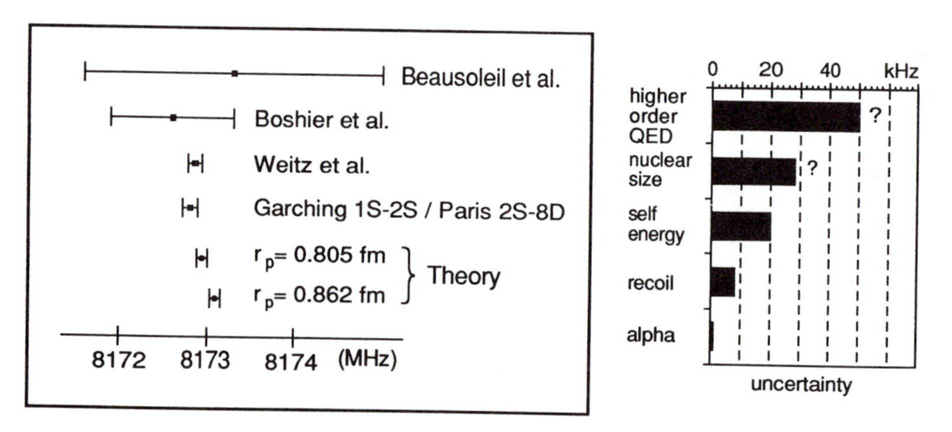

Fig. 8a: Comparison of this measurement with other recent measurements of the hydrogen 1S Lamb shift[34,35,11] and with the theoretical predictions assuming two different proton charge radii as given in Refs. 28 and 29. Fig. 8b: Uncertainties of the theoretical prediction of the hydrogen 1S Lamb shift.

have been done by electron scattering using an accelerator. Laser spectroscopy of myonic hydrogen should provide a more accurate charge radius in the future.[33] In this exotic atom due to its very small Bohr radius ($m_{myon} \approx 207\ m_e$), nuclear size effects have a much larger influence on the energy levels and govern the theoretical uncertainty. A precise measurement of the $2S_{1/2}$-$2P_{1/2}$ splitting in myonic hydrogen should be able to yield an uncertainity of 1 part in 10^3 with the proton charge radius.

The simple hydrogen atom thus permits unique comparisons of spectroscopic experiment with quantum electrodynamic theory. Precise comparisons and measurements of its transition frequencies can provide stringent tests of basic laws of physics and also yield accurate values of fundamental constants. Dramatic advances in resolution have recently become possible, to a large extent due to the development of well stabilized laser sources from the infrared to the ultraviolet spectral region. The Ti:sapphire laser has played an important role in these developments.

The authors thank C. Zimmermann, D. Leibfried, A. Huber and S. Zimmermann for much help and W. Vassen for his important contributions during the early phase of this project. We are grateful to L. Julien and F. Biraben for calculation of the theoretical lineshapes, and J. Hall for helpful discussions.

LITERATURE

(1) *The Hydrogen Atom*, G. F. Bassani, M. Inguscio, and T. W. Hänsch (Ed.) (Springer, Berlin, 1989).

(2) W. E. Lamb and R. C. Retherford, Phys. Rev. **72**, 241 (1947); W. E. Lamb and R. C. Retherford, Phys Rev. **79**, 549 (1950); W. E. Lamb and R. C. Retherford, Phys Rev. **81**, 222 (1951).

(3) T. Kinoshita (Hrsg.), *Quantum Electrodynamics*, (World Scientific, Singapore, 1990).

(4) S. R. Lundeen and F. M. Pipkin, Phys. Rev. Lett. **46**, 232 (1981).

(5) F. Schmidt-Kaler, T. Andreae, W. König, D. Leibfried, L. Ricci, M. Weitz, R. Wynands, C. Zimmermann, and T. W. Hänsch in *Proceedings of the Thirteenth International Conference on Atomic Physics* (American Institute of Physics, New York, in press).

(6) C. Zimmermann, R. Kallenbach, W. Vassen, F. Schmidt-Kaler, M. Weitz, D. Leibfried, and T. W. Hänsch, in *Proceedings of the Twelfth International Conference on Atomic Physics* (American Institute of Physics, New York, 1991).
(7) C. Zimmermann, R. Kallenbach, and T. W. Hänsch, Phys. Rev. Lett. **65**, 571 (1990).
(8) T. W. Hänsch, S. A. Lee, R. Wallenstein, and C. Wieman, Phys. Rev. Lett. **34**, 307 (1975).
(9) M. Weitz, F. Schmidt-Kaler, and T. W. Hänsch, Phys. Rev. Lett. **68**, 1120 (1992).
(10) M. Weitz, Ph. D. thesis, LMU-München (1992, unpublished).
(11) F. Nez, M. D. Plimmer, S. Bourzeix, L. Julien, F. Biraben, R. Felder, O. Acef, J. J. Zondy, P. Laurent, A. Clairon, M. Abed, Y. Millerioux, and P. Juncar, Phys. Rev. Lett. **69**, 2326 (1992).
(12) T. Andreae, W. König, R. Wynands, D. Leibfried, F. Schmidt-Kaler, C. Zimmermann, D. Meschede, and T. W. Hänsch, Phys. Rev. Lett. **69**, 1923 (1992).
(13) C. Zimmermann, R. Kallenbach, T. W. Hänsch, and J. Sandberg, Opt. Comm. **71**, 229 (1989).
(14) P. F. Moulton, J. Opt. Soc. Am. B **3**, 125 (1986).
(15) E. S. Polzik and H. J. Kimble, Opt. Lett. **16**, 1400 (1991).
(16) L. Julien, private communication.
(17) M. W. Hamilton, Comtemp. Phys. **30**, 21 (1989).
(18) See for example: J. Hough, D. Hils, M. D. Rayman, Ma L.-S., L. Hollberg, and J. L. Hall, Appl. Phys. B **33**, 179 (1984).
(19) R. W. P. Drever, J. L. Hall, F. V. Kowalski, J. Hough, G. M. Ford, A. J. Munley, and H. Ward, Appl. Phys. B **31**, 97 (1983).
(20) R. C. Dorf, *Modern Control Systems*, (Addison-Wesley, Reading, 1980).
(21) W. Vassen, C. Zimmermann, R. Kallenbach, and T. W. Hänsch, Opt. Comm. **75**, 435 (1990).
(22) T. L. Boyd and H. J. Kimble, Opt. Lett. **16**, 808 (1991).
(23) G. Camy, D. Pinaud, N. Courtier, and H. C. Chuan, Rev. Phys. Appl. **17**, 357 (1982).
(24) J. L. Hall and T. W. Hänsch, Opt. Lett. **9**, 502 (1984).
(25) J. L. Hall, in *Proceedings of the International Conference on Lasers*, P. R. China (Hrsg.) (Wiley, New York, 1983).
(26) J. L. Hall, private communications.
(27) J. C. Garreau, M. Allegrini, L. Julien, and F. Biraben, J. Phys. France **51**, 2263 (1990), 2275 (1990), 2293 (1990).

(28) L. N. Hand, D. G. Miller, and R. Wilson, Rev. Mod. Phys. **35**, 335 (1963).

(29) G. G. Simon, Ch. Schmitt, F. Borkowski, and V. H. Walther, Nucl. Phys. A **333**, 381 (1980).

(30) F. Schmidt-Kaler, D. Leibfried, M. Weitz, and T. W. Hänsch, to be published.

(31) M. Weitz, A. Huber, S. Zimmermann, and T. W. Hänsch, Verhandlungen der DPG, in press.

(32) H. R. Telle, D. Meschede, and T. W. Hänsch, Opt. Lett. **15**, 532 (1990).
R. Wynands, T. Mukai, and T. W. Hänsch, to be published.
S. Zimmermann, A. Huber, D. Leibfried, F. Schmidt-Kaler, M. Weitz, and T. W. Hänsch, Verhandlungen der DPG, in press.

(33) D. Taqqu, private communication.

(34) R. G. Beausoleil, D. H. McIntyre, C. J. Foot, E. A. Hildum, B. Couillaud, and T. W. Hänsch, Phys. Rev. A **35**, 4878 (1987).

(35) M. G. Boshier, P. E. G. Baird, C. J. Foot, E. A. Hinds, M. D. Plimmer, D. N. Stacey, J. B. Swan, D. A. Tate, D. M. Warrington, and G. K. Woodgate, Phys. Rev. A **40**, 6169 (1989).

INDEX